BUILD, OPERATE, TRANSFER

BUILD, OPERATE, TRANSFER

Paving the Way for Tomorrow's Infrastructure

SIDNEY M. LEVY
Frank Mercede & Sons, Inc.
Stamford, CT

JOHN WILEY & SONS, INC.
New York · Chichester · Brisbane · Toronto · Singapore

This text is printed on acid-free paper.

Library of Congress Cataloging in Publication Data:

Levy, Sidney M.
Build, operate, transfer : paving the way for tomorrow's infrastructure / Sidney M. Levy.
p. cm.
Includes index.
ISBN 0-471-11992-X (cloth : alk. paper)
1. Infrastructure (Economics) 2. Privatization. I. Title.
HC79.C3L48 1996
363—dc20 96-10268

Printed in the United States of America

10 9 8 7 6 5 4 3 2 1

CONTENTS

PREFACE

The Industrial Revolution that occurred in the eighteenth century had a dramatic effect on the role of government as agrarian societies evolved into industrial centers and new demands were made of the government to provide the services required by the emerging economy. As people left the farms for the city, the obligation of government grew to include the development of waste disposal facilities and potable water distribution systems. The steam engine powering John Kay's flying shuttle in 1733 and James Hargreave's spinning jenny in 1764 brought other economic forces into play. Coal had to be mined to fuel the engines, roads and bridges had to be built to bring raw materials to the factories housing the engines, and transportation networks had to be constructed to cart the finished products to the marketplace. New methods of taxation had to be formulated to provide the revenue that these government-sponsored projects required.

The term *Infrastructure*—the essential elements of a structure—became the soul mate of this new activity called economic growth. The infrastructure became the foundation for economic development and was critical to sustaining the development in the nineteenth century.

In the twentieth century, and most certainly into the twenty-first century, infrastructure development, or the lack thereof, will separate the "have" from the "have-not" nations as emerging nations and developing nations attempt to enter the mainstream of world commerce.

Even the United States, the richest nation in the world, has yet to provide all of its citizens with a comprehensive, cost effective infrastructure system. Highways and bridges in the United States are in need of repair and water and waste treatment facilities must be expanded and upgraded. The United States devotes a lower percentage of its gross domestic product (GDP) to infrastructure upkeep than do Japan, Italy, Germany, France, Canada, and the United Kingdom.

Although the last several decades have shown vast improvements in the quantity and quality of infrastructure in emerging and developing countries, much remains to be accomplished. Of the current world population of approximately 5.7 billion people, fully two billion lack adequate sanitation systems and one billion people are without electric power. Those countries that have set infrastructure priorities and goals in the next century and meet their milestones may well be responsible for dramatic shifts in global wealth.

The twenty-first century will see the emergence and growth of several economic blocs—some old and some new.

The North American Free Trade Agreement (NAFTA), which now includes only the United States, Canada, and Mexico, will probably expand to include several South American countries and might even change its name to NOSAFTA—the North and South America Free Trade agreement. On the other side of the globe the Association of South East Asian Nations (ASEAN), comprising Brunei, Indonesia, Malaysia, Philippines, Singapore, and Thailand have only to team up with Japan, Korea, China, and India to create the biggest economic power in the world. And—in between —the European Union has embarked on a grandiose scheme known as the Trans-European Networks (TENS) to forge a unified transport, telecommunications, and energy network to span and unite member countries of the European Economic Community into a single economic dynamo.

Except for isolated examples in the last two centuries, infrastructure development has been the responsibility of public agencies, and taxes collected by local and central governments have provided the funds by which infrastructure projects have been built. However, the traditional role of government as the sole provider of railroads, bridges, tunnels, telecommunications networks, and power generation and transmission facilities is no longer feasible.

Private developers have access to vast amounts of investment capital and, in many cases, they have better management skills than government workers do, and so they are beginning to challenge government's traditional role. And, faced with severe budgetary restraints, governments are starting to encourage private investors to participate in a number of infrastructure projects. They are enacting legislation to create new public/private partnerships in these fields, utilizing the Build-Operate-Transfer concept or one of its derivatives.

The Build-Operate-Transfer trend, begun in earnest in the 1970s and 1980s, has gained momentum in the 1990s. Still, some projects that were expected to start with a bang opened with a whimper.

The Channel Tunnel, stuck fast to a financial tar baby, struggles to break free.

The first of the new public/private toll road ventures to become operational in the United States was the Dulles Greenway in Fairfax County, Virginia. The Greenway opened to the public on September 29, 1995, and this project in which the public pays to avoid congestion on existing highways has been attracting an average of 10,500 drivers daily—far short of its projected 34,000 per day for the first year.

In California, State Road 91, an express lane toll road built in the median strip of existing SR91 opened in late December 1995. California Private Transport, the project owner, has not divulged ridership numbers but it has been offering commuters a money-back guarantee of a delay-free ride. Computers connected to the electronic toll-collecting equipment automatically increase toll charges as traffic volume builds. The toll increases may cause drivers to elect to forego a hasslefree commute and use the congested old toll-free highway system.

In another arena of public/private partnerships, the Minneapolis-based Education Alternatives Inc. (EAI) a private corporation that contracts to manage local school systems, had a tough go in 1995. In late December, the last of three top corporate officers resigned and a lucrative contract with the City of Baltimore was cancelled. In Dade County, Florida, school board officials elected not to renew a contract with EAI, and the company is in the midst of a financial dispute with the city of Hartford, Connecticut, where it has a contract to manage the secondary school system. It appears that the concept of an *educational* public/private partnership may be doomed to failure.

Most of the public/private infrastructure partnerships created in the 1970s and 1980s have yet to yield the hard evidence of a series of successfully completed projects since few of the typically 20- to 40-year concession agreements are near maturity. These new public-private partnerships are challenging long-standing institutions though and institutions are slow to change. Perhaps too much has happened too soon; the public may need more time to absorb these institutional changes before it can accept them, or perhaps certain segments of the public/private partnership business need to mature. This book explores a series of Build-Operate-Transfer projects, highlighting the challenges and the methods by which the public and private sectors have responded to these challenges.

Will the public ultimately benefit from this new project delivery system of Build-Operate-Transfer or will the system go the way of the flying shuttle and the spinning jenny? The twenty-first century holds the answer.

BUILD, OPERATE, TRANSFER

1

THE GLOBAL MOVE TOWARD PUBLIC/PRIVATE PARTNERSHIPS

Privatization is a word frequently used during the first half of the 1990s. But what does it mean? When describing events taking place in the former Soviet bloc, or in South America, Africa, and Asia, where previously state-owned industries are being sold to private investors, the term has a readily understood meaning. However, there is another form of privatization, more aptly referred to as "limited term privatization" or perhaps a "public/private partnership," taking place in all parts of the globe today. It involves the investment of private risk capital to design, finance, construct, operate, and maintain a project for public use for a specific term during which a private investment consortium is able to collect revenue from the users of the facility. When the consortium's limited term of ownership expires, title to the project reverts to the government at no cost. By then, the consortium should have collected enough revenue to recapture its investment and turn a profit on the investment. Both forms of privatization have resulted from an acknowledgment by governments that the private sector can in many cases produce a much more cost-effective end product than can the public sector.

The end of the Cold War in the 1980s dramatically illustrated how a centralized dictatorial government maintaining tight control over all phases of economic endeavor, shutting out the entrepreneurial spirit, will eventually fail. The former USSR may have been able to produce weapons of war at a phenomenal rate, but it was unable to produce much else effectively and economically. China's attempt to enter the mainstream of world economics during Mao Ze-dong's Great Leap Forward of 1958 produced nothing but a catastrophic failure resulting in tens of millions of deaths due to mass starvation. Governments the world over, whether they administer emerging, developing, or developed nations, are beginning to embrace the concept that the private sector, spurred on by the profit motive and unencumbered by layers of bureaucracy, can perform certain tasks more efficiently than they can. Timely implementation of projects in the private sector can reap the benefit of greater efficiencies and productivity. The private sector's enhanced decision-making capability allows for more effective use of available human resources.

As the world accelerates into the twenty-first century, social, economic, and political changes as profound as any encountered in the previous 100 years are taking place almost daily. The disintegration of the Russian monolith, the unification of Germany, and the European Economic Community's chrysalis all point to the potential for dramatic social and economic changes in the coming century. In Asia, Japan grows more comfortable with its role as a world power and actively seeks to establish a working relationship with its longtime adversaries, the People's Republic of China and the Republic of South Korea. New "Asian Tigers," such as Taiwan, Thailand, Indonesia, Vietnam, Malaysia, and the Philippines, are on their way to collectively creating an Asian economic powerhouse. The North American continent is seeking to become one giant trading partnership, as evidenced by the implementation of the North American Free Trade Alliance of 1994 (NAFTA). The maturing of South America and the recovery of its economy proceeded slowly but steadily through the 1990s. Only Mexico's devaluation of its peso in the closing moments of 1994 cast a shadow over Latin American economic stability. The emerging market in debt securities, which plunged sharply for several days after Mexico's announced peso devaluation, dramatically revealed how the economy of a nation in one part of the world can have a serious effect on the economy of other nations, some of which halfway round the world.

THE INFRASTRUCTURE CHALLENGE

One characteristic common to mature, developing, and undeveloped nations in today's global economy is the necessity to construct, repair, refurbish, and modernize their infrastructures. Economic infrastructure is made up of public utilities such as power plants, telecommunications, piped water supply, sanitation and sewage facilities, arrangements for solid waste collection and disposal, and gas pipe lines. Public works such as roads, dams, canals, urban and interurban rail systems, ports, waterways, and airports round out the investments known as infrastructure. The composition of a country's infrastructure changes according to income levels as shown in Figure 1.1.

The World Bank, in its *World Development Report 1994,* lists the physical measure of worldwide infrastructure grouped by low-income, middle-income, and high-income economies.

In February of 1995, speaking at the World Economic Forum in Davos, Switzerland, Jerome Monod, Chairman of the European Roundtable of Executives, identified five potential infrastructure growth areas around the world:

1. Emerging markets, such as Vietnam and Malaysia
2. Strong markets with fragile economies, such as Mexico

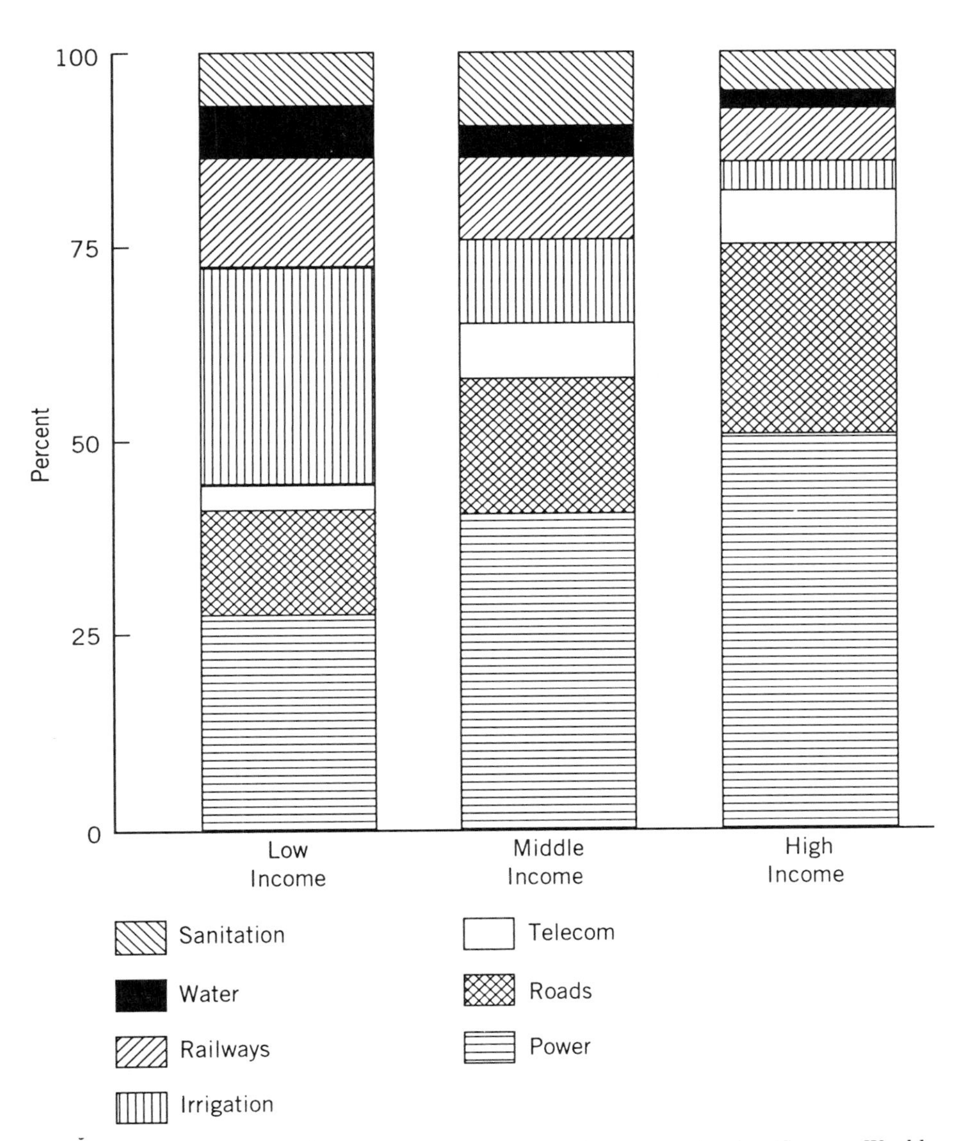

Figure 1.1. Composition of a country's infrastructure by income levels. *[Source: World Bank.]*

3. Reconstructed economies, such as the Eastern European countries
4. Potentially giant economies, such as China, India, and some South American countries
5. Africa, but only if developed nations initiate action, without which very little infrastructure growth activity will occur

Many emerging nations are pursuing an aggressive infrastructure construction program to provide

- Sufficient, reliable power to enable the growth of an industrial base and provide a higher standard of living for their citizens
- Clean water and adequate sewage disposal systems for industrial growth and protection of their countries' populations, but not at the expense of the environment
- Roads, airports, port facilities, and public transportation systems to get both worker and product to market faster and more efficiently
- A link-up of the country via a greatly expanded telecommunications system for both mercantile and residential usage

In middle- and low-income nations, infrastructure expansion has made significant strides in the past three decades, as witness the growth rates shown in Figure 1.2.

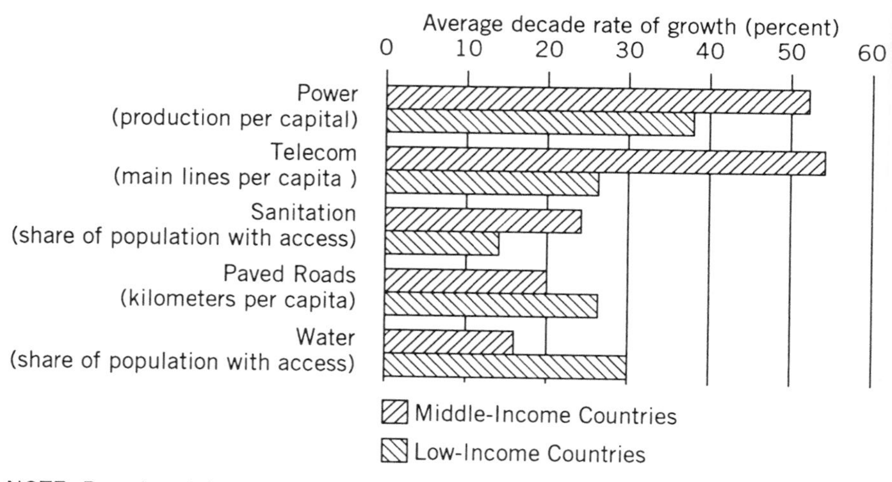

Figure 1.2. Infrastructure expansion in low- and middle-income nations in past three decades. In recent decades, infrastructure has expanded. *[Source: World Bank.]*

Developed nations, faced with the need to upgrade and revitalize their infrastructures in order to compete in the aggressive global marketplace in which they operate, must

- Rehabilitate existing infrastructure and expand the construction of new roads, bridges, tunnels, port facilities, and rapid transportation facilities to speed workers to and from their jobs and reduce the time and cost of transporting goods
- Create a healthy environment, provide clean air and water, and dispose of waste materials efficiently and effectively
- Establish a sophisticated telecommunications system to make optimum use of the worldwide information superhighway

Today, more than 1 billion people in the developing world lack access to clean water; 2 billion don't have adequate sanitation facilities. Electric power has yet to reach 2 billion people, and, in many countries where power is available, its quality is so undependable as to render it almost useless for sustained growth. Approximately 40 percent of all power generated in developing countries is unsuitable for industrial production.

The Need for Electric Power

Today's world is a world of "haves" and "have-nots." Electric power consumption per capita in the United States is 11,333 kilowatts per year (kW/year), the French consume 7,177 kW/year, and the citizens of Hong Kong use 4,418 kW/year. These figures sharply contrast with those for Thailand (711 kW/year), China (562 kW/year), and India, where power usage per capita is only 332 kW/year—not enough to meet the minimum power requirements for a small studio apartment in the United States for more than two or three months.

However, some estimate that energy consumption in developing countries will exceed that of all of Western Europe by the year 2010. And, as the economies and populations of developing countries expand, sources at the Industry and Energy Department of the World Bank have estimated that by the end of the 1990s $100 billion per year will have been spent by these countries to expand their power plants—for a total of $1 trillion. Anthony A. Churchill, Director of World Bank's Industry and Energy Department, estimates that only $150 billion will come from his institution, or from bilateral or other lending agencies with an interest in power plant investments, leaving an $850 billion shortfall in capital funds waiting to be raised by innovative methods. The private sector can be counted on to formulate and implement strategies to take on and complete such formidable tasks, if adequate return on investment is in the cards.

Enron Corporation of Houston, Texas, is one private-sector risk taker betting on reaping profits from public/private–sector partnering in the power generating field. Enron, a company with $11.5 billion in assets, has a number of irons in the fire:

- A $130 million 150-megawatt (MW) Build-Operate-Transfer (BOT) project in the Chinese province of Hainan, off China's southeast coast
- A $1.5 to 2 billion joint venture BOT gas pipeline project with Yacimentos Petroliferos Fiscales Bolivanos (YPFB), the Bolivian state-owned oil and gas company, to export Bolivian natural gas to Brazil and Paraguay
- A 200-MW, $200 million cogeneration plant in Cali, Columbia, and another 200-MW gas power plant and pipeline in Temovaelle, Columbia

One of Enron's major projects, a 2,015-MW joint venture liquefied natural gas power plant in Dabhol, India, valued at $2.5 billion, was abruptly terminated by the government of India in August 1995 ostensibly because the opposition party upon gaining power argued that the project and the electricity it would generate were too costly. The ensuing legal action will cast a pall over future projects in India and create more nervous investors throughout Asia.

Switzerland's Asea Brown Boveri Ltd. (ABB) and U.S. power generating equipment manufacturers Westinghouse and General Electric have entered the field of public/private power generation and are roaming the world searching for opportunities to join with other consortiums. The $60 billion General Electric Company, coupled with its GE Capital affiliate, has become the number one power plant investor for the past five years. Gunnar Johnannesson, ABB's president, said that customers are not merely buying turbines these days; they are buying a saleable product, kilowatt hours.

A Healthy Economy Requires Good Transportation Systems

Transport infrastructure is another key element in a country's ability to achieve economic prominence. Examples from around the world illustrate how transportation infrastructure, or the lack thereof, plays a critical role in determining how well a developing country's economy is capable of prospering. In the former USSR, farmers were able to produce substantial harvests of grain, but it was still necessary to purchase vast amounts of grain from the United States because most of the Russian harvest rotted due to lack of adequate storage facilities and the Russians' inability to transport it rapidly to consumer markets. During the 1980s, India's ports were slow to convert to containerization. This factor, combined with regulatory restrictions imposed by the Indian government, considerably slowed the passage of goods through Indian ports. The costs of many products shipped through these ports to the United States were 33 percent higher than those shipped from Bangkok or Singapore as a result. In China, transport shortages affected the distribution of coal, the fuel consumed by 73 percent of the country's power plants; by 1989, China experienced a power shortfall of 20 percent which, by some estimates, was responsible for reducing that country's gross national product by approximately 1 percent. Several steps have since been taken to alleviate these problems: Two 810-kilometer (500-mile) coal slurry pipelines are under construction in 1995 at a combined cost of U.S.$1.78 billion; new highway construction is now one of China's priorities, with 4,541 km (2,800) miles of roadways in various stages of planning. The cost to construct the extensive highway transportation system in the United States in the 1950s has been justified because 75 percent of the government's investment resulted in reductions in trucking costs, benefiting industry and consumers alike and also contributing to the gross national product.

Telecommunications—Joining the World Together

Telecommunications statistics developed during the early 1990s indicated the disparities among worldwide telephone access:

Country	Number of Telephone Lines per Person
China	0.72
Japan	46.06
Peru	2.70
Thailand	2.80
United States	55.00
Venezuela	8.20

Clearly widespread telecommunications systems are needed not only to provide better voice communications in developing and emerging nations but also to provide the modern data transmission facilities necessary for sustained economic growth.

The private sector investment can be a key to bringing modern service to developing nations. When previously state-owned telecommunications systems have been turned over to private operators, dramatic changes often take place. Venezuela, for example, recognized the need to lift its country out of the Middle Ages of telecommunications, and in 1991 contracted with a private operator who was granted an exclusive franchise for local, long distance and international phone service. By 1993 private investment in that country's telephone system had increased by U.S.$1.1 billion and 850,000 new and replacement lines had been installed.

INFRASTRUCTURE MANAGEMENT AND MAINTENANCE

Any existing infrastructure must be efficiently managed and properly maintained. Such management is often lacking in developing and emerging nations, and that lack of timely and quality maintenance can have a profound effect on the life span of the infrastructure.

An expenditure of U.S.$1 million to curb power generation line losses in some

African countries during the 1980s would have saved U.S.$12 million in generating capacity during that decade. The efficiency of irrigation systems in developing countries is only 25 to 30 percent whereas "best practices" in other countries result in efficiencies in the 40 to 45 percent range. One-third of the U.S.$13 billion expenditure on roads in sub-Saharan Africa has been frittered away because of lack of proper maintenance over the past 20 years. Review of the availability of locomotives in the world's railway systems reveals that 60 percent are available for service in Latin America, 70 percent are available for service in the Middle East and North Africa, while 90 percent are available for service in North America. Forty-two percent of the income from the operation of water and sewer treatment systems in Bogota, Columbia, is lost due to poor maintenance of the distribution system. In Costa Rica inadequate maintenance of the national water system in the early 1990s resulted in an annual loss of income equivalent to 24 percent of the investment planned for the years 1994–1999. By the end of the 1980s, Mexico City's neglect of the water system, along with the failure to increase tariffs to compensate for increased costs, required a federal subsidy equal to 0.6 percent of the country's gross domestic product to keep the system in operation.

Such inefficiencies as these place more strain on the already stretched financial resources of many countries.

Outsourcing Infrastructure Services

Efficient operation and maintenance of publicly owned infrastructure facilities is essential and requires experienced, technically trained staffs. Many public officials, in an attempt to achieve maximum efficiency at the least cost to their agencies, are beginning to look to the private sector for assistance.

The involvement of private concerns in the operation of public infrastructure operation and maintenance is growing rapidly in South America, Europe, and the United States. Fifty percent of Italy's motorways are operated and maintained by a private corporation, Autostrade S.p.A. In 1982, Lyonnaise des Eaux S.A., a $17 billion conglomerate headquartered in Paris, purchased the Delaware-based General Waterworks, a corporation that services 1 million people in 14 states, thereby gaining a foothold in the U.S. water treatment market. Lyonnaise des Eaux had already obtained lucrative contracts to provide drinking water for a large portion of Mexico City; Guangzhou, China; and Buenos Aires, Argentina. In 1993, a Lyonnaise des Eaux–backed consortium was awarded a $72 million, five-year contract to manage the Indianapolis, Indiana, waste treatment facilities. Waste management is also a growing field: in mid-1994 Wheelabrator Technologies reported revenues of $1.14 billion and net income of $163 million from its long-term operation and management of resource recovery facilities throughout the country.

The Mercer Group, a management consulting company based in Atlanta, Georgia, surveyed 82 cities in 34 states in 1987, 1990, and 1995 to determine the extent of municipal outsourcing. A portion of its report containing the percentages of various services contracted to private companies is set forth in Table 1.1.

TABLE 1.1. Percentage of Services Outsourced

Service	1987	1990	1995 (Projected)
Solid waste management	30%	39%	50%
Janitorial services	50	61	70
Building maintenance	35	39	42
Security services	26	33	40
Street repairs and maintenance	18	20	33
Park maintenance	18	25	30
Data processing	15	20	32

Private Operation of Public Schools and State and Federal Prisons

As municipal and state governments throughout the country continue to search for new ways to transfer more services to the private sector to increase quality and lower the costs, private firms are becoming involved in the management and operation of public schools and even prisons. When transferring government-operated facilities to the private sector, a public agency must negotiate a price that serves the public interest while providing sufficient incentives to the private investor to make anticipated rates of return on investment attractive to that investor.

Public School Solutions Many public schools throughout the United States are discovering the advantages of engaging private contractors to provide certain necessary services. For example, a New Jersey school superintendent posted a 62 percent savings on a $4 million budget in 1995 by contracting the district's school bus operations to a private concern. The district saved an additional $500,000 by contracting with a private food service company to run the cafeteria program.

Other public school boards have decided to experiment with even broader forms of outsourcing—by contracting out entire schools. One such private firm that manages schools is Educational Alternatives, Inc., founded in Minneapolis, Minnesota, in 1986. When Educational Alternatives became a publicly held corporation in 1991, it began to acquire contracts with municipal school systems.

Educational Alternatives has managed private schools in Eagan, Minnesota, and Paradise Valley, Arizona, as well as maintained a five-year consulting agreement with Dade County, Florida, to develop and implement a management plan for south Pointe Elementary School.

In the fall of 1994, the Board of Education of the City of Hartford, Connecticut—prompted by financial projections that the yearly educational budget of $171 million was insufficient to deal with student dropout rates and falling student test scores—voted to turn the management of 32 of its secondary schools over to Educational Alternatives. Educational Alternatives planned to invest $1.6 million in the City of Hartford school system during the first year of its contract and to spend another $14 million for teaching and administration technology upgrades. The contract called for the company to keep 50 percent of all savings effected through any created efficiencies.

The jury is still out on Educational Alternatives' business, and it will be several years before any definitive analysis of its modus operandi can be made.

The Private Prison Industry America's federal and state prison population rose from approximately 50,000 inmates in 1910 to slightly more than 200,000 in 1970 before skyrocketing to 1 million in 1994. State and federal corrections officers, unable to cope with the burgeoning prison population, began contracting prison management services to private concerns in the late 1980s. By 1989, slightly more than 10,000 prison beds were under contract to private corrections companies, a number has grown steadily—15,000 in 1991, 20,000 in 1992, and 32,000 in 1993. By the end of 1994, the number of prisoners incarcerated in privately run prisons operating in 14 states had increased by 51 percent to 49,000. A higher percentage of private prisons have won accreditation from the Amercan Correctional Association than have government-operated facilities.

In the area of controlling costs, the private sector has been far more capable than government agencies. A private prison has operating costs ranging from 15 to 30 percent lower than a comparable publicly operated facility. And the private sector has the ability to construct a prison more quickly and more economically than the government. According to the figures made public in 1994 by the Reason Foundation, private firms can build a prison in two years versus the four-year time frame required by the government; the private firm cost of $50,000 per cell is 60 percent less than the $80,000 average cost incurred when government construction is involved.

One of the leaders in this field is Corrections Corporation. This private prison operator receives a per diem fee for each prisoner under its control and derives profits by operating the prison less expensively than its state or federally provided budget. The *New York Times,* on August 14, 1994, reported that Corrections Corporation had 1993 income 30 percent higher than the previous year. An analyst from Equitable Securities predicted that the income derived from the 13,000 beds under Corrections Corporation's control could increase by 85 percent between 1994 and 1996 and its profits could double. So it would appear that crime does pay.

ROADBLOCKS TO PUBLIC PLACEMENT OF INFRASTRUCTURE

Infrastructure services—transportation, power, telecommunications, clean water, and hygienic disposal of wastes—are central to the development and well-being of a country. Although it may be the obligation of governments throughout the world to ensure that their citizens are provided with adequate infrastructure, many factors enter into a government's inability to do so.

First of all, areas of rapidly expanding population in some parts of the world strain a government's ability to build new or expand existing infrastructure facilities while preserving the environment. For example, China has experienced an economic growth unmatched by any other country—10 percent average annual growth over the last decade. But China's phenomenal growth has been at the expense of its environment. Sun-fed algae blankets many of the country's large lakes, air pollution over parts of Manchuria is so thick that satellite photos can't penetrate to the ground, and parts of Shanghai are sinking because the water table has been depleted. China, now recognizing that it has abused its environment, will be diverting much-needed funds from infrastructure construction to cleaning up the mess that years of neglect have created. Companies or consortiums capable of providing up-front money to purchase the technology and equipment to clean up the environmentally despoiled areas will have a field day in China once they are assured that they can turn a decent profit and easily get their money out of the country.

Second, growing world debt and the resources available to financial institutions throughout the world will divert investment money to those markets that appear to offer the lowest risks and yield the highest returns. Thus, the world's leading financial institutions do not look favorably upon loans to the "Silics" (severely indebted, low-income nations), the bulk of which exist in sub-Saharan Africa, and which, according to a United Nations report, already owe $179.4 billion, mostly in the form of government-to-government loans.

Third, citizen resistance is increasing to the imposition of added taxes as a means of obtaining more money for a variety of government projects. And citizen reaction to government waste, whether real or apparent, is also becoming more vituperative. Political officials throughout the United States are acutely aware of voter dissatisfaction and opposition to tax increases. California's experience is a case in point. The passage of Proposition 13 the 1978 property tax slashing initiative in California presaged the voter's unwillingness to stand idly by while politicians continued to call for increased taxes. An even stronger grass-roots insurgency took place in that state in June of 1994, when California voters defeated approximately $6 billion in bond issues intended for previously popular purposes such as school renovations, acquisitions of parklands, earthquake recovery, and state college improvements. California voters told the politicians in Sacramento that they were fed up with mediocre services at high costs, and they no longer gave carte blanche spending approval. Unsurprisingly, many U.S. elected officials won their seats in 1995 by promising to maintain or even improve certain government services lowering or not increasing taxes. One such high-profile, newly elected official is Christine Todd Whitman, the Republican governor of New Jersey. On January 23, 1995, she announced her plan to reduce residents' tax burdens, as she had promised during her campaign: she would reduce taxes on families with an annual income of $85,000 or less by laying off 3,400 state employees whose jobs were no longer necessary due to the privatization of their agencies.

The Skunk Works Paradigm In his recent book published by Little, Brown, Ben Rich, former CEO of Lockheed's famous Skunk Works, relates a classic example of government bureaucracy adding considerable cost to a project. Rich's company was involved in building super-secret airplanes such as the U-2 spyplane, the F-111 Stealth fighter-bomber, and the SR-71, still considered the fastest aircraft in the world. According to Rich, General Electric's jet engine plant in Evendale, Ohio, sold its engines to commercial airlines for approximately 20 percent less than those sold to the United States Air Force (USAF). The USAF would station 300 inspectors at GE's plant to inspect the engines it had purchased; commercial buyers had no representatives at the manufacturer's plant as those customers expected to receive the quality for which they paid. Mr. Rich also discusses expenditures for servicing the supersonic Air Force SR-71 Blackbird: When two of these aircraft were stationed in England in the 1970s, they were serviced by a Skunk Works' maintenance crew of 35—but the USAF assigned 600 people to handle the two SR-71s stationed on Okinawa.

PUBLIC PROCUREMENT ALTERNATIVES

When times are tough, capital spending for infrastructure projects is an early victim of government cutbacks. Repair, maintenance, and plant upgrades are close behind, omissions that will probably result in much higher costs in the future. In fact, the World Bank reported that timely maintenance of some existing highway systems in Africa would have cost U.S.$12 billion over the past decade whereas reconstruction costs will amount to U.S.$45 billion. The number of bridges in the United States in critical need of repair will continue to rise unless $50.7 billion is spent on repairs and improvements. A review of power generating plants in 51 developing countries by the World Bank revealed that older plants consume between 18 and 44 percent more fuel per kilowatt hour than do power plants operating at best practices levels.

Mature economies, developing countries, and emerging markets all seek, each in its own way, methods to achieve common goals—economic growth, political independence, and more opportunities and higher standards of living for their citizens. To these ends, more and more countries are looking to the private sector to assist them in constructing the necessary infrastructure. There are movements afoot in all parts of the world to cast aside the old ways of doing business and get on with the task of providing people with not only the basics for survival, but also opportunities for improvement of the individual's well-being. Developing and emerging nations, eager to join the bandwagon of economic growth, are desperately seeking funds so that they can begin to climb the long road to political and economic stability. To this end, partnership development activity on the needed infrastructure projects has been accelerating rapidly.

According to *Public Works Financing* (PWF), the authoritative newsletter on public/private infrastructure partnerships, from the mid-1980s to late 1993, 115 of these new types of transportation, power, and water treatment projects had been financed, were under under construction, or had been completed. The total value of these projects was estimated to be $58.4 billion. But when *PWF* published its survey of public/private infrastructure projects in October 1993, it revealed that 319 projects with a value of $227.6 billion were under development (see Table 1.2.). Just one year later, PWF's survey included 788 power, transportation, and water projects in various stages of development with a total value of $534 billion (Table 1.3).

TABLE 1.2. New Public/Private Infrastructure Projects Reported in 1993

	Projects Underway					
	Transportation Number	Value (U.S.$ millions)	Power Number	Value (U.S.$ millions)	Water Number	Value (U.S.$ millions)
North America	35	$ 9,981	2	$ 643	12	$1,222
Caribbean and Latin America	9	3,032	5	450	1	3,000
Europe	9	17,184	2	1,200	1	240
Mideast	0	0	1	2,000	1	700
Central Asia	0	0	0	0	0	0
Far East	9	6,206	3	4,950	1	25
Pacific Rim	13	4,062	7	3,112	4	444
Total	75	$40,466	20	$12,355	20	$5,631

	Projects Under Development					
	Transportation Number	Value (U.S.$ millions)	Power Number	Value (U.S.$ millions)	Water Number	Value (U.S.$ millions)
North America	59	$ 23,445	6	$ 6,775	8	$ 75
Caribbean and Latin America	44	10,420	15	6,106	7	3,316
Europe	50	49,182	7	8,140	2	25
Mideast	5	2,420	5	10,740	2	450
Central Asia	9	4,094	7	7,333	0	0
Far East	44	47,194	16	15,107	3	3,090
Pacific Rim	17	20,890	8	5,583	5	3,226
Total	228	$157,645	64	$59,784	27	$10,182

Source: *Public Works Financing*, (Westfield, NJ) October, 1993.

TABLE 1.3. Public-Private Infrastructure Projects as of October, 1994

	Power		Transportation		Water	
	Number of Projects	Cost (U.S.$ millions)	Number of Projects	Cost (U.S.$ millions)	Number of Projects	Cost (U.S.$ millions)
North America						
United States of America	6*	$ 8,793*	47	$ 16,599	16	$ 650
Mexico	4	1,680	69	13,617	29	2,963
Canada			8	4,765	5	766
Puerto Rico	1	500	5	1,123	1	300
Total	11	$ 10,973	129	$ 36,104	51	$4,679
Financed as of 10/94	2	$ 643	73	$ 17,415	25	$2,345
Latin America and the Caribbean						
Brazil	7	$ 5,995	12	$ 5,445	3	$ 3,006
Argentina	6	861	30	4,370	4	4,350
Costa Rica			3	2,724		
Chile	4	895	12	671		
Colombia	6	1,220	3	256	1	50
Uruguay			1	1,000		
Panama	1		3	675		
Venezuela	2	135	3	387	2	100
Ecuador	1	220	1			
Jamaica	1	181				
Guatemala	1	92	1	69		
Trinidad & Tobago			1	80		
Belize	1	60				
Honduras	1		1	50		
Peru					1	
Total	31	$ 9,659	71	$ 15,727	11	$ 7,506
Financed as of 10/94	9	$ 901	24	$ 4,010	1	$ 3,000
Europe						
United Kingdom	4	$ 1,385	27	$ 27,214		
France			8	23,881		
Czech Republic			4	20,680		
Italy	2	1,000	1	18,000		
Germany			6	16,178	1	240
Spain			10	11,069	1	82
Greece	2	4,500	9	5,883		
Poland	1	550	14	7,000	1	364
Sweden			5	6,900		
Portugal	3	1,560	3	3,749		
Hungary			7	3,115	1	25
Romania			1	3,000		
Holland	2	620	3	720	1	40
Ireland	1	460	2	52		

TABLE 1.3. *(Continued)*

	Power		Transportation		Water	
	Number of Projects	Cost (U.S.$ millions)	Number of Projects	Cost (U.S.$ millions)	Number of Projects	Cost (U.S.$ millions)
Yugoslav Republic			1	330		
Bulgaria			3	290		
Iceland			1	60		
Albania			1	44		
Estonia			1	30		
Belarus			1			
Total	15	$ 10,075	108	$148,195	5	$ 751
Financed as of 10/94	2	$ 1,200	20	$ 21,589	2	$ 280
Africa & Mid-East						
Turkey	5	$ 2,308	5	$ 3,900	2	$ 700
Israel			2	2,120	3	
South Africa			1	952		
Qatar			1	350	1	450
Mali, Mauritania, and Senegal	1	746				
Oman	1	240				
Iran			1	100		
Total	7	$ 3,294	10	$ 7,422	6	$ 1,150
Financed as of 10/04	1	$ 1,005	1	$ 952		
Central Asia						
Pakistan	29	$ 24,700	8	$ 1,047	1	$ 96
India	10	8,581	10	1,471		
Russia			9	10,050	2	
Nepal	2	120				
Sri Lanka	1					
Total	42	$ 33,401	27	$ 12,568	3	$ 96
Financed as of 10/94	1	$ 1,700	1	$ 2		
East Asia						
China	38	$ 28,992	82	$ 45,363	5	$ 956
Taiwan	1	22,000	1	17,000		
Hong Kong	1	3,900	11	16,997	1	3,100
South Korea			1	13,200		
Thailand	1	615	6	7,850	1	250
Japan			1	7,700		
Vietnam	4	1,405	4	1,357		
Laos	1	700				
Macao			1	577	1	25
Myanmar	3					
Total	49	$ 57,612	107	$110,044	8	$ 4,331
Financed as of 10/94	5	$ 5,340	13	$ 8,038	2	$ 198

TABLE 1.3. *(Continued)*

	Power		Transportation		Water	
	Number of Projects	Cost (U.S.$ millions)	Number of Projects	Cost (U.S.$ millions)	Number of Projects	Cost (U.S.$ millions)
Pacific Rim						
Malaysia	8	$ 18,799	14	$ 7,426	4	$ 3,676
Indonesia	8	6,568	7	4,473		
Philippines	24	7,150	6	3,472	1	40
Australia	6	2,105	14	6,504	4	286
New Zealand					1	
Total	46	$ 34,622	41	$ 21,875	10	$ 4,002
Financed as of 10/94	5	$ 5,063	15	$ 4,339	5	$ 476
Worldwide						
Total	201	$159,636	493	$351,935	94	$22,515
Financed as of 10/94	25	$ 15,852	147	$ 56,345	35	$ 6,299

Source: Public Works Financing, October 1994.

* *Pumped storage IPP projects only*

There are many innovative project procurement methods being considered today that may suit the needs of government and voters alike.

- *Super Turn-key.* In addition to the conventional "turn-key" process, the proposed developer receives real estate development rights along the projected right-of-way of the proposed project. The developer is expected to contribute in some manner to the financing requirements of the project in exchange for these rights.
- *Contract Service.* A private company provides services for a public project, such as the operation of a waste water treatment facility.
- *Developer Financing or Proffer.* In exchange for a real estate development entitlement, a private developer will finance the construction of a new public facility, or finance the expansion of an existing one.
- *Build, Operate, Transfer.* BOT variations are explained in the next section.
- *Design, Build, Finance, Operate.* DBFO is explained in the next section.

THE BUILD, OPERATE, TRANSFER APPROACH

The BOT approach—sometimes referred to as BOOT (Build, Own, Operate, Transfer)—involves the assembling of private sponsors, usually a consortium of private companies, to finance, design, build, operate, and maintain some form of revenue-producing infrastructure project for a specific period. At the end of this concessionary period, when it has been estimated that all investment costs have been recouped from user fees and a profit turned, title to the project passes from the private consortium to the host government. The BOT theme has several variations:

BOO Build, Own, Operate (without any obligation to transfer ownership)

BTO Build, Transfer, Operate (a method of relieving the consortium of furnishing high-cost insurance required by the project during the operation of the facility)

BRT Build, Rent, Transfer

BOOST Build, Own, Operate, Subsidize, Transfer

DBFO Design, Build, Finance, Operate (similar to BTO, the government will retain title to the land and lease it to the private concern over the life of the concessionary agreement)

In a BOT project, the private-sector sponsors generally provide equity financing in the amount of 10 to 30 percent of the total project cost and seek debt financing for the balance of the investment. The host government may, on occasion, furnish a portion or all of the land required for the project or possibly grant partial tax relief in some form or another. However, many BOT projects are structured without any financial or other form of assistance from the host government.

When the downturn in international construction took place in the late 1970s, many globally operating construction companies saw an opportunity to obtain construction contracts through the innovative BOT method. By teaming up with an engineering firm that was also seeking work, a construction company could approach investors and developers with an eye to both forming a consortium and presenting a plan for a proposed infrastructure project to a receptive government. In many other cases, requests for submission of BOT proposals emanated from the host government to relieve pressure on tight budgets. Because the risks associated with these projects were high, so were the potential rewards. And, even as BOT consortiums were creating their own construction projects from which normal profits could be earned, they were opening the door to other ancillary projects—and corresponding profits—for investors who were quick to recognize the potential. Some agreements with government agencies allowed developer return on investments to exceed 25 percent.

The World Bank, international financial institutions, commercial banks, and some government agencies such as Japan's Overseas Economic Cooperation Fund (OECF) and Official Development Assistance (ODA) agency were willing to provide loans to many Asian neighbors for the purpose of conducting feasibility studies for potential infrastructure projects. In the case of Japan, government and industry work closely together. One or more Japanese engineering or construction firms with backing from a Japanese government agency would conduct feasibility studies for another government's projects. As a matter of course, these same designers or contractors might well end up with a BOT contract should the host government agree to advance that project.

The Role of the U.S. Overseas Private Investment Corporation and BOT

The Overseas Private Investment Corporation (OPIC) was created in 1971, under the provisions of Foreign Assistance Act of 1961, to facilitate U.S. private investment in developing countries and emerging market economies. OPIC is a quasi-government agency somewhat similar to the Tennessee Valley Authority (TVA) and is subject to financial decisions and management controls of the Office of Management and Budget (OMB). The OPIC's mission is to facilitate American investments in international markets through several means:

1. Financing direct business loans or loan guarantees for new ventures or for the expansion or modernization of existing operations. It can provide direct loans in the $2 million to $30 million range to small businesses, and loan guarantees in the $10 million to $200 million range for larger projects.
2. Insuring investments against a broad range of political risks such as currency inconvertibility, expropriation, and political violence.
3. Providing a wide variety of investment services, such as conducting investment missions to selected countries, acting as liaison between U.S. business executives and government officials in those countries, and bringing foreign officials to the United States to meet with American business people.

Although there has been criticism leveled at OPIC for using taxpayer money to assist big business expand overseas, OPIC is self-sustaining and turns a profit. Its 1994 annual report reflects revenues of $220 million, expenses of $53 million, and net income of $167 million—a $7 million increase over 1993's $160 million net income.

OPIC's clients include a host of large companies such as Air Products; American Express; Bechtel Group; Louis Berger Group; Crown, Cork and Seal; Del Monte Foods; Enron Corporation; FMC Corporation; GTE; ITT; and a number of construction and engineering firms operating internationally. In 1994, OPIC financed two power plants in Turkey, one involving an $85 million loan to allow a BOT project to proceed and the other a $545 million cofinanced loan to Enron Corporation and Wing International for a 500 MW combined-cycle gas turbine facility near Istanbul. In 1993, OPIC provided $435 million in financing and $100 million in political risk insurance to K&M Engineering to structure a BOOT natural-gas–fired generating facility in Cartagena, Columbia, along with several other infrastructure projects in Jordan, Nicaragua, and St. Petersburg, Russia.

THE HISTORY OF BUILD, OPERATE, TRANSFER

Although the term Build, Operate, Transfer is relatively new, the practice of permitting private concerns to develop and operate infrastructure projects has been around for several centuries. In Europe, such projects were called "concessions" whereby the government would establish the major objective of a particular project and assume the role of defender of the public interest, but would allow a private company or consortium to design, finance, construct, and operate the particular project for a certain concessionary period. The concessionaire would assume responsibility for the completeness of design, any risks associated with construction, and the control of operational costs, all of which would be reimbursed by the collection of revenue from the public making use of the finished project. The concessionaire would be granted a specific concessionary period, after which time the contract might be renewed at the option of the government, or title would be transferred from the concessionaire to a government agency.

The First BOT Project in the Modern World: The Suez Canal

In 1834, British engineer Galloway Bey proposed linking the Mediterranean Sea with the Red Sea via a railroad line. The English, who at that time were world leaders in railroad construction, found this scheme appealing. Over the next several years, various French and English business officials, working with their respective government officials, discussed methods by which the two seas could be linked for commercial exploitation, and both groups sought to enhance their respective positions with the Ottoman Empire that ruled the Middle East.

Frenchman Barthelemy Prosper Enfantin formed the Société d'Études du Canal du Suez in 1847, a triparte group of French, British, and Austrian members charged with investigating the feasibility of constructing a canal to link the Mediterranean and Red seas. This canal, to be called the Suez Canal, was to be the world's first great international enterprise of the nineteenth century. It was to be financed by European capital with Egyptian financial support, and a concession to design, construct, and operate this revenue producing facility was expected from the Egyptian ruler at that time, Pasha Muhammad Ali.

Ferdinand de Lesseps was given the responsibility for forming the company to build and operate the canal, and on November 30, 1854, his Compagnie Universelle du Canal Maritime de Suez obtained the concession from the Egyptian government. The concessionary period was for 99 years. The land was donated by the Egyptian government which relaxed many of the taxes generally levied on imported materials and equipment. The Compagnie Universelle du Canal Maritime de Suez was to pay all costs for design and construction. The schedule of tolls for the canal would be set by the company after consultation with Viceroy Muhammad Ali, and there was to be no preferential rate treatment—all nations would be required to pay the same schedule of rates. The government of Egypt was to receive 15 percent of the company's annual profits from the canal's operation, 10 percent of the profits would be paid out to the initial stockholders, and the balance of 75 percent would accrue to the company. Also, in what might have been the first affirmative action plan, the concession agreement was modified in January 1856 to include, among other provisions, a requirement that at least 80 percent of all workers employed on the project must be Egyptian.

Construction began in 1859 and was scheduled for completion in 1864 at a total cost of £8 million, a sum that grossly underestimated the real cost of the project. In

1863 British engineer Sir John Hawkshaw projected the eventual total cost would be £10 million when the project was to be completed in 1868—four years later that originally anticipated. The substantial cost overruns would eventually result in the title to the project being transferred from French to British hands.

The Viceroy of Egypt accumulated so much debt in the course of financing a substantial portion of the canal that by 1875 the interest on the debt alone exceeded the gross income of his entire country—and he went looking for a buyer. The Dervieu Brothers of Paris and Alexandria, and a group associated with French concern Credit Foncier sought assistance from the French government to purchase the canal. The French government, showing some lack of foresight, saw no need to assist the Dervieu Brothers in their quest to purchase the canal. The British used the canal considerably more than the French, so English interest in ownership was high; the British government stepped in and bought the Egyptian Viceroy's shares. Shortly thereafter, the remaining shares from Ferdinand de Lesseps' company were also purchased by the British.

Although the Suez Canal ultimately cost £18 million when the 10-year construction time ended, it proved to be a bargain. By the end of the first decade of the twentieth century, Great Britain's investment had increased in value tenfold.

Privatization Fever

In the United States, the construction of privately built and operated toll roads was commonplace during Revolutionary days, as were privately owned and operated ferries and even jails. But "privatization fever" truly hit the United States in the 1980s and 1990s when newspapers began to announce the privatization of prisons and school systems. And, on a larger scale, Build, Operate, Transfer projects encompass the full range of infrastructure projects—roads, bridges, tunnels, power generating plants, and water and waste treatment plants—and are in process in all parts of the globe.

The U.S. federal government today acknowledges BOT and encourages the implementation of BOT-related transportation projects. In a speech to a group of contractors attending the Associated General Contractors of America convention in Las Vegas, Nevada, in March 1993, John Randolph, Jr., of the Major Projects Branch of the U.S. Department of Commerce made the following comments about this concept:

> I don't have to tell you BOT (Build-Operate-Transfer), sometimes expanded to BOOT (Build-Own-Operate-Transfer), is the buzzword of the nineties.
>
> I first heard the term used in connection with some larger power plants Turkey wanted to build in the mid-1980s. With the emphasis on privatization, BOT, or versions of it, is the wave of the future. Private sector construction and ownership of power plants and toll roads are nothing new. Water purification and sewage treatment may be next.
>
> Despite all of the attention to BOT, putting these projects together is an inexact science. Exposure to risk is greater, particularly in countries with weak currencies and governments. BOTs are even hard to put together in the United States. Again, financ-

ing is critical. Who's going to take the risks? Most official export credits are based on sovereign country risk. A government either borrows the money or provides a guarantee of repayment. The whole purpose of BOT and priviatization is to reduce government borrowing and exposure.

Many if not most major international projects are beyond the scope of any one contractor or even country. They require multinational consortia with a wide range of finance and insurance options. One acronym I've seen for the international contractor of the future is AFLEEECS—an organization of architects, financiers, lawyers, economists, engineers, environmentalists, and constructors.

Commerce Secretary Ronald Brown saw BOT as a primary way to make U.S. business more competitive on an international basis. During a trip to China in 1994, Brown was quoted by the *New York Times,* on August 9, as stating that he hoped to persuade the Chinese to make broader use of the "build, operate and transfer method" to finance major construction projects such as airports and power plants. According to the *Times* article, Brown said that the United States could use BOT to its advantage and that American companies could walk away winners. The United States has indeed taken a number of steps to provide direct and indirect entities to support the privatization movement.

U.S.-Supported Financial Institutions

Augmenting the Overseas Private Investment Corporation (OPIC) is the U.S.-supported World Bank and its branches. Seventeen percent of the World Bank's capitalization is supplied by the United States. The World Bank's mission is to assist developing countries in raising their standards of living. To accomplish this goal, the Bank has four aims:

1. International Bank for Reconstruction and Development (IBRD) This bank borrows money in the international capital market and loans this money to commercial banks in developing countries.
2. Export-Import Bank (EXIM) This bank assists in the financing of sales and goods and services overseas. It guarantees working capital loans to exporters and provides U.S. companies with credit insurance against nonpayment by foreign purchasers of the goods and services sold by these companies.
3. International Monetary Fund (IMF) This organization strives to correct economic imbalances such as inflation and plunging currencies.
4. European Bank for Reconstruction and Development (EBRD) This arm of the World Bank promotes privatization of state-owned businesses and promotes technology advancement in Europe.

Where the Market Is

Continuing economic progress depends upon the development, growth, and preservation of a country's infrastructure. According to World Bank statistics, developing countries invest $200 billion each year in new infrastructure. The Asian market is where most of the big dollars will be spent on infrastructure development in the next two decades. South Korea plans to spend $400 billion on infrastructure, Taiwan about $245 billion, and China $92 billion between 1995 and the end of the century. India has embarked on a $12 billion, 10,560 MW power generating plant. Another study indicates that East Asia will require at least $1 trillion in infrastructure investment by the year 2000.

Even the United States with its mature infrastructure is not immune to the spreading movement toward a private/public alliance. With the U.S. Department of Transportation's announcement that it will need to spend $51.6 billion annually just to maintain the condition and performance of the federal highway system, and another $16.7 billion each year to improve the system, it is no wonder that the Intermodal Surface Transportation Efficiency Act allows state governments to enter into contracts with private consortiums to design, build, finance, operate, and maintain roads, bridges, and tunnel projects within their respective sovereign domains for specific times. The vehicle for many of these private initiatives may be found in the BOT concept.

Developed and undeveloped nations are embracing the BOT concept for different reasons, but the end result is the same—private consortiums assembling extremely complex construction projects for public usage, anticipating that the very nature of these complicated projects could yield high returns on investment over the life of that investment. But can the private sector produce a complex project in less time, with less money, and can the public be the beneficiary of its efforts? Where these completed projects have opened for public use in all parts of the world, the answer has been a resounding Yes!

BASIC CHARACTERISTICS OF A BUILD, OPERATE, TRANSFER PROJECT

A BOT project requires a financially viable project, receptive host government, private sponsors, local partners, and a group of experienced professionals.

The Host Government The host government must be fully committed to the project and, in most cases, enact legislation that will permit the BOT project to form and operate within its assigned jurisdiction. The host government must be able to provide bureaucratic support for the project, not only at conception but throughout the concessionary period. Government negotiators must posses the necessary skills or contract with consultants who have the requisite skills and knowledge to comprehend, analyze, and evaluate a complex project set out before them. In some cases, the host government will be the sole purchaser of servces provided by the BOT consortium and must therefore be able to enter into a long-term contract in good faith with the consortium. In the case where the public, such as users of a toll road or toll bridge, will be the ultimate purchaser of the services provided by the BOT team, the host government must have ample resources to take over the project in case of default.

Private Sponsors Historically, private sponsorship of a large BOT project has been an organization composed of a large international construction company or several of these companies, an international engineering firm or firms, as well as several specialized engineering firms. Oftentimes a capital equipment manufacturer, such as General Electric, Asea Brown Boveri, or Seimens, will participate, particularly when power plant construction is involved. Lending institutions, insurers, and other types of equity investors—private and government-backed—will round out the team.

Local Partners Some host countries require the use of local labor, local contractors, and local products if practical, and BOT sponsors also recognize the value of having a respected local participant or participants on their team. A strong proponent of this view is Toichi Takenaka, president of Takenaka Corporation of Tokyo, Japan, a 385-year-old firm that is the world's oldest continuously operating construction company. His company's philosophy is "Think globally, Act locally," a concept that has worked well for his firm for almost four centuries. If the local team member happens to be politically well connected, that can only be viewed as a major plus.

Construction Consortiums A typical BOT project is generally rather large and complex; it usually requires participating construction companies to assume some degree of the project's risk. Civil engineering plays a major role in these projects; depending upon where the project is located, a local civil engineering firm that is knowledgeable in local site conditions, equipment, and manpower availability will find ready acceptance as a member of the design and construction team. If the construction is of a specialized nature then a construction company experienced in that particular field will be employed. Large, multinational, multidisciplined projects require international contractors with the experience, know-how, and organizational and administrative skills to control these processes.

Financing

BOT projects are not the arena in which untested or untried technologies should be applied; state-of-the-art technology is welcome, but unproven technology will make investors skittish. Unless the financial viability of the project over its entire life can be clearly demonstrated, equity investors and other long-term investors will be unwilling to provide the amount of funding required at competitive interest rates. The integrity of the cost of construction must be sound and the revenue stream must be realistically calculated.

Equity financing is generally provided by the members of the BOT project whereas debt financing is obtained from commercial banks, international financial institutions, or bilateral government lenders. Typically, equity financing ranges from 10 to 30 percent of the total project cost. Nonrecourse debt ensures that lenders will have no recourse to the project developers; instead, they must rely solely on the revenue generated by the project as their source of repayment. If a project is well structured, international banks will provide funding for 15 years and beyond. Commercial banks are a more flexible source of funds, but they generally require early repayment and may not be the most appropriate for use on these long-term projects.

The Advantages of a BOT Offering

To the Host Government:

- BOT allows a government to obtain a much-needed infrastructure project at little or no cost to taxpayers.
- The government will incur little or no risk as there are generally sufficient bonds in place and sufficient letters of credit in hand to insure completion of the project in the event that the sponsors default prior to project completion.
- Because the private sector can usually move the preconstruction activities along more rapidly than government agencies, the project will probably progress from concept to construction faster.
- Since the construction activities will be those of a private-sector operation rather than a public-sector one, the construction cycle will be more rapid.
- The sponsors must operate and maintain the facility for periods of time generally exceeding 20 years, so the chances are good that initial quality will be high.

To the Citizens of the Host Country:

- General taxes will not have to be increased, nor will revenue bonds have to be issued to pay for the project.
- When projects receive income from tolls, only the users of the BOT facilities will be required to pay for them. Citizens can elect to use alternate routes if toll rates are not reasonable, which exerts pressure on the consortium to keep toll rates as low as possible.

To the BOT Consortium: The construction and engineering firms and equipment manufacturers have created their own market for their services and products and anticipate making a profit. Some or all of the consortium's members may receive profits from the operation of the project; if their market analysis was correct, these profits could be substantial. Also, by purchasing the land adjacent to the project, these concessionaires can make the most of that land's increased value.

With reward come the risks to which no BOT project is immune. Any number of problems can arise before, during, or after construction of the project:

- Political instability in the host country is a concern at all stages of a BOT project. Since most concessions range from 20 to 40 years, political stability must be equally long range. The cancellation of a power plant, for example, will cost the developers millions of dollars if they cannot collect damages from the new regime.

- Cost overruns, if significant, may change a project's entire pro forma. If additional financing is required and not forthcoming in a timely fashion, project progress may grind to a halt or end in default. The Channel Tunnel is a classic example of cost overuns: An initial budget of $9.7 billion ballooned to almost $16 billion. The project has been kept alive as much by the determination of the British and French governments as by the commitments made by private developers and financial institutions.
- Unfavorable exchange rate fluctuations can place an undue burden on the consortium which must pay back loans with devalued revenue. The 20 percent devaluation of the Mexican peso in December 1994 reverberated through many markets—a warning that these sorts of things don't happen only in third-world countries.
- Consumers may balk at paying the infrastructure usage fees set by the host government, in conjunction with the concessionaire. Toll rates for concession type highways, set by the Mexican government, for example, were about eight times higher per mile than comparable rates in the United States, which resulted in increased toll jumping. The San Jose Lagoon Bridge in Puerto Rico also suffered from a paucity of users because tolls were set too high.
- Drastic changes in demographics over the concessionary period may substantially affect revenue, as a BOT consortium's source of revenue is based upon a specific number of consumers who are projected to use the facility. Although such changes can be favorable—England's Dartford River Crossing is generating much higher revenue than projected, other changes can be disasterous—a Hong Kong tunnel project is suffering from undersubscription.

Mexico's devaluation reverberated through many markets near and far. As the 1994 bankruptcy of Orange County, California—one of the richest counties in the United States—has made clear, no government agency, be it local, state, or central, is an absolutely secure partner. Risk is an intrinsic part of the BOT concept.

Faced with limited sources of funding, developed and developing nations alike are seeking innovative ways to promote the creation of much needed infrastructure projects. Governments the world over are realizing that "profit" and "reasonable return on investment" are not dirty words, but are the lifeblood of the private sector and are necessary to convince investors to assume the risks that go with reward. The BOT vehicle is attractive enough to have gained the attention of scores of countries and developers throughout the world, and projects are underway in Great Britain, France, Portugal, and the United States, to name a but few mature economies. This book takes an in-depth look at public/private projects from all parts of the globe, revealing both the strengths and weaknesses of what is still a relatively untested, but promising, venture.

2

TOLL ROADS IN AMERICA— AN OLD CONCEPT UPDATED

The concept of the toll road in the United States dates back to the late 1700s. By the middle of the 1800s, approximately 15,000 miles of toll roads had been built. Many of these roads were constructed with some federal assistance in the form of land grants or subsidies, while others were built by wealthy businessmen who formed stock-based companies where investors could receive returns on their investment from collected tolls. During the first four decades of the nineteenth century, more than 450 chartered toll roads were built throughout New England the area of major mercantile and financial activity at that time. The federal government was also engaged in road construction during that time, and, between private and public road construction, a fledgling highway system emerged to link communities together.

Alternative modes of transport were also under development, however. In 1825, the first experimental steam locomotive, built by John Stevens in Hoboken, New Jersey, ushered in the era of the railroad. Coincidentally, the opening of the Erie Canal in that same year provided merchants and travelers with yet another mode of transportation. The canal barge, pulled by teams of horses or mules on tow paths adjacent to the canal, cut travel time between Buffalo and New York City by one-third and was instrumental in opening up the Great Lakes to commerce. During the later part of the nineteenth century railway and canal construction took precedence over highway construction, and it was not until the commercial production of the internal combustion engine and Henry Ford's introduction of the Model T in 1908 that the idea of building a highway system throughout the country would capture the attention of the American public.

The first Federal Highway Act in 1916 established a federal highway policy. By 1920, the Federal Bureau of Public Roads was in operation and a federal standard for road construction dictated the specifications to be followed by states seeking federal aid in developing their highway systems. States requesting federal assistance in the construction of their intrastate road systems began to levy taxes on gasoline to help pay for this construction and for the maintenance of these roads. Although toll charges were allowed on tunnels and bridges, state roads built in the mid-1930s with federal aid were prohibited from collecting tolls except in special situations. The construction of the Merritt Parkway in Connecticut in 1937 became one such special situation: Traffic on this road—once proclaimed the most beautiful road in America—was limited to passenger car traffic only. The densely populated Northeast Corridor was thus the spawning ground for a interconnecting system of toll roads, but the sparsely populated western part of the country relied mainly on toll-free roads, logically dubbed freeways, financed by gasoline and other motor vehicle related taxes.

When World War II created an awareness of the need to create a high-speed, national highway system linking strategic parts of the country together for passenger, freight, and military travel, the Federal-Aid Highway Act of 1956 came into being. The Highway Trust Fund was created whereby proceeds from a federal tax on gasoline, tires, and other motor vehicle parts would be deposited with the government to be used solely for highway construction. The collection of tolls on the new interstate highway system was not allowed except for those portions of the system that had already been built prior to the enactment of the Highway Act.

With the institution of the federal grant system of 90–10, 90 percent federal participation to 10 percent state participation, the construction of toll roads decreased dramatically. From 1960 to 1980, only around 1,500 miles (2,430 km) of toll roads were built in this country. Additionally, at the beginning of 90–10 period some states removed existing tolls to comply with a federal requirement that would make states eligible for reconstruction and widening funds.

On October 15, 1966, the Department of Transportation was established and one year later, in 1967, the Federal Highway Administration (FHWA) came into existence to administer the nation's federal highway system.

THE FEDERAL FUNDING CRISIS OF THE 1980s

Prior to the passage of legislation creating the Highway Trust Fund in 1956, the federal dollars necessary to pay for obligations incurred by the Federal-Aid Highway Program came from the Treasury's general fund as the revenue from taxes on motor vehicle fuels and products bore little relationship to the amount of money actually being spent on highway construction. The Highway Trust Fund changed that: A companion act, the Highway Revenue Act of 1956 increased user taxes and established new ones, and all the revenues now would be credited to the Fund which would be dedicated to finance a larger highway program. The act was due to expire in 1971, but after several extensions, was rescheduled to expire on September 30, 1999.

Revenues generated from the Fund still proved woefully inadequate to deal with either the expansion of the program or the costs of repairing and maintaining a severely neglected infrastructure system. Even federal and state gasoline tax increases during the 1980s were not enough to cover the high cost of new construction and of rehabilitation and repairs to the rapidly deteriorating roads and bridges. The United States was caught in an infrastructure crisis, heightened by a growing budget deficit and increased revenue demands for various nonhighway-related federally funded programs. As Representative Norman Y. Mineta (D-California), Chairman of the House Public Works and Transportation Committee, pointed out in a speech to the American Road and Transportation Builders' Association in 1993, decades of erosion in infrastructure spending beginning in the 1960s had left the country with a serious long-term problem.

The need to reconstruct and repair neglected, aging highways, bridges, and tunnels was only one of the many demands being placed upon overburdened federal and state treasuries. Thus, finding innovative methods of providing funds for new highway construction became a necessity. Federal legislation had prohibited states from establishing toll roads on highway projects utilizing federal highway aid, but the U.S. Congressional Budget Office in 1985 reconsidered the government's position on toll road construction. The Budget Office concluded that financing a portion of highway construction could be accomplished with revenues collected via tolls, but that only a limited amount of urban highway in densely populated areas was capable of generating enough revenue to suport toll road construction. In 1987, Congress authorized a pilot program that allowed the states to construct toll roads with a maximum of 35 percent federal aid. Only new or dramatically improved roads in a noninterstate system would be eligible to participate in this program, and several states took advantage of this opportunity.

As the decade of the 1980s drew to a close, some of the funding difficulties for new highway construction had been alleviated, but the problem of what to do with the deteriorating and inadequate existing highway system went basically unanswered.

THE INTERMODAL SURFACE TRANSPORTATION EFFICIENCY ACT (ISTEA)

In the 1990s searching for answers to aiding highway maintenance and construction, the federal government implemented the Intermodal Surface Transportation Efficiency Act (ISTEA) on December 18, 1991. This act was one of the significant factors responsible for accelerating new toll road construction and, more importantly, for speeding up the process of creating a public/private partnership for the construction of much-needed roads and bridges. ISTEA (pronounced just like "ice tea") is comprised of eight parts:

Title I. Surface Transportation
Title II. Highway Safety Act of 1991

Title III. Federal Transit Act Amendments of 1991
Title IV. Motor Carrier Act of 1991
Title V. Intermodal Transportation
Title VI. Research
Title VII. Metropolitan Washington Airport Act Amendments of 1991
Title VIII. Surface Transportation Act of 1991

The new ISTEA legislation permitted states for the first time to collect tolls and use the revenue to supplement any taxes collected from the sale of motor vehicle fuel and vehicle taxes for the construction of roads receiving federal funding assistance. In addition, ISTEA expanded toll facility eligibility for federal aid to include a variety of activities:

1. Initial construction of toll facilities (except for on the interstates)
2. "Four Rs" (resurfacing, restoring, rehabilitating, and reconstructing) work on toll facilities
3. Reconstruction or replacement of free bridges or tunnels and conversions to toll facilities
4. Reconstruction of free highways (except interstate roads) to convert to toll status
5. Preliminary studies to determine the feasibility of the above work

Federal aid by project type is summarized in Table 2.1.

ISTEA not only provided that federal aid to reimburse states for expenditures on eligible projects, but it also expanded the types of expenditures eligible for reimbursement. For instance, ISTEA allowed states to use federal funds to pay up to 50 percent of the cost of building or expanding a noninterstate toll facility with the federal share increasing from 35 percent. To take advantage of these ISTEA programs each state must pass its own enabling legislation.

THE PATTERN FOR PUBLIC/PRIVATE PARTNERSHIPS IS SET

Involvement of the private sector in the construction of new highways has many obvious benefits. Such cooperative ventures

- Allow the construction of highways that, even though recognized as important by state and federal governments, could not be funded because available funding had to be committed to projects of a higher priority
- Transfer risk of delays and construction cost overruns from the government—and ultimately from the taxpayer—to the private developer
- Permit the efficiency of a profit-motivated entity to complete design and con-

TABLE 2.1. ISTEA Federal Aid by Project Type

	Federal Aid Share (percent)			
	Interstate		Noninterstate	
Activities Eligible for Toll Financing and Public/Private Partnerships	Roads	Bridges/ Tunnels	Roads	Bridges/ Tunnels
Initial construction (except in the Interstate system) of a toll highway, bridge, tunnel, or approach thereto	NA	NA	50	80
Reconstruction of an existing toll highway, bridge, tunnel, or approach*	50	80	50	80
Resurfacing, restoring, and rehabilitating of a toll highway, bridge, tunnel, or approach*	50			
Reconstruction or replacement of a free (noninterstate) highway, or a toll free bridge or tunnel on or off the interstate, and conversion to a toll facility	NA	80	50	80
Preliminary studies to determine the feasibility of the aforementioned toll construction activities	50			

**Note:* An exception to the 50 percent share is that highway facilities under existing federal toll agreements are eligible for 80 percent federal aid share until the expiration of the existing toll agreement.

struction in the most cost effective manner, thereby delivering the project to the public for its use as quickly as possible

- Assure the highest quality construction because the private sector would be responsible for the project's maintenance over a 30 to 40 year period
- Create land development opportunities adjacent to and in the vicinity of the highway, thereby stimulating the local economy

Executive Order 12893

President Clinton signed Executive Order 12893 on January 28, 1994, to reaffirm the federal government's commitment to the public/private infrastructure partnership. In the preamble to EO 12983, President Clinton states that

> a well functioning infrastructure is vital to sustained economic growth, to the quality of life in our communities, and to the protection of our environment and natural resources. To develop and maintain its infrastructure facilities, our Nation relies heavily on investments by the Federal Government.

The goal of Executive Order 12893 was to stimulate interest in several forms of public/private programs, permitting the federal government to divert more federal funds from a sorely needed infrastructure revitalization program to other politically sensitive programs. Thus, paragraph 2(c) of Executive Order 12893 deals specifically with private-sector participation:

> Agencies shall seek private-sector participation in infrastructure investment and management. Innovative public/private initiatives can bring about greater private-sector participation in the ownership, financing, construction and operation of the infrastructure programs referred to in Section 1 of this order (sic. transportation, water resources, energy and environmental protection). Consistent with the public interest, agencies shall work with the State and local entities to minimize legal and regulatory barriers to private sector participation in the provision of infrastructure facilities and services.

Federal agencies were directed to submit initial plans to implement these principles to the Director of the Office of Management and Budget (OMB) by March 15, 1994. By the beginning of fiscal year 1996, budget submissions to OMB from each agency were to incorporate the principles set forth in EO 12893 to justify major infrastructure investment and grant programs. Major programs were defined as those with annual budgetary resources in excess of $50 million.

IMPLEMENTING ISTEA

The Intermodal Surface Transportation and Efficiency Act of 1991 made significant amendments to Title 23 of the United States Code, and while ISTEA did not change the form of federal aid, it did change the way it was administered.

Expanded Project Eligibility Construction of new toll facilities (except for highways on the interstate system), reconstruction, resurfacing, rehabilitation, and conversion of some facilities to toll facilities (except on the interstate system)—all were added to toll projects as eligible for federal aid. And tolls did not have to be removed once the state had achieved recovery of costs.

Public/Private Partnerships ISTEA allowed the comingling of federal, state, and private-sector funds and the sharing of responsibility between public and private sectors. Both sectors were allowed to design, finance, construct, and operate new highway facilities and to participate in the repair and expansion of existing facilities.

Loans under ISTEA Section 1012(a)

Because the funds can be either loaned or granted to a toll project, the funds are not returned to the federal government, and ISTEA allows federal aid funding to be recycled. If the funds were loaned, the repayments (including interest) are made to the sponsoring state, where they remain to be used for other transportation projects. These repayments can then be used to fund other transportation projects eligible under Title 23, free of the federal requirements that originally accompanied the federal aid funds.

State Revolving Funds These funds are a natural extension of an ISTEA loan and involve the following three steps:

1. The state provides an initial grant or loan to the revolving loan fund agency.
2. The revolving loan fund agency lends the money and receives loan repayments.
3. The revolving loan fund agency uses the loan repayments to make new loans.

Construction grants acting as seed money can stretch available federal aid and be looked on as new principle to be applied to state revolving loan funds for toll projects.

State Matching Requirements Matching funds can derive from state appropriations, proceeds from revenue bond sales, general obligation bonds, or private funds.

Project Eligibility Requirements To be eligible for a public/private partnership under ISTEA, the public entity must have negotiated a contract with the private entity. This agreement is to contain the roles and responsibilities of each party. Each state must enact is own enabling legislation setting forth the ground rules for the roles and responsibilities of the public sector and private entity.

Reconstruction Opportunities ISTEA permits funding of previously nonrevenue-producing facilities, including federal aid highways, bridges, tunnels, and approaches to these facilities. These previously nonrevenue-producing facilities may be converted to tolls in order to cover that portion of the project not funded by federal aid.

COMPONENTS OF A MODEL STATE ORDINANCE

The Department of Transportation issued publication no. FHWA–PL–93–015 with guidelines to assist states implementing a program incorporating ISTEA toll provisions. The leaders in enacting legislation paving the way for the establishment of private infrastructure projects have been the states of Arizona, California, Florida, Illinois, Minnesota, Washington, South Carolina, and Virginia. The Federal Highway Administration publication No. FHWA-PL-93-015, includes excerpts from various sections of existing state legislation to assist other state legislators to formulate their own program. The following components and illustrative portions of existing state laws may prove helpful for that purpose.

What a State Should Include in Its New Legislation

1. The Legislative Declaration An introductory section to legislation should clarify the state's goals by seting forth the state's purpose for proposing a public/private partnership to create an efficient, cost effective transportation system. California's Assembly Bill 680 offers an excellent example.

> The Legislature hereby finds and declares all the following:
>
> (a) It is essential for the economic well-being of the State and the maintenance of a high quality of life that the people of California have an efficient transportation system.
>
> (b) Public sources of revenue to provide an efficient transportation system have not kept pace with California's growing transportation needs, and alternative funding sources should be developed to augment or supplement available public sources of revenue.
>
> (c) One important alternative is privately funded Build-Operate-Transfer (BOT) projects whereby private entities obtain exclusive development agreements to build, with private funds, all of a portion of public transportation projects for citizens of California.

2. Reference to ISTEA Legislation should cover the relevant ISTEA toll and public/private partnership provisions. Chapter 26 of the State of Arizona's Title 28 contains the proper verbiage:

> Apply for, receive, and accept, from any Federal agency or any other government body, grants for or in aid of the design, construction, reconstruction, resurfacing, restoring, rehabilitation, replacement, maintenance or operation of toll and non-toll highways, bridges and tunnels, or study of the feasibility of such activities, and enter into any contracts with the granting body, or any other governmental body, and with private entities as may be required to qualify for such grant. The Department may transfer or lend the proceeds of any such grant, or utilize such proceeds credit enhancement, to public agencies or private entities, on terms and conditions complying with applicable Federal and State law and otherwise acceptable to the Department.

3. Identification and Empowerment of Implementing Agency States will need to empower a specific agency to implement the public/private partnership project. For example, Florida House Bill 175, Section 3, clearly identifies the agency and its authority:

> (1) A person may not construct or operate a private toll roadway unless the Florida Transportation Commission has issued a certificate authorizing such construction or operation. A person may not extend or enlarge a private toll roadway for which a certificate has been issued unless the commission issues a certificate authorizing such extension or enlargement.

4. Grants, Loans and Other Aid to Private Entities Laws will be needed to authorize grants or loans to private developers of transportation projects. Often amendments to state laws or changes in the transportation agency policies that allow the state to award a grant or loan are required. Federal Department of Transportation cites Arizona state law as an example of such legislation:

> In addition to the agreements of the type described in Chapter 26, the Department many enter into agreements, on such financial and other terms as the department may deem to be in the public interest, with private entities for the private development and operation of toll transportation facilities and for private participation in public toll highways financed pursuant to this chapter. Subject to the determination being made in the final sentence hereof, such agreements may provide for one of more of the following:
>
> (a) Sale or lease to the private entity by the Department of undeveloped real property to be utilized in connection with the private toll facility.
>
> (b) Sale or lease to the private entity by the Department of an existing State highway to be utilized in connection with the private toll facility.
>
> (c) Lending to a private entity the proceeds of bonds issued pursuant to this chapter, to be utilized by the private entity to finance the construction of a private toll facility.
>
> (d) Imposition of tolls on currently existing State highways, and making toll revenues available for the financing of privately constructed toll facilities.
>
> (e) Engaging private entities to operate and/or maintain toll highways constructed by the Department pursuant to this chapter.

5. Allowable Models for the Public/Private Relationship How will the private entity be structured? The legislation may define a specific joint venture model to be used for projects or options open for negotiation on a project-by-project basis. ISTEA allows for various models to be incorporated in the enabling legislation, including Build, Own, Operate (BOO); Build, Operate, Transfer (BOT); Buy, Build, Operate (BBO); Lease, Develop, Operate (LDO); and temporary privatization.

Implementing the Grant and Loan Provisions of ISTEA

1. The Basic Project Loan Federal aid funds remain an apportionment to the state which may loan the proceeds to projects eligible under Title 23 provided that the state has the legislative framework in place to make loans and grants to private-sector entities.

2. The State Revolving Loan Fund (SRF) Chapter 26, Title 28, Section 28–305 of Arizona's Title 28 legislation dealing with the creation of a state revolving fund (SRF) contains the following language:

> (a) In general. There is hereby established a Revolving Loan Fund. The Fund shall be maintained and administered by the Board in accordance with the provisions of this subtitle and such rules as the Board may from time to time prescribe. The Fund shall be available in perpetuity for the purpose of providing financial assistance in accordance with the provisions of this Section. Subject to the provisions

of any applicable bond resolution governing the investment of bond proceeds deposited in the Fund, the Fund shall be invested and reinvested in the same manner as other State funds. Any investment earnings shall be retained to the credit of the Fund.

(b) The proposed provision also included statements regarding the use of the SRF funds.

(c) Purposes. Amounts in the Fund may be used only:

(1) To make loans for the construction, reconstruction, resurfacing, restoring, rehabilitation or replacement of public or private toll transportation facilities within the State, or study of the feasibility thereof;

(2) To guarantee, or purchase insurance for, bonds, notes, or other evidences of obligation issued by the developer of a public or private toll facility for the purpose of financing all or a portion of the cost of such toll facility, if such action would improve credit market access or reduce interest rates;

(3) To earn interest on Fund accounts; and

(4) For the reasonable costs of administering the Fund.

3. Loan Agreements ISTEA requires loan repayments to begin within five years after the project is completed and opened for use. An amendment to Section 1720 of California's State Labor Code contains the basic language to establish this requirement:

> Loans made from the State Transportation Revolving Fund shall bear interest at the average rate of interest earned by the State Pooled Money Investment Fund over the preceeding twelve months. Loan repayment shall begin no later than 5 years from the date that the facility is opened to toll traffic and shall be completed by no later than 30 years from the time the loan was obligated. The amount loaned may be subordinated to other debt financing for the facility other than loan from any other public agency. The Department may charge reasonable origination fees for any loan pursuant to this article.

State-Private Agreements The authority to enter into franchise agreements must be established in enabling legislation by the states. Illinois Proposed Act I, Section 100-8 deals with this issue in a straightforward manner:

> The Authority shall have the power:
> (g) To acquire, or to enter into long-term contracts to acquire, privately constructed roadways, or portions or parts thereof, pursuant to Article II of this Act. Upon acquisition, such roadways may become a part of the system of toll highways, operated, owned and controlled pursuant to Article I of this act.

The State of California's authority to enter into agreements with private developers is clearly defined in Section 143, California Streets and Highways Code:

(a) The department may solicit proposals and enter into agreements with private entities, or consortia thereof, for the construction by, and lease to, private entities of four transportation demonstration projects, at least one of which shall be in Northern California and one in Southern California.

(b) For the purpose of facilitating these projects, the agreements may include provisions for the lease of right-of-way in, and the airspace over or under, State highways, for the granting of necessary easements, and for the issuance of permits or other authorizations to enable the private entity to construct transportation facilities supplemental to existing State-owner transportation facilities. Facilities constructed by a private entity pursuant to this section shall, at all times, be owned by the State. The agreement shall provide for the lease of those facilities to the private entity for terms up to 35 years. In consideration thereof, the agreement shall provide for completion reversion of the privately constructed facility to the State of California at the expiration of the lease at no charge too the State.

Available State Aid, and Services and Requirements for Public/Private Ventures

According to the provisions of ISTEA, the state government must provide assurance to the private entity that some responsibilities are either shared or limited, such as—who will assume responsibility for damages and to what degree. There may be limits placed upon the tort liability imposed on the private entity. A state may wish to establish its right to reclaim a concession if there are compelling public interests to do so, but it must compensate the private entity in an equitable manner. The state should consider addressing a range of issues:

- Clarification of the exercise of state powers on behalf of the developer.
- Provisions to escrow sales taxes levied on the privately developed transportation project.
- The establishment of special financing districts for tax purposes to solicit taxes from private owners who will benefit directly from the new project.
- Government credit support providing limited support for private financing of for privately developed projects relating to the transportation project.
- Local government financial support, if desirable. Most private toll road laws prohibit the use of state funds for private transportation projects but often don't mention local government assistance.
- Provisions for the comingling of tax exempt and taxable debt to finance the project in its enabling legislation. Comingling of these two debts may be accomplished by obtaining an allocation on the state's volume cap on private-purpose tax-exempt debt.
- Clarification of the right of eminent domain. When will the government exercise or deny its right of eminent domain on the part of the private entity?

Powers and Duties of the Operator State legislatures must address the rights of the franchisee to toll, operate and develop a project. Arizona used the following language to outline the rights of the franchisee:

A. The Operator may:
   1. Operate the roadway and charge tolls for the use of the roadway to receive reimbursement of costs and a reasonable return on investment.
   2. Subject to applicable permit requirements, cross any right-of-way if the crossing does not unreasonably interfere with the use of the right-of-way.
   3. Classify traffic on the roadway according to reasonable categories for the assessment of tolls.
   4. With the consent of the department, adopt and enforce reasonable rules including rules:
      (a) Prescribing maximum and minimum speeds that conform to Department and State practices.
      (b) Excluding vehicles or cargoes or materials from the use of the roadway if exclusion of such vehicles, cargoes or materials by this State or a local authority is authorized by the laws of the State.
      (c) Establishing high occupancy vehicle lanes for use during all or any part of a day and limiting the use of these lanes to certain traffic.

Toll Collections State legislation generally does not include provisions for the levying and collection of tolls by private entities; therefore, this kind of legislation will be required along with any incentives to allow operators to vary rates from time to time to reward off-peak usage, discourage rush-hour traffic, and attempt to relieve congestion.

Priorities on the Use of Revenue Although developers may be required to pledge first lien on their toll revenue toward the operation and maintenance fund, once this reserve is fully funded toll revenues can be used to retire debt; if that requirement is met, excess revenues will be diverted to a state highway fund.

Duration of the Public-Private Franchise Agreement Typically a toll road project will require at least 20 years and usually no more than 40 years to achieve the developer's required return on investment. The State of California stipulates a term of 35 years, Virginia allows the Dulles Greenway 40 years, and ISTEA requires 30 years to repay loans obligated with federal aid highway funds.

Quality of Facilities/Service Requirements The federal DOT refers to Arizona legislation as a model for facility quality requirements. Arizona's Title 28, Chapter 26, Section 28–3066 B has the following.

> If the Board approves the project, project design and connections of the roadway, the Department shall enter into an agreement with the applicant providing that the Department, on a reimbursable basis, shall:
>
> (1) Review the plans and specifications for the roadway and approve them if they conform to State practices.
>
> (2) Inspect and approve construction of the roadway if it conforms to the plans and specifications and State construction and engineering standards.
>
> (3) Throughout the life of the roadway project, monitor the maintenance practices of

the operator and take action as appropriate to ensure the performance of maintenance obligations.

(4) Perform other necessary services that the private entity is unable to perform, including project development and environmental impact statements.

Provisions for Default and Agreement Modification Florida addressed the default issue in their House Bill 175, Section 14.

> Default—If an operator commits a material and continuous violation of, or fails to comply with, the terms of its contract with the department, the commission may revoke the operator's certificate, declare a default in the construction or operation of the private toll roadway, and make or cause to be made claims under any completion or performance bonds or may take other appropriate action. Prior to revoking an operator's certificate, the commission shall provide the operator notice and an opportunity to be heard. Upon revocation of a certificate, the operator to whom it was issued ceases to have the authority to construct or operate the project that is the subject of the certificate.
>
> The department may close the project, or otherwise dispose of the project in a manner that the department determines is in the public interest. The operator shall, at the discretion of the department, grant the department, without cost, all its rights, title, and interest in the project, real property acquired as rights-of-way for the project, and facilities of the project.

On the other hand, the state can provide the private entity with certain guarantees; the two most common are debt repayment and traffic guarantees. The debt guarantee obligates the state to assume all or part of the debt of the private operator should the project go into default. Depending upon how this debt guarantee is worded it might decrease the private operator's incentive to operate the facility profitably.

In those cases where the state provides traffic projections to the private entity and the developer's proposal is based upon those studies, some form of guarantee may be forthcoming if traffic flow falls below these projections. The franchise period may be extended or provisions included that allow the state to buy back the facility from the private entity.

THE EMERGENCE OF THE TOLL ROAD CONCEPT IN THE 1990s

In October 1993, Rodney Slater, in his first meeting with the press after being appointed Administrator of the Federal Highway Administration by President Clinton, affirmed the federal government's position on public/private ventures. The private sector is not only going to be encouraged to form partnerships with the public sector, but he believes these partnerships are essential to maintain an effective transportation program. According to Slater, as reported in the *New York Times* October 1, 1993 edition, new highway construction and maintenance was estimated at $200 billion with an additional $90 billion required for bridge repairs, but restraints placed upon the federal budget preclude allocating all of the funds that will be required. Slater went on to state that, although the creation of public-private highways and the application of new tolls or the raising of existing ones may come about, the pressure of the market will keep these tolls at reasonable level.

Some doomsayers in Congress have warned that public outcry will ensue if toll booths start cropping up all over, but Robert Poole, head of the Reason Foundation, a Beltway thinktank, is convinced of the benefits: the advent of new toll roads "means that cars will be increasingly paying their own way in the transportation system." Even Lester Lamm, President of the Highway Users Foundation which represents the trucking industry, was quoted in the *New York Times* as stating that the private sector operation of some interstate highways is "something that probably should be considered."

Several situations occurred during the early 1990s that reawakened interest in toll roads in general, and public/private toll road partnerships in particular. ISTEA and other federal legislation were certainly factors. The outgoing 103rd Congress had reacted to the budget squeeze by trimming the transportation appropriations for the fiscal year 1995. Highways, the largest segment of the program at $19.7 billion, would be trimmed by $74 million. After the November 1994 congressional elections placed a Republican majority in both the House of Representatives and the Senate, the Republican goal of "less government" placed immediate attention on the distribution of power from the federal government to the states. The term "privatization" was heard a great deal during the first 100 days of this new Republican-dominated congress, and it has continued to be a topic inside and outside of the Beltway.

In the meantime, major cities throughout the country were experiencing more and more traffic congestion problems, particularly during the morning and evening commuting hours. In some cases the costs of expanding some existing highways were prohibitive; for example, lane widening for an existing bridge that spaned six lanes of traffic and had no usable median strip is extremely expensive. Thus, public agencies were searching for new ways to smooth out the flow of traffic, even as former defense-oriented industries were envisioning using technology previously developed for the military to create new transportation concepts. Private developers, always on the lookout for an innovative competitive edge, picked up on both the public's needs and the industry's technologies—and the intelligent highway was born.

THE INTELLIGENT HIGHWAY

According to the Intelligent Transportation Society of America (ITS), highway traffic has increased 200 percent since 1960. More that 2 trillion vehicle miles are traveled each year in the United States. According to ITS Executive Director James Constantino, traffic delays will cost the public an estimated $120 billion annually. One potential way of dealing with traffic congestion on the roads is to create more "intelligent" highways. Is the intelligent highway just another cyberspace hype, or is it a practical way to assist motorists in traveling from Point A to Point B more quickly, safely, and with less wear and tear on both motorist and machine?

A number of transportation systems in the United States and throughout the world have already successfully employed various components of the intelligent transportation system, and others are experimenting with more sophisticated aspects of ITS. An intelligent highway system is composed of an alphabet of components:

ETC—electronic toll collection

ETM—electronic traffic management

AVI—automatic vehicle identification

ETTM—electronic toll and traffic management systems (a marriage of ETC and ETM)

The idea behind all of these systems is to speed motorists on their way and monitor and alert them to conditions that may affect their routing and safety.

ETC Electronic toll collection systems can be as simple as providing a coin deposit chute at a toll booth. A motorist throws in a selection of coins totaling the required toll charge, waits for a barrier to lift, and then proceeds through. When operated properly, this system is more efficent than a manual toll collection operation. Labor costs account for about 80 percent of all collection expenses in tolled ticket systems; fully automatic ETC lanes cost 90 percent less to operate than attended lanes, so the ETC is more cost effective and speeds up the transaction.

ETM Electronic traffic management involves the use of highway surveillance systems to search out and report accidents or other disruptions to traffic flow and alert motorists so that they may elect alternative routes to avoid the congested area. The ETM system usually involves the use of variable message signs along the highway to visually display either traffic alert problems or a highway advisory radio system (such as signs that display "Tune in on 85.6AM to Receive Highway Information").

AVI An automatic vehicle identification system uses vehicle-mounted transponders' signals to record traffic and/or toll information.

ETTM The electronic toll and traffic management program works best when ETC and ETM are combined with a third component, automatic vehicle identification (AVI). AVI can function as a stand-alone system, but it has a synergistic effect on the ETTM process.

How a Typical AVI System Functions

Conventional toll road ticket booths and collection stations slow down the passage of traffic; AVI systems allow a vehicle to pass through entrance and exit stations at speed, thus eliminating two major areas of traffic congestion. The AVI system, such as that produced by MARK IV of Amherst, New York, has three basic components.

1. A small transponder about the size of a credit card which mounts on the inside of a vehicle's windshield (Figure 2.1) or a slightly larger unit which mounts over or under the front license plate of the vehicle (Figure 2.2).
2. Roadside antenna—which are either mounted on poles or on overhead gantries (Figure 2.3) as flat panels spanning the highway—that pick up the signals from the vehicle-mounted transponder.
3. The reader which establishes a two-way communication between the vehicle-mounted transponder and the highway antenna (Figure 2.4).

Upon entering the toll road, a vehicle with a transponder passes the highway antenna. The transponder, which is implanted with the individualized vehicle code, emits a signal that is passed on by the antenna to the reader (in effect, "I am entering the toll area"). When the vehicle exits the highway at an antenna-equiped ramp, the vehicle's transponder sends out another signal to the reader ("I have just left the toll road—please calculate the amount of my toll and charge my account").

States considering AVI/ETC systems must establish the method by which motorists can apply for transponders. Generally, a driver opens an account by submitting an application to the relevant state agency along with a minimum monetary deposit. Thereafter the motorist's account is debited accordingly with each trip on the AVI-equiped highway. When a driver's account nears the point when it should be replenished, some transponders activate a small yellow light. Other systems use a conventional "traffic light" system at the point of entry: The green light glows and the barrier opens for a properly funded transponder; the signal glows yellow and the barrier still rises for a low account; but when the red light activates, the gate will not rise for a depleted account. The toll authority can decide if these individual motorist accounts can be replenished automatically by tapping into the individual's bank account or credit card, or if sufficient notice should be given to motorists to allow them to remit payment by check or credit card.

Educating the Public to AVI Use

When existing highway systems are retrofitted for AVI or new roadway construction incorporates these systems, all components must function as designed. The motorist who is to use the road is one such component, so educating the public in the proper use of AVI and controlling those who will attempt to abuse the system must be taken into consideration. California's bold initiative to introduce ETTM and AVI systems on several new highway projects has provided much useful information on public interaction with AVI systems.

The Foothills Transportation Corridor (FTC) of Orange County, California, is part of a 68-mile (110-kilometer) highway that will ultimately extend from Camp Pendleton north of San Diego to Interstate 91 via the Eastern Transportation Corridor. Phase 1 of FTC, a 3.2-mile (5.2-kilometer) stretch known as Backbone I, opened in 1993; Phase 2, the southern portion known as Backbone II, opened in April 1995. The FTC started using a toll collection system employing AVI in 1992.

Short Range Vehicle-to-Roadside Communications Equipment for Intelligent Transportation Systems Applications

FLATPACK TRANSPONDER

The ***ROADCHECK™*** Flatpack Transponder functions as a vehicle's short range communications device. Communication between vehicle-mounted transponders and roadside mounted ***ROADCHECK™*** readers occurs at 500 Kbits per second, permitting data transfer between the roadside and vehicles traveling at highway speeds.

ROADCHECK™ Flatpack Transponders are suitable for every type of vehicle and application where portability is desired. Transponders are designed to be mounted securely on the interior surface of the vehicle's windshield, yet can be removed if desired.

Installation of the Flatpack Transponder is quick and simple. Using adhesive-backed material, the transponder is easily installed in the correct position on the windshield. Markings on the case guide proper orientation.

The transponder is a half-duplex device which uses the same frequency and modulation scheme for both up and downlinks. Its receiver is a simple AM detector while its transmitter is a single stage on/off unit. This elegant design has consistently demonstrated excellent performance in testing and field operations.

ROADCHECK™ semi-active technology employs a lithium battery as the transponder's source of power, giving the transponder a minimum life of 10 years, regardless of the number of interrogations it undergoes. The transponder casing is made of durable impact-resistant molded plastic and is available in different colors. MARK IV ***ROADCHECK™*** transponders are highly suitable for Intelligent Transportation Systems (ITS) applications such as electronic toll collection, traffic monitoring and commercial vehicle operations.

The Flatpack Transponder has "read/write" capabilities and can store fixed, pre-programmed data as well as variable data added in real time as the vehicle passes a reader antenna at highway speeds. Partitioning between fixed, preprogrammed data fields and reprogrammable fields can be altered to suit the needs of the client. Closed toll systems operators can write variable point of entry data into the transponder for subsequent toll

ROADCHECK™ Flatpack Transponder
This compact transponder can be installed on any kind of vehicle quickly and easily. Available in different colors for different vehicle types.

The transponder mounts on the interior surface of the windshield, behind the rearview mirror. This maintains a clear view of the road for the driver, and communicates with antennas mounted above or alongside the road.

MARK IV

Figure 2.1. Windshield-mounted transponder. *[Courtesy MARK IV Industries, Inc.]*

ROADCHECK™

Short Range Vehicle-to-Roadside Communications Equipment for Intelligent Transportation Systems Applications

EXTERIOR TRANSPONDER

The exterior transponder, when used with MARK IV's ***ROADCHECK™*** Vehicle-to-Roadside Communications (VRC) systems, acts as a vehicle's electronic communications device. In response to an interrogation pulse from the ***ROADCHECK™*** reader equipment, the transponder transmits its information packet at 500 Kbits per second. While Type 1 transponders are capable only of transmitting data, Type 2 transponders can store new data received from roadside Readers. Type 2 transponders can also be operated in a read-only mode. The portion between the fixed and reprogrammable data fields is flexible.

ROADCHECK™ exterior transponders can be adapted for any type of vehicle. Although exterior transponders are usually mounted on license plates, they can be positioned anywhere on the outside of a vehicle as long as there is a clear line of sight to the antenna. (Consult the factory for detailed mounting instructions.)

Installation is quick and easy using commonly available license plate mounting hardware.

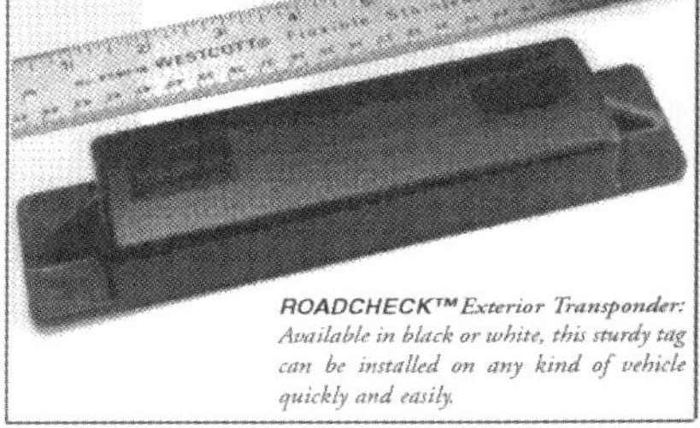

ROADCHECK™ *Exterior Transponder: Available in black or white, this sturdy tag can be installed on any kind of vehicle quickly and easily.*

Installing the transponder is simple: it is mounted on either the front or rear license plate using commonly available license plate mounting hardware. On vehicles with protruding bumpers, the transponder can be secured under the bumper of the vehicle using double-sided adhesive tape. All transponders within a particular ***ROADCHECK™*** system will usually be mounted in the same position.

The transponder is a semi-active, half-duplex device which uses the same frequency and modulation scheme for both up and downlinks. Its receiver is a simple AM detector while its transmitter is a single stage, on/off unit. This elegant design has consistently demonstrated excellent performance in testing and field operations.

A lithium battery gives the transponder a minimum life of 10 years, regardless of the number of interrogations it undergoes. The transponder casing is made of impact-resistant molded plastic available in either black or white. (Consult the factory for custom color availability.)

All MARK IV Type 1, 2 and 3 transponders are compatible with ***ROADCHECK™*** Readers. They can be integrated with Intelligent Vehicle Highway Systems (IVHS) applications such as confidential transaction systems, or serve as an interface between on-board computers and roadside readers.

MARK IV

Figure 2.2. Exterior transponder that can be mounted on automobile's license plate holder. *[Courtesy MARK IV Industries, Inc.]*

ROADCHECK™

Short Range Vehicle-to-Roadside Communications Equipment for Intelligent Transportation Systems Applications

OVERHEAD LANE KIT
with *ROADCHECK™* Flat-Panel Antenna

Each lane monitored by the ***ROADCHECK™*** VRC system requires an RF (Radio Frequency) Lane kit consisting of an RF module, an antenna, and adapter cables. For channelized toll lanes, the MARK IV Flat-Panel Antenna provides highly suitable capture zone definition.

The RF module, via the Flat-Panel antenna, sends out a periodic trigger pulse, establishing a lane based capture zone. Any transponder passing through this capture zone transmits its information packet to the antenna which sends it to the Reader via the RF module.

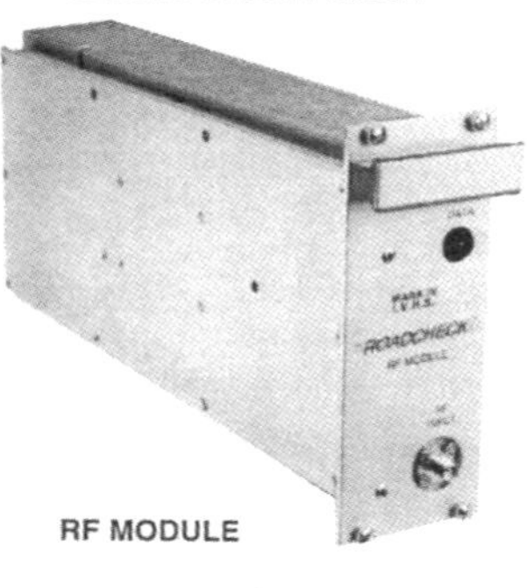

RF MODULE

The link between the ***ROADCHECK™*** Reader and transponder, the RF module slides into one of the 8 slots on the RF rack located with the Reader unit. One Reader with multiplexed antennas can simultaneously monitor up to 8 lanes of high speed traffic. Both the Reader and the RF Rack are enclosed in a weather-proof cabinet.

When desired antenna mounting locations are less than 30 meters (100 feet) cabling distance from the RF Module, RG-8U or RG-214U cable can be used. Consult the factory for longer distance cabling configurations.

FLAT-PANEL ANTENNA

This ruggedized patch array antenna is installed on a suitable overhead structure, such as the underside of a toll plaza canopy, using stainless steel fasteners. Optional mounting brackets are available from MARK IV.

For a typical channelized toll lane, the Flat-Panel antenna is centered over the lane to be monitored and tilted slightly towards the flow of on-coming traffic. To ensure adequate clearance for tall vehicles, the Antenna is typically installed approximately 4.9 meters (16 feet) above the road surface.

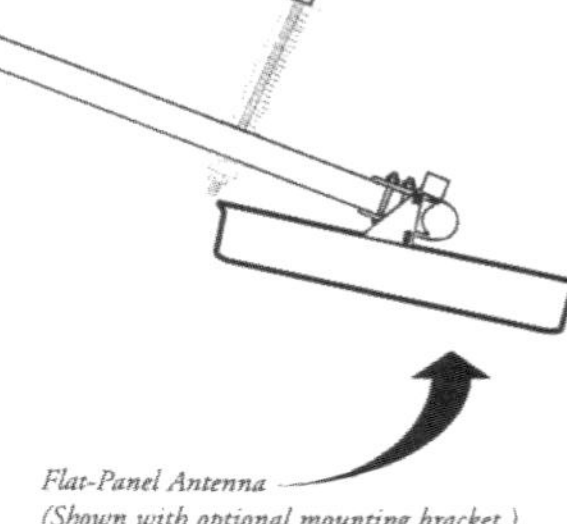

Flat-Panel Antenna (Shown with optional mounting bracket.)

I.T.S. APPLICATIONS

The Overhead Lane Kit with Flat-Panel Antenna is suitable for Intelligent Transportation Systems (I.T.S.) applications which need to employ a mixture of both internally and externally mounted VRC transponder styles.

Both internally and externally mounted MARK IV transponders can be read and reprogrammed using the Overhead Lane Kit.

The MARK IV Overhead Lane Kit achieves outstanding capture performance and excellent lane discrimination using high data rate, [semi]active, VRC technology.

MARK IV

Figure 2.3. Flat plate overhead-mounted antenna. *[Courtesy MARK IV Industries Inc.]*

ROADCHECK™

Short Range Vehicle-to-Roadside Communications Equipment for Intelligent Transportation Systems Applications

BASIC READER

At the heart of MARK IV's ***ROADCHECK™*** system is the Reader. It establishes high speed, two-way communications between vehicle mounted transponders and a Host Computer system.

The Reader, using multiplexed antennas, can simultaneously monitor up to eight lanes of traffic moving at speeds of up to 100 mph.

Linked with other Intelligent Transportation Systems (ITS) technologies, the ***ROADCHECK™*** system is used in situations where a fast two-way communications link between vehicles operating at highway speeds and fixed roadside equipment is required. These applications include electronic toll collection, commercial vehicle operations and parking revenue control.

The ***ROADCHECK™*** system operates automatically and unattended. An extremely high level of accuracy and reliability is assured through the use of Cyclic Redundancy Checks (CRC) and built-in self test routines.

The **Basic Reader** is made up of two independent units linked in a Master-Slave configuration. The Master unit is comprised of a Power Supply Module, a CPU Board, an RF Control Board and two Communications Boards. It also features a Master Failsafe Module. The Slave unit is identical to the Master unit but features a Slave Failsafe Module. Closely located Basic Readers can be synchronized via their SYNC Boards.

Readers are installed in weatherproof cabinets, typically located beyond the shoulder of the roadway.

ROADCHECK™ *Basic Reader for high performance, failsafe and reliable VRC applications.*

MARK IV transponders mounted on passing vehicles, travelling at speeds of up to 100 mph, transmit their data at 500 kbits per second in response to the antenna's trigger pulse and in turn receive new programming from the Reader. Transactions take place accurately and reliably in milliseconds.

Each Roadcheck Reader, using multiplexed antennas, can monitor up to eight lanes of traffic simultaneously with no loss of performance. Multiple Roadcheck Readers can be coupled together to expand coverage where there are more than eight lanes of traffic such

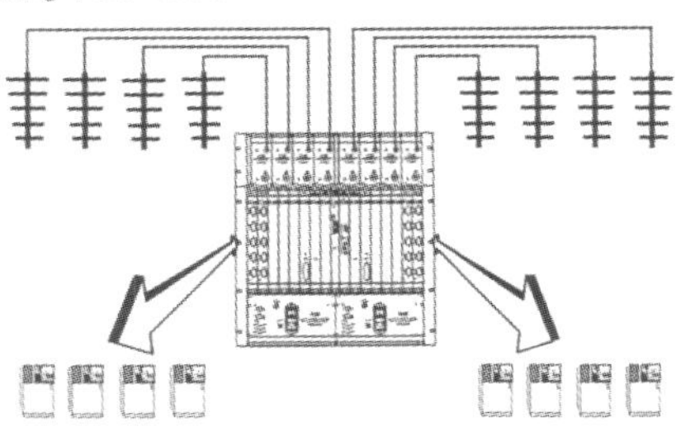

Each ***ROADCHECK™*** *Basic Reader can simultaneously link up to 8 lane based antennas and up to 8 lane controllers, plus a host computer.*

MARK IV

Figure 2.4. Mark IV's basic reader. *[Courtesy MARK IV Industries Inc.]*

This system was found to be not in compliance with Title 21 of the State of California Code for automatic vehicle identification systems and was replaced in 1994. Lockheed Martin Information Systems, working with the state's Transportation Corridors Agency (TCA), selected MFS Network Technologies and Texas Instruments to supply various subsystems to ensure compliance of the Foothills' system with the prevailing state codes. Testing of the new system began in early 1995 and DeLeuw, Cather & Company, a leading transportation engineering firm was brought in by Lockheed and TCA to provide oversight engineering during the certification period required by the state.

The AVI system used by MFS and Texas Instruments called for windshield-mounted backscatter transponders and a one-in lane antenna. A backscatter transponder is one in which the AVI system's reader sends out a carrier signal that is reflected back by an antenna in the transponder tag. The tag interprets the reader's trigger. The formal AVI test period began on January 18, 1995, and ended on March 14, 1995, during which time it was certified that an equipment accuracy reading of 99.99 percent had been achieved. The FTC system incorporated video surveillance as part of the revenue collection portion of the AVI program. Video cameras photographed the license plates of toll jumpers so that the appropriate law enforcement agencies could be notified and the tolls collected. During the testing program, however, the video system was also used to evaluate the types and nature of patron abuse or misuse.

As part of the testing proceedure, Lockheed test drivers were used to provide a control sample. They worked the system the same as the average driver would and completed 9,000 AVI transactions. Only three test drivers generated any systems violations, and these occurred over a two-day period. Upon investigation, it was determined that these violations occurred because of hardware failure, so the test results confirmed an equipment reliability factor of 99.99 percent.

The patron test group revealed real world conditions: The over 49,000 transactions resulted in approximately 490 system violations. Many of these violations occurred because the patron test drivers either had not installed their windshield-mounted transponders or had installed them improperly. The list of patron misuses that was compiled indicated that in order for AVI systems to succeed (1) operating instructions must be easy to understand and (2) a certain amount of driver education is required. The following categories of transponder tag abuse were assembled during this period:

- Tags Not Installed or Not Visible in the Vehicle
  - Transponder not in the car
  - Transponder left in the pocket of the car door
  - Transponder shut in the glove compartment
- Tags Installed in the Wrong Place
  - Tags not installed low enough on the driver's side of the windshield
  - Tags installed on the sun visor
  - Tags installed on the dashboard of the car

- Tags Not Installed at All
    Tag displayed in the driver's hand on entering the system
- Tag painted by owner to match car color.

The DeLeuw, Cather report arrived at the following conclusions after the complete comprehensive AVI testing period:

1. AVI technology is as good as it promised and is able to deliver with a 99.99 percent performance factor.
2. Toll facility operators must not expect the system to function at 99.99 percent efficiency under real world conditions and should anticipate that many violations will occur during the initial period of operation primarily because of patron misunderstandings. Operators should plan on implementing a continuing patron education program in the use of AVI equipment.
3. Expect Murphy's Law to be in force at all times.

Other Innovations in the ETTM System

In order to install most ETTM systems some form of toll structure must be built to house the required electronics. Other structures on which antennas can be mounted to pick up transponder signals from vehicles must be built at various places on the highway. If these antennas are the flat panel type, they can be mounted at the toll structure or on gantries. The structures for new highway construction where AVI equipment will be installed add to the total project cost, but generally these costs are not significant. When existing highways are being retrofitted for AVI and ETM, however, not only must structure costs be considered but additional land on which to build the various structures may be required. If more rights-of-way must be acquired, the costs can be considerable, and the process of acquiring these rights can be lengthy and arduous. An innovative system that does not require any substantial investment in roadside infrastructure is the Global Positioning System (GPS), which utilizes all in-vehicle-mounted equipment.

Germany's ROBIN System The German Ministry of Transportation is conducting comprehensive tests on advanced freeflow ETTM technology, and one of the systems under test is the Road Billing Net (ROBIN) developed by Mannesmann Pilotentwicklungsgesellschaft mbH. ROBIN uses a global navigation satellite in combination with a compact in-vehicle unit (IVU) which allows for identification, billing, and collection of toll charges exclusively from inside the vehicle. Infringements are detected and enforced from outside the vehicle by spot checks from portable control stations using a dedicated communication link.

The in-vehicle ROBIN equipment consists of a GPS receiver, a central processing unit, a card reader, and an in-board digital display. A GPS receiver with external antenna delivers the geographic position of the vehicle to a central processing unit (CPU) where coordinates are compared to the highway and toll data

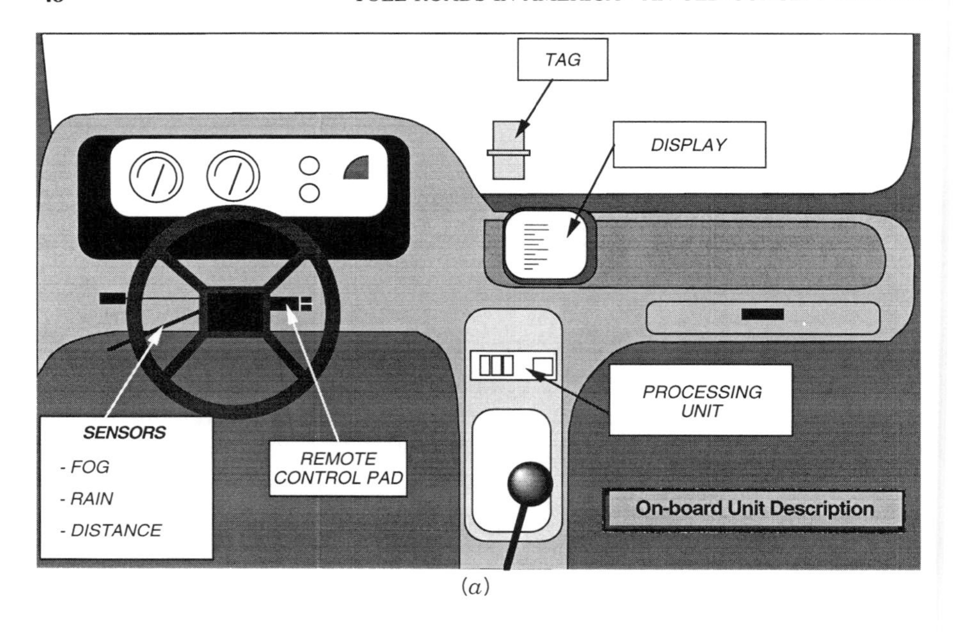

(a)

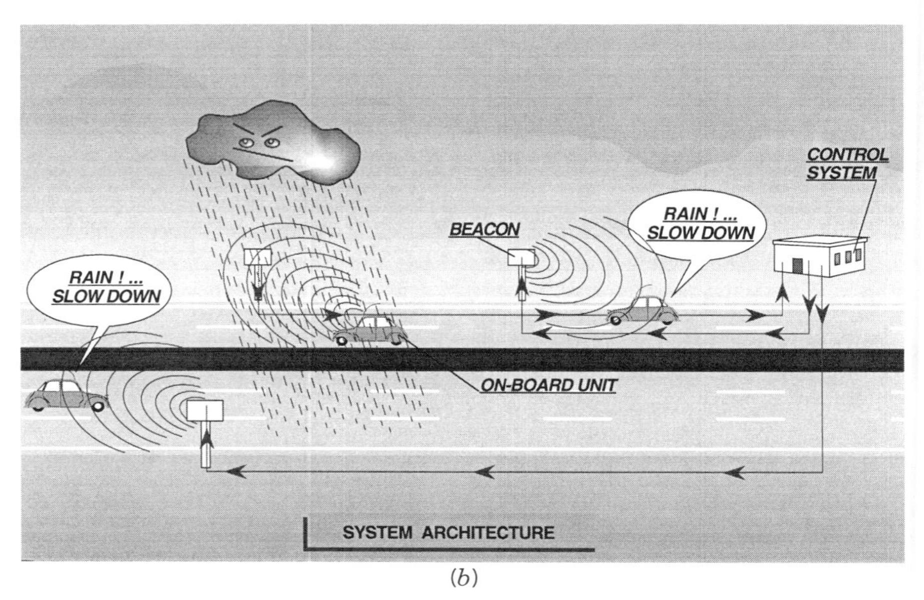

(b)

Figure 2.5. (a) Schematic of an automobile dashboard outfitted with Cofiroute's ADAMS system. (b) Schematic of ADAMS operating system. *[Courtesy Cofiroute, Sèvres, France.]*

stored in the unit's memory via an alogrithm. The CPU receives all of this information and calculates the toll. A smart-card reader stores the driver's portable account information; it is inserted into the smart-card reader before the driver departs on a journey. As the car enters and leaves the toll road, the CPU makes the correct deducts on the smart-card and a digital display advises the driver of the remaining funds. Prepaid smart-cards are sold at various terminals along the highway, and these terminals also allow motorists to replenish the balance on their current card. Since ROBIN does not depend upon highway mounted antennas, it is a mobile system that can be used with any toll highway, bridge, tunnel, and in urban areas where existing toll facilities are installed.

Cofiroute and Renault's ADAMS System Although the ETTM systems are expected to survey the highway; identify areas of construction, accidents, and traffic congestion; and report any problems to drivers either visually or by radio, an onboard reporting system is under development by Cofiroute and Renault in France. This system is called the Automatic Debiting Application for new Motorway Services, or ADAMS.

The ADAMS system is not just for toll collection, its on-board CPU, a transmitter and receiver, and dash-mounted display provides drivers with helpful real-time information. A schematic of the ADAMS system is shown in Figure 2.5. The heart of ADAMS is its information-gathering capability. Emergency messages can be transmitted directly to the vehicle via its dashboard-mounted digital display. "Accident ahead," "Construction in the right lane 2 miles ahead," "Heavy thunder showers headed this way," "fog rolling in around Exit 25"—all kinds of warnings can be instantaneously transmitted to the driver. Personalized information can also be received via the in-board receiver-transmitter. ADAMS can provide such information as the distance to the next exit, estimated time of arrival at a driver's destination, distance to the next gas station, and location of rest areas and even leisure activities or other tourist information.

Advanced electronic systems such as ROBIN and ADAMS will play a vital role in the design and development of future highway systems. And with the private sector being increasingly invited to participate in the design, development, construction, and operation of future transportation infrastructure projects, the government and the public will share in the benefits derived from these advanced technologies.

3

THE DULLES GREENWAY

It has been 130 years since the last privately financed toll road, the Little River Turnpike built in the 1860s that ran west from Alexandria for about 34 miles, was built in Virginia. The renewal of road privatization in Virginia came about in 1988 when the Virginia legislature passed a law authorizing the creation of private toll roads in the Commonwealth. The new Dulles Toll Road extension, known officially as the Dulles Greenway, has only the collection of tolls in common with its toll road ancestors.

THE HISTORY BEHIND THE GREENWAY

As the facilities of the National Airport were being taxed to the limit, the federal government recognized that another airport would be needed to service the Washington D.C. area and the adjacent Virginia and Maryland countryside. Consequently, a new facility, the Dulles International Airport, was finished in 1962 to relieve congestion for travelers and permit expansion of service in and out of the area. The Dulles Airport planners had envisioned easy access to this new facility. The Dulles Access Road was built to connect the airport with the Washington Beltway, I-495, and Interstate 66, which ran between downtown Washington D.C. and the Beltway. No interchanges were built on the Dulles Access Road to prevent its being clogged by local traffic; however the federal government had anticipated that another road might be required in the area for local and nonairport traffic and had purchased rights-of-way for that inevitability.

The Dulles Corridor is an area that extends from Tysons Corner in Fairfax County to Leesburg, the seat of Loudon County. This area saw great growth in recent years as the allure of relatively low-cost land, affordable housing, and close proximity to the nation's capitol resulted in a great deal of residential and commercial construction. Several Fairfax County communities boomed in the early 1970s, and small towns like Tysons Corner and Falls Church experienced frenetic growth. New towns such as Reston sprang up, adding to the population increases. Loudon County was also experiencing considerable growth, with population climbing almost 100 percent from 57,000 in the mid-1970s to slightly more than 100,000 by 1994.

The sustained community growth brought home to Virginia that its transportation network that linked the road systems with those of Maryland and the District of Columbia had to keep pace with continued development. Fairfax County officials, faced with the substantial costs of increasing all forms of infrastructure for their rapidly expanding communities, saw no alternative to charging tolls on a new highway. And after this roadway was built, the fairly reasonable toll rate of 7 cents per mile and the high volume of expected traffic could not only meet the construction costs of $57 million but could produce a small profit for the Virginia Department of Transportation (VDOT). Thus, in the 1970s VDOT leased the Dulles Corridor right-of-way from the federal government and began planning the construction of a toll road to service the increased traffic generated by the rapid growth of nearby communities.

THE GENESIS OF THE DULLES GREENWAY PROJECT

Two men were primarily responsible for the idea that eventually evolved into the Dulles Greenway. John Miller had been the guiding force behind the Municipal Development Corporation (MDC), a small New York–based company that was an affiliate of Catalyst, a developer of cogeneration plants. The mission of MDC and John Miller was to look for and develop privatized infrastructure projects. Bill Allen, a top executive in the transportation engineering firm of Parsons Brinckerhoff had worked on the Dulles Toll Road project and was also involved in consulting work at the Dulles International Airport. Allen was familiar with the growth potential of the area and first brought the concept of a toll road extension to Miller's attention.

At about the same time that Miller and Allen were examining a toll road's potential, the governor of Virginia, Gerald Baliles, had as one of his top priorities increasing the quality of transportation. To better formulate policy, the Commission on Transportation for the 21st Century was established. The report issued by the Commission in August 1986 identified the transportation needs of Virginia and put a price tag of $7 billion on the program. These monies would have to be raised by a combination of bond issues and state gasoline tax and sales tax increases. The state legislature eventually reached a compromise between providing what was actually needed in revenue as projected by the Commission and what revenue enhancing legislation the lawmakers thought they could enact. John Miller saw an opportunity to step into the breach; Miller and a consultant, Steve Pearson of Hazel and

Thomas, presented a legislative draft proposing a privately funded toll road to the Commission. The Commission in turn recommended that the Virginia state legislature explore the potential for privately funded and operated roadways.

Miller and Pearson's arguments in favor of legislation permitting the private financing, construction, and operation of toll roads within Virginia were compelling and resulted in a bill introduced by state senator Charles Waddell, the Chairman of the Senate Transportation Committee. In early 1988 the privitization legislation was enacted, resulting in the creation of the Highway Corporation Act.

THE HIGHWAY CORPORATION ACT OF 1988

The Virginia Assembly announced its desire to invite developers to participate in a renewed highway expansion program as the following policy:

> The General Assembly finds that there is a compelling public need for rapid construction of safe and efficient highways for the purpose of travel within the Commonwealth and that it is in the public interest to encourage construction of additional, safe, convenient, and economic highway facilities by private parties, provided that adequate safeguards are provided against default in the construction and operation obligations of the operators of the roadways.
>
> The public interest shall include without limitation the relative speed of the construction of the project and the relative cost efficiency of private construction of the project.
>
> The General Assembly further finds that the use of public funds for the purposes set forth in this section is in the public interest.
>
> Accordingly, the General Assembly finds that this chapter is necessary for the public convenience, safety and welfare.

The Highway Act and the Dulles Greenway Project

The Virginia Highway Corporation Act of 1988 established the prerequisites for construction and operation of a toll road in the Commonwealth of Virginia. The Act would provide the guidelines not only for the Dulles Greenway, but for any other public/private transportation projects to be built in Virginia. The Act set forth the powers of the Commission. It also contained information regarding the submitting applications, obtaining a certificate of authority, and dealing with eminent domain issues, as well as the powers and duties of the roadway operator; various inspection, insurance, and approval processes; what constitutes default and how it is to be addressed; the issue of police powers; conditions where termination of the certificate could be effected; and the provisions of the Improvement Fund.

Powers of the Commission The Commission was granted the authority to regulate the applicant as a public service corporation and was charged with the responsibility of supervising and controlling the operator in its performance. The Commission would also have the responsibility of approving or revising the toll rates charged by the operator. Initial rates would be approved if they appeared reasonable in relation to the benefits received and appeared not to discourage use of the roadway. A reasonable rate of return to the operator would be determined by the Commission, but after acting on a complaint about toll rates or its own initiative the Commission may require the operator to set the tolls at a rate deemed more reasonable in relation to the benefit received. The Commission would charge the developer a small annual fee to cover the cost of reviewing and inspecting its operations.

The Application for a Certificate of Authority The Commission set a reasonable fee to cover the cost of reviewing and processing, approving or denying an application to design, build, operate, and maintain a private toll road. The Act specified that the following items must be included in the developer's application:

1. The geographic area to be serviced and a topographical map indicating the route of the proposed roadway
2. A list of property through which the proposed roadway will traverse and the names of the property owners
3. The method by which the applicant plans to secure the necessary rights-of-way
4. A comprehensive plan for all counties, cities, and towns through which the proposed roadway will pass and an analysis to reflect that the roadway conforms to that plan
5. The operator's plan to finance the project, including proposed tolls, anticipated traffic, and detailed plans to show distribution of funds collected
6. A plan for the operation of the proposed roadway
7. A list of all permits and approvals required for the project
8. An overall description of the project, the design, and the proposed interconnections with state, interstate, and secondary highways and local streets
9. A list of public utilities to be crossed and/or relocated
10. Certification that the proposed roadway will be designed to meet VDOT standards and built in accordance with a proposed timetable
11. Completion and performance bonds in a form and amount satisfactory to the Commission

The Power of Eminent Domain The power of eminent domain was not to be used by a project's proposer for the acquisition of land for the project. This was in keeping with the Act's concept of noninvolvement in the private developer's land acquisition activities.

Powers and Duties of the Roadway Operator The operator of the toll road would be vested with the authority to operate the roadway, collect tolls, and pledge any revenue net of operational expenses to repay obligations incurred during construction. Financing of the project was to be at the discretion of the operator; repayment would be effected by the collection of tolls, and the state would have no responsibility to assume any financial obligations of the operator.

In operating the roadway, the operator may

- Classify traffic according to reasonable categories for the assessment of tolls
- Set and enforce maximum and minimum speed limits, and exclude undesirable vehicles or cargoes from using the roadway, with the consent of the Department
- Establish commuter lanes for use during the day or any part of the day, with the consent of the Department
- Do anything that is deemed reasonable and proper in the operation of the roadway provided the practice is reasonable, nondiscriminatory, and meets with the Department's approval

The operator was vested with the following duties over and above those listed above:

- The operator must file and maintain at all times with the Commission an accurate schedule of rates charged to the public and a statement that those rates would apply uniformly to all users within the assigned classification.
- The operator will construct and maintain the roadway in accordance with the appropriate standards of the Department and allow for periodic inspection of any construction or enlargement of the project. The operator must cooperate fully with the Department in establishing any interconnection with the roadway that the Department may make.
- The operator must contract with the state for the enforcement of traffic and public safety laws and can also contract with local authorities for those portions of the roadway within their local jurisdictions.

Department Approval and Inspection Agreements The state Board of Transportation has the authority to approval or deny application for a certificate. The Board approves the project and its interconnections if there is a public need for the project and if the project is compatible with the existing road network. Construction costs must be deemed to be reasonable, and project authorization is conditional upon receiving all required approvals.

If the project is approved, the Board enters into a comprehensive agreement with the operator which will include reviews of all plans, specifications, and proposed maintenance practices; reimbursement of all Department direct costs; and assurance that the operator will establish and fund an account to meet all of their financial obligations including establishing reasonable reserves for contingencies and maintenance replacement costs. The costs for which an operator must reimburse the Department include any services performed by the Department on behalf of the operator such as project development review costs and the cost of any necessary environmental impact statements that the Department prepared for a private developer.

Insurance and Sovereign Immunity The operator must obtain public liability insurance in the form and amount that satisfies the Commission. The state does not waive its right of sovereign immunity with respect to its participation in or approval of any portion of the proposed roadway application or operation, including but not limited to the interconnection of the roadway with the existing state highway system. Counties, cities, and towns through witch the proposed roadway would pass will also retain their sovereign immunity with respect to the proposed roadway construction and operation.

Utility Crossings and Relocations The operator and any public utilities whose works are to be crossed or relocated must cooperate with each other and do what is necessary for the project to proceed. If the operator and public utility owner cannot agree on the terms and conditions of the crossing or relocation, the Commission could be called upon to review the situation and render a decision in the matter. The Commission can employ experts to review a particular dispute and both the applicant and the public utility will share in paying the Commission's costs.

Default If construction has not begun within two years after the issuance of a certificate, the Commission can hold a hearing to review all facts involved in the delay, and can revoke the certificate of authority if necessary. Appropriate claims could then be made against applicable bonds in effect at that time. The Department would receive the full proceeds of any payments due to claims against the bonding companies. The Department would take into account any costs incurred in connection with the completion of fulfillment of unperformed applicant obligations and return any remaining funds to the applicant.

Police Power The Highway Act provides that state police officers will patrol the roadways even though portions of the roadway may lie within the corporate limits of other jurisdictions. The operator and the Department of State Police must agree on reasonable terms and conditions for the policing of the roadway. The traffic and motor vehicle laws of Virginia apply to all persons and vehicles traveling on the roadway, and officers in charge of policing the roadway are under the exclusive control and direction of the superintendent of state police.

Termination of Certificate The operator must provide the Commission with a statement of full disclosure of all financing arrangements including the terms of all bonds within 90 days of the completion and closing of the original permanent financing. The operator must notify the Commission of the term of the original financing and its termination date. the authority and the duties of the operator cease and all highway assets and improvements revert to the state when the certificate of authority terminates.

The Improvement Fund The Act requires that the state's Board of Transportation establish a fund from a portion of the toll revenues to provide for transportation improvements related to the toll road. A percentage of each toll that exceeds the amount necessary to enable the operators to meet their obligations and earn a reasonable return must be committed to the improvement fund.

The Qualifying Transportation Facilities Act of 1994 The Qualifying Transportation Facilities Act, effective July 1, 1995, elaborated on the requirements of a private operator's application for approval. Section 56–562 requires a private operator to obtain a certificate of public convenience verifying that there is a public need for the construction, enlargement, or acquisition of the planned transportation facility. Another proviso in the Act prevented an operator imposing tolls or user fees on any interstate highway or existing road, bridge, tunnel, or overpass without the consent of the affected local jurisdiction.

The Toll Structure In February 1995 Virginia's Secretary of Transportation, Robert E. Martinez, affirmed the state's stance on toll facilities: "It's always nicer not to pay tolls. But if the alternative is an increase in the overall tax level or deferral for several years of projects with merit, tolls and user fees are preferable." VDOT paragraph 56.542 conveyed to the State Corporation Commission the power to approve and revise the toll rates proposed by an operator:

> Initial rates shall be approved if they appear reasonable to the user in relation to the benefit obtained, not likely to materially discourage use of the roadway and provide the operator no more than a reasonable rate of return as determined by the Commission.

State regulators considered a reasonable rate of return to the Dulles Greenway investors as beginning at 30 percent and reduced to 15 percent once toll revenues exceeded debt service. After the latter point is reached, the return on investment would decline to 14 percent and remain at that level until the end of the contract period, 42.5 years.

THE DULLES GREENWAY PROJECT BEGINS

The Public Hearings

The Virginia Department of Transportation began to hold public hearings on the proposed 14-mile Dulles toll road extension in the spring of 1988. Through the hearings the residents of Fairfax and Loudon Counties became aware of the state's search for a private corporation to build, finance, operate, and maintain this new highway, the title to which would revert back to the government after a specific term. This road would be a toll road with revenues going to the developer, and even though toll rates would be established by the private developer, these rates would have to pass the Commission's test of being "reasonable user fees."

The toll rate finally accepted by the Commission was $1.75 for the Greenway. Travel on the government-owned Dulles Toll Road was $0.85, so the ride from Leesburg to the District of Columbia, a distance of about 35 milles, would cost commuters $2.60. The alternative route—while less expensive—would probably be less desirable to most daily travelers, considering the 23 traffic lights that dot the way to downtown Washington. Both alternate State Routes 7 and 28 and the route of the toll road with interchanges are shown in Figure 3.1.

With hearings out of the way and legislation in place, on July 20, 1989, the Board approved an application for the Dulles Toll Road extension submitted by the Toll Road Corporation of Virginia.

The Project Parameters

The toll road, referred to as the Dulles Greenway, would extend about 14 miles from the Dulles International Airport northwest to Leesburg, the Loudon County seat. The roadway would connect to the existing Dulles Toll Road and, combined with that highway, would provide a limited access route through the Loudon–Fairfax County corridor.

The Dulles Greenway was to be a four-lane limited-access highway within a 250-foot right-of-way. An 88-foot median would allow for the creation of an additional two lanes while still retaining an easement for future mass transit development. There would be nine interchanges along the route, seven of which were to be constructed as the highway was being built; the other two would be future projects. The Greenway would fulfill several needs:

1. Providing a key link in the east-west Fairfax–Loudon County roadway system
2. Relieving congestion along several existing routes
3. Accommodating the increased growth of Dulles Airport and environs
4. Acting as an economic stimulus to Northern Virginia
5. Providing an estimated 4,500 jobs during construction

The Dulles Greenway would also employ the latest Automatic Vehicle Identification (AVI) equipment to save travelers from having to come to a complete stop at toll collection points.

The Principal Members of the Toll Road Corporation of Virginia (TRCV) Team

The Primary Private Financial Backer The driving force behind TRCV was Magalen O. Bryant, heiress to the family fortune amassed by her father, George Ohrstrom. Ohrstrom had enjoyed a successful career as a stockbroker, and had also built a sizable conglomerate which included the elevator manufacturer Dover Corporation and the Carlisle Companies, world leaders in single-ply elastomeric mem-

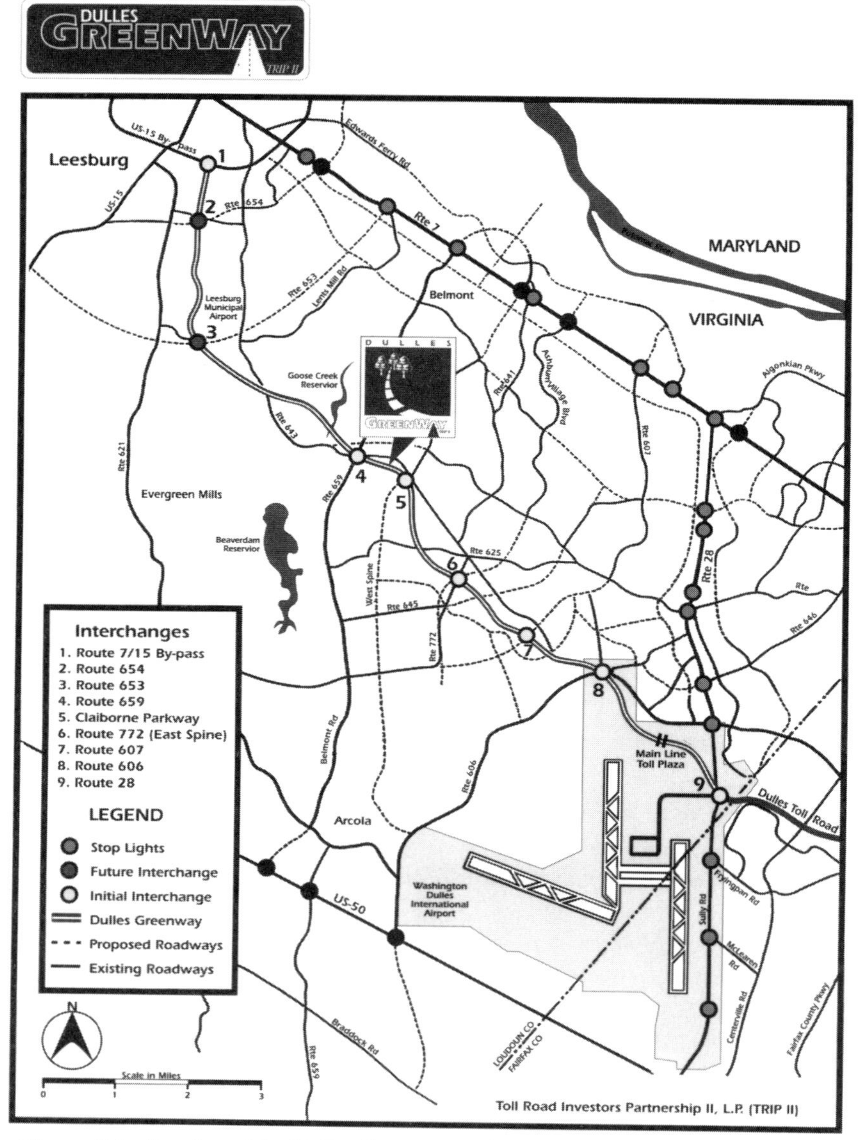

Figure 3.1. Routing of the Dulles Greenway. *[Courtesy Toll Road Investors Partnership II, L.P. (TRIP II).]*

brane roofing systems, as well as several other corporations. Bryant owns Locust Hill Farms, a horse breeding estate located near Middleburg, Virginia. She is an avid environmentalist, and is vitally interested in the welfare of her state and the need to increase Virginia's transportation system while preserving its scenic open spaces. As an astute investor, she is dedicated to ensuring that any investments made by her family-owned investment company, Lochnau Limited, will yield satisfactory rates of return. Bryant saw an opportunity to accomplish all of these goals by investing $68 million in the Dulles Greenway project where it was conceivable that a 30 percent return on investment, the maximum allowed by the newly enacted legislation of the Commonwealth of Virginia, could be attained.

The Contractor Initially Kiewit Eastern Company, a division of Peter Kiewit Sons of Omaha, Nebraska, had been proposed as the general contractor for the Dulles Greenway project, but their executives decided the risks were too great to become involved in the construction. According to an article in the November 15, 1993, issue of *Engineering News Record* magazine, when Kiewit pulled out it had already invested $5 million in time, money, and materials in the project. When Brown and Root stepped into the picture it was awarded a $145 million fixed-fee contract and also became a 13.8 percent investor in the limited partnership that was created.

Brown and Root was established in 1919, and it has evolved into one of the world's top design and construction firms. It operates worldwide, and employs approximately 10,000 technical people and 30,000 office and support personnel. Founder George Brown had been one of the great men of Texas politics back in the 1940s and had counted among his best friends U.S. Congressman Lyndon Baines Johnson. This Texas background led the company into energy-related construction projects. In later years, Brown and Root would construct the world's largest offshore pipeline which stretches 670 miles from Kalsto, Norway, to Zeebrugge, Belgium. The company also designed, built, and installed the world's first commercial offshore oil and gas production platform in the Gulf of Mexico in 1947.

The Operations and Maintenance Entity Autostrade International S.p.A., based in Rome, Italy, is a constructor, concessionaire, and operator of extensive roadway systems in its home country. Autostrade International became a general partner in the Greenway corporate entity, and will employ an AVI system on the Dulles Greenway project.

Autostrade International got its start in 1956 with the construction and concession assignment for the Autostrada Del Sole running from Naples through Bologna and Florence to Milan. Today the company operates more than 1,800 miles of roadways in Italy and its activities include toll collecting, traffic control and regulation, and maintenance of these roadways. The company also acts as a general contractor and subcontracts all phases of work necessary for highway construction.

Autostrade has a fair amount of experience in the installation and operation of AVI equipment as well. All the Italian routes it services employ the VIACARD AVI system, in which a current account or prepaid motorway credit card can be inserted like an ATM card at the toll booths. It also provides a service known as VIACARD PLUS, which allows the cardholder to charge other services available on the motorway such as fuel, restaurants, and lodging. The system TELEPASS allows roadway users to enter and exit the highway without stopping by providing these users with a small on-board transceiver. A signal sent from this windshield-mounted transceiver communicates with ground-stationed electronic equipment to pick up entry and exit and to charge the tolls to the VIACARD holder.

The Construction Process Manager General Charles E. Williams retired from the U.S. Army Corps of Engineers in 1989 after directing the Engineering Division's $2 billion of design and construction work. Williams was program manager for the largest Corps construction project since World War II, the $1.3 billion Fort Drum complex in New York. Upon leaving the service, he became president and chief executive officer of the New York School Construction Authority which had a $4.13 billion public school building program. As chief operating officer of the Toll Road Investors Partnership, Williams became the key element in getting the Dulles Greenway project moving, an ability he demonstrated by commencing the clearing and rough grading one day ahead of schedule.

Financing the Project

The projected total cost of the project was estimated at $326 million, and raising additional equity capital was no easy task at the time—wary investors had been burned by previous deals, and the country was still trying to recover from a recession. To attract new financing sources the Toll Road Corporation of Virginia (TRCV), the entity that had originally obtained a certificate of authority from Virginia, decided to change its structure from a corporation to a limited partnership. The State Corporation Commission approved the transfer of the certificate of authority to this new entity, the Toll Road Investors Partnership II (referred to hereafter as TRIP II). The strategy proved successful, but Bryant's Lochnau Ltd. remained the largest investor with a 57.04 percent portion. Of the initial $68 million investment, $22 million was for equity financing and the remaining $46 million provided access to various lines of credit that would serve as guarantees against project risks. Autostrade International, Italy's largest toll road operator, bought a 19.16 percent interest and the right to operate the Greenway.

Long-term financing in the amount of $202 million was provided by a consortium of 10 lending institutions including John Hancock Insurance Company, CIGNA Investments Inc., Barclays Bank, NationsBank, and Deutsche Bank. Deutsche Bank also provided a $40 million revolving credit instrument. Prudential Power Funding Associates purchased $58,333,000 in senior fixed-rate notes due 2022 as part of its function as co-lead investor in institutional financing for the construction and operation of the roadway.

As an expression of confidence in the project, many of the owners of property that would have been designated as right-of-way donated their land to TRIP II with the expectation that other portions of their properties that were contiguous or near the right-of-way would increase in value substantially, making future land sales or development highly lucrative.

Early Setbacks

The passage of the Dulles Greenway from concept to reality was a slow and arduous one. The acquisition of rights-of-way, inextricably tied to the acquisition of equity and debt financing, brought the project close to disaster at several points along the way.

On September 1, 1992, after delays in obtaining short-term financing and the necessary rights-of-way the original contractor for the job, Kiewit Eastern, decided to mothball its field operations until all financing was in place. The plan had been to start construction on July 1, 1992, but this was pushed back to August 3—after which it was further extended to September 1.

In October 1992 TRCV announced that it was "furloughing" a number of officers and company employees in order to minimize the continuing development costs resulting from delays in completing the project's financing arrangements. TRCV had been waiting for additional commitments from some other participating banks, but these commitments never materialized. At that point it appeared that the entire toll road project would collapse.

TRCV was to have obtained all necessary rights-of-way by January 1, 1993, but it became doubtful that the deadline could be met. On December 14, 1992, a public hearing was scheduled at the Loudon County Board of Supervisors to consider a request by TRCV to extend its application for another six months. In that meeting Supervisor Zurn, Republican from Sterling, Virginia, as reported in a local publication the EIR News on Jan. 7, 1993, said that as far as he was concerned, this latest request for renewal of TRCV's special exception would be the group's last: "I'm not going to do any more extensions. Someone has got to tell them, 'Put up or shut up.' " TRCV responded by announcing that it was optimistic about a spring 1993 start. But the apparent black cloud hanging over the project continued to plague the developers: The December 14 meeting was invalidated because the 7:30 P.M. meeting time had been incorrectly published as 7:00 P.M.

The fourth extension to the special exception was finally granted and carried with it an expiration date of June 1, 1993, but a joint hearing was set for July 7, 1993, at which time the developer was scheduled to submit an entirely new application which would ostensibly require another public hearing. A Catch 22 had developed—no financing could be finalized unless TRCV received approval for an extension of the special exception beyond June 1, 1993, and the extension was required before lenders would sign off on their commitments. In January 1993, contractor Kiewit Eastern had voiced concerns about the new financial arrangements and, according to a January 7 newspaper article, a spokesman for TRCV was quoted as saying that negotiations were underway with other contractors. Even though Kiewit Eastern had not yet officially withdrawn as the project's construction contractor, negotiations already had begun with Brown and Root which was also expected to join the team as an equity investor. On July 7, 1993, the Board of Supervisors gave

TRCV a final 90-day extension and it appeared that TRCV was finally going to wrap up all financial considerations. TRCV had yet to finish negotiating a right-of-way agreement with a partnership that owned one 190-acre parcel; in early August of 1993 the partnership decided to file a lawsuit against the developer. On Sept. 24, 1993 the Leesburg Today newspaper reported that the recalcitrant landowners finally agreed to sell the necessary right-of-way and withdrew their lawsuit. Financing was completed by the third week of September 1993. At that time TRCV also announced that a construction contract had been signed with Brown and Root. Thus the way was cleared for the ground-breaking ceremony on September 29.

PUBLIC RELATIONS AND THE DULLES GREENWAY PROJECT

Even though most of the local residents had been wholeheartedly in favor of the Greenway project, there were some vocal opponents. For example, one article published in the local paper called the project a "boondoggle" that would allow banks and wealthy note holders to steal money from the pockets of residents. Another article assured readers that the "bloodsucking debt service guaranteed that the transit system would be plunged into a total state of collapse and decay." The Greenway project required a sensitive public relations effort to reduce adverse public opinion to a minimum. The public needed assurances that the developers were not only aware of its concerns, but were doing everything in their power to lessen the impact of the Greenway's construction on local quality of life and the surrounding environment.

The Greenway and the Community

Suzanne Conrad, Director of Public Affairs and Marketing, displayed an unusual sensitivity to the needs of the communities in and around the Dulles Greenway construction area. An early program was the Greenway-sponsored contest for 1,200 Loundon elementary school children from kindergarten to 6th Grade to name the Greenway mascot. Contest entry forms included a black-and-white outline of a cute chipmunk wearing a hard hat and a Dulles Greenway T-shirt standing on the Greenway construction site (Figure 3.2). Students were asked to color in the drawing and name the mascot. The winning entry would receive a $100 savings bond and a ride in the opening day parade. The response was immediate and enthusiastic, and Scooter the Commuter was born. Scooter would be used in a number of other public relations announcements (Figure 3.3). Conrad also invited the Loudon Healthcare Association to host a 10K run on the Greenway in September 1995 as one of the kick-off ceremonies for the Greenway Grand Opening in November 1995. A Tour de Loudon bike race was also scheduled for September.

The announcement that construction workers would be working double shifts for a period of time commencing January 1994 was one of the project's greatest public relations challenges. Although TRIP II officials indicated that the noise generated would be well below the 75 decibel level allowed by county ordinances, the sound

Figure 3.2. The coloring sheet for naming the Dulles Greenway mascot distributed to area school children. *[Courtesy Toll Road Investors Partnership II, L.P. (TRIP II).]*

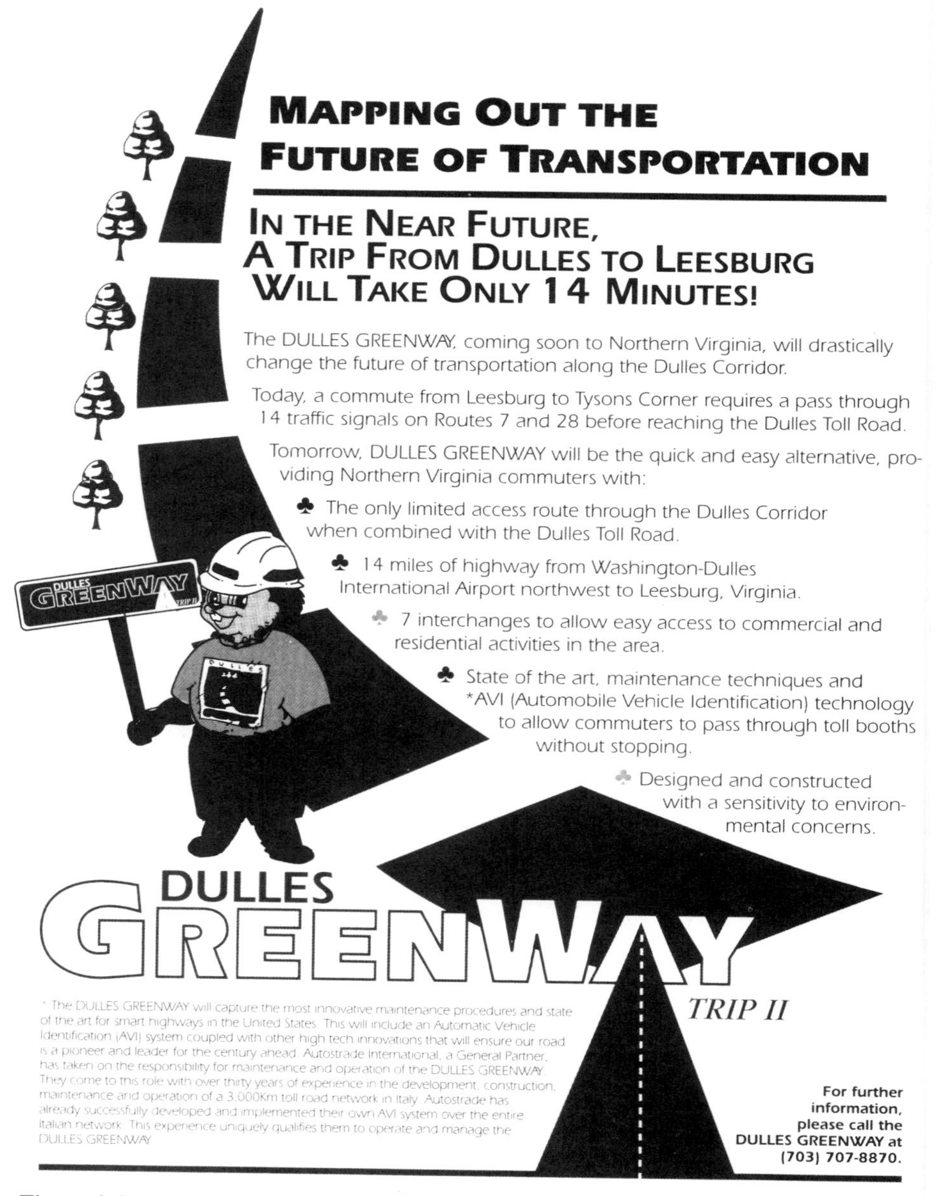

Figure 3.3. Scooter the Commuter, the Dulles Greenway mascot. *[Courtesy Toll Road Investors Partnership II, L.P. (TRIP II).]*

of earthmoving equipment operating near adjacent homesites until 4:00 A.M. each morning for 45 to 60 days would undoubtedly prove unnerving to homeowners. The double-shift operation began on Wednesday, January 5, but project officials were ever sensitive to community relations and modified the work schedule just a few days later. Conrad notified Asburn Farm, near-neighbors to the construction activity, that no equipment would again work between the hours of 11:30 P.M. and 4:00 A.M. With a potentially severe winter in store and a huge debt service facing TRIP II if timely completion was delayed, the work schedule modification was a significant public relations compromise. Additionally, residents were assured that decibel level checks would be conducted frequently during the day.

Keeping homeowners apprised of planned construction activites was an important part of Conrad's daily routine. She was also responsible for a special telephone hotline to handle construction-related complaints regarding noise, glare and lighting intensity, public safety, and so on. As of April 1995 when construction was more than 80 percent complete, she had received only 29 complaints through the hotline, nine of which came in when the double-shift work took place. Residents appreciated the project's responsiveness; a spokesperson for Smokehouse Court residents said that "to their credit toll road officials have been very nice and cooperative."

The Greening of the Greenway

The Dulles Greenway project's approach to environmental issues reflected the influence of Magalen Bryant, the owner of Lochnau Ltd., the project's primary investor. Bryant has a long history of activism in the field of environmental concerns. She served on the Board of Directors of the National Fish and Wildlife Federation, and founded D.E.L.T.A, a nonprofit organization for the conservation and restoration of 5 million acres of the lower Mississippi River delta ecosystem. Bryant also is the founder of Tara Wildlife Management and Services Inc., a firm dedicated to the continuation and enhancement of wildlife.

Construction of the Dulles Toll Road extension would result in the loss of about 64 acres of federally protected wetlands. However, the Greenway was allowed to obtain an Army Corps of Engineers 404 Permit to replace forested wetlands on a 2:1 ratio and emergent wetlands on a 1.5:1 ratio. The new wetlands would be monitored over a five year period to ensure that the re-created natural habitat would sustain wetlands plants and wildlife. In order to ensure the survival of these wetlands, a performance bond in the sum of $2.5 million was submitted to the Corps of Engineers.

Temporary and permanent storm sewer management ponds were established in the right-of-way whenever possible, and any trees cleared outside of the right-of-way were to be replaced on a 1:1 acreage replacement schedule. Large areas outside of the right-of-way would be reforested and approximately 622 two- to three-year-old seedlings per acre would be planted. More than 23,000 native hardwoods, shrubs, and grasses were individually planted by TRIP II in April 1994.

The preservation of Goose Creek was critical to the environmental planners of the Greenway. Along with the standard erosion control measures taken by the contractor, the piers of the bridge spanning Goose Creek were located on-shore rather than in the water (see Figure 3.4). Further safeguards against creating sedimentation problems while constructing these bridge piers would be accomplished by building cofferdams around the pier construction areas. Water infiltrating into the cofferdams would be pumped into sediment traps a distance from the creek. Turbidity monitoring of the Goose Creek Bridge was conducted twice daily during the active construction period. Provisions to guard against oil and gasoline spills into Goose Creek were numerous, and included the installation of two emergency call boxes alongside the creek to report spills immediately to the proper authorities.

A CLOSE LOOK AT THE CONSTRUCTION OF THE DULLES GREENWAY

The Greenway's construction started a day early and Charles Williams, the CEO of TRIP II, kept it at a fast pace. As of September 1994 the project was ahead of schedule; half the work activity had been completed in the first 11 months of the 30-month project schedule. Williams' complete understanding of the construction process and his unique ability to view a problem through the eyes of the other parties in the construction process were responsible for the rapidity with which the Greenway construction proceeded—as of April 1995, it was seven months ahead of schedule. His methods and relationship with the other members of TRIP II can serve as inspirations not only to large road building operations but to any type of vertical construction.

In an April 1995 interview Williams shared some of his management techniques that result in a successful construction cycle:

> The management of this project represents the function of three things:
>
> 1. Understanding the myriad of interplays between the construction execution processes. You must have knowledge of that in order to focus the execution apparatus. And you must set in place resolve and teeth into the management apparatus for the owner. Protect the owner.
> 2. And then, of course, once that happens, you must then control the process. And you have to go beyond paper—you have to physically do things to control it.
> 3. And then what you do with this control element is where we start, is a combination of, first of all disciplining the process. That is, put the general contractor in its rightful role—and that is to manage the CONSTRUCTION not the process.
>
> We sprinkle all of that on the grid and with a good communications system, that, in my view, is a paradigm shift.

Williams draws an analogy between the bakery and the construction businesses. Just as to open a bakery needs someone who knows how to bake bread, operating a construction company requires someone who knows construction. And thus, operating a development company needs someone who either understands all of the

(a)

(b)

Figure 3.4. (a) The span across Goose Creek. (b) The bridge piers spanning Goose Creek to mitigate environmental concerns.

aspects of development or is surrounded with a cadre of experts who do know and understand all aspects of the development project being considered:

> Following the bakery concept, no one on this job can know the engineering, can know the numbers, and sort the cost parameters any better than the owner. Otherwise I (the owner) can't manage it. So what I did was go through the thing (refering to the project), engineered it, priced it out. Then I issued and invitation to bid. It was private, so I didn't have to issue an RFP (Request For Proposal). I called five of the top road builders.

Williams recognizes that contractors are willing to assume risks—for a price; so he decided it made good business sense to reduce the risk exposure of the Greenway contractors in order to extract the best possible price for construction. First, he advised the contractors that his office would assume responsibility for obtaining all of the permits required for the project, a process that is often a major headache for a contractor not thoroughly familiar with the myriad regulations of governmental agencies. In a project of the scope of the Dulles Greenway, processing all of the required permits would have been a real problem for anyone but Williams, who was already intimate with the process.

Second, he recognized that insurance is a significant element in a general contractor's cost estimate, and some of these insurance requirements are imposed on subcontractors as well, adding even more costs to the insurance package. So Williams announced that TRIP II would assume responsibility for all the insurance requirements for the Greenway project. And because TRIP II was paying the insurance premiums, he instituted a very effective safety program which ensured that workers' compensation claims were held to a bare minimum.

Charles Williams' view of the construction world provides several valuable lessons:

1. Know what you are doing—knowledge of product is essential in every field of endeavor.
2. As leader of the team, be knowledgeable in all aspects of the process.
3. Learn to share risks.
4. Be fair in your dealings—don't just emphasize penalties, but offer incentives for initiative and hard work.
5. Don't be a creature of habit. Look for new solutions to problems. Be creative. Look for new paradigms.
6. If you're going to open a bakery, you'd better know how to bake bread!

THE DULLES GREENWAY AVI SYSTEM

Syntonic, a pioneer in automatic vehicle identification and electronic toll and traffic management systems headquartered in San Diego, California, provided the AVI and ETTM systems for the Greenway.

Syntonic, founded in 1938 as a two-way radio service company in the wireless communication field, was acquired by Science Applications International Corporation (SAIC) in 1994. SAIC was founded in 1969 as a consulting company to the defense industry and has grown rapidly into a full systems provider of traffic control and management systems.

Syntonic's Dulles Greenway installation was its Advanced Revenue Collection System (ARCS), which provides windshield-mounted transponders for vehicles. Highway-mounted antennas pick up the vehicles' signals as they pass through an AVI lane and debit motorists accounts. The need to slow down, stop, and drop coins into a chute has been eliminated for commuters on this toll road.

The Dulles Greenway is faced with a special problem because it connects with the previously existing Dulles toll road. The AVI system for the Greenway is compatible with the Dulles Toll Road's existing system in that certain lanes are available to handle cash and coin collection in the conventional manner. When Greenway AVI customers continue down the Dulles Toll Road to the District of Columbia both tolls are debited to their accounts; a portion is payable to the Greenway and a portion to the Virginia Department of Transportation.

The Greenway ARCS system must also deal with AVI drivers who have insufficient funds in their accounts after traversing the Greenway to proceed on the VDOT toll road. After patrons leave the Greenway and enter VDOT's Dulles Toll Road, 15 to 20 minutes elapse before they pass through the VDOT toll plaza. During this time their account status must be updated and transmitted to the VDOT computer. The situation is being addressed in three ways:

1. The AVI transactions are transmitted from the lane controller to the plaza computer to the tag store and back to the lane controller in "real time" using leased telephone lines or fiber-optic lines linking all components of the toll collection system together. If an account status changes from valid to invalid at a single point, the change in status is immediately passed on to the next toll plaza before the driver reaches it.
2. Two account thresholds are established to alert drivers and toll authorities as the patron's account dwindles. One threshold will advise both parties that there is enough money for a few more trips, and the second threshold will advise both parties that funds are no longer available for further trips. When the second threshold is reached the driver must make payment manually at the toll booth or be classed as a violator subject to law enforcement procedures.
3. The third method provides some sort of automatic replenishment of the driver's account either by credit card or automatic withdrawals from the driver's bank account. Replenishment would take place after the first threshold is reached, thereby saving both driver and toll booth attendant from holding up traffic.

The ARCS system should be able to perform the following functions:

1. Interface with a third-party AVI customer service center whose other participant is the FASTOLL system operated by VDOT, thereby providing Greenway drivers with a single point of contact for the VDOT Dulles Toll

Road, the Dulles Greenway, and any other AVI systems initiated later by VDOT in Virginia
2. Establish semivariable message signs over each toll lane to indicate traffic problems ahead
3. Accommodate multiple toll schedules in order to collect VDOT tolls and provide audit functions to separate and dispense the proper share of revenue to VDOT
4. Maintain a reversible lane system at the main toll plaza
5. Remotely program the operational mode of each toll lane from the control center in the plaza headquarters

The Dulles Greenway AVI system will permit an estimated 1,400 cars per hour to pass through each AVI-dedicated lane, as opposed to an estimated 600 to 800 cars passing through conventional toll collection facilities.

WHAT WILL THE DULLES GREENWAY DO FOR VIRGINIA?

On September 30, 1995, amidst great fanfare, the Dulles Greenway was opened to the public. Over 3,000 people attended the opening ceremony including Virginia's Governor Allen and U.S. Secretary of Transportation Peña. Public Relations and Marketing Director Suzanne Conrad had survived a hectic week: "It was an impressive event, and I felt like Steven Spielberg before the planning was over and the event held." The celebration marked the end of the long race to become the first operational public/private highway project utilizing an advanced AVI system in the United States—quite an accomplishment.

Without this private venture, it is doubtful that Virginia alone would have been able to obtain sufficient funding for the Dulles Greenway. With this new roadway in full operation, travelers to and from Dulles Airport and commuters from all parts of Fairfax and Loudon Counties have a first-rate highway addition, and the only people who pay for this service are those who use the road.

According to CEO Williams, the operation and maintenance costs of $1.2 billion can be paid entirely out of toll revenues, and the Dulles Greenway business entity will pay more than $1.3 billion in federal and state income taxes over its 42.5 year franchise period. An additional $20.2 million will be paid to the Metropolitan Washington Airport Authority in rental fees. During construction of the project, 456 new jobs were created, and small businesses—from local hardware stores and stationary stores to bottled water suppliers and xerox repair services—all shared in the increased business that inevitably piggybacks on large-scale construction activity. As for the future, the additional taxes that will be paid to local communities and the Commonwealth of Virginia from the creation of new businesses and new residential communities and from increased business activity generated by existing concerns may be more difficult to quantify but not to difficult to envision.

The Dulles Greenway was conceived by visionaries with an eye to the practical.

Hard work, perserverance, and faultless preparation were the cornerstones of bringing this public/private concept to fruition. The willingness of one women to risk a fortune on an idea whose time had come allowed this project to advance from inception to completion. And Bryant's opinion was shared by some of the world's leading financial institutions who were willing to provide the required additional financing. The experience, dedication, and just plain grit of the operations team and its partners made it all happen. The Dulles Greenway project can serve as an ideal model for future highway Build, Operate, Transfer projects in the United States.

4

CALTRANS AND THE PUBLIC/ PRIVATE SECTOR PROJECTS

From its ambitious beginning, the transportation privatization program in California has experienced several ups and downs and a few major upheavals. Six years after California Governor George Deukmejian signed Assembly Bill 680 into law in 1989 giving the green light to private sector participation in the state's transportation system, two of these projects are in a holding pattern, one is bogged down in environmental issues, but the fourth is barreling along and is scheduled to be completed by the end of 1995.

Federal cutbacks in defense budgets hit the California economy hard, and no one ever could have predicted that Orange County, one of the richest counties in the United States, would file for protection from creditors under Chapter 9 of the Federal Bankruptcy Code in late 1994 after suffering a reported $2.5 billion loss in its investment portfolio. But Caltrans, the California Department of Transportation, has forged ahead in its pilot program to enhance the state's transportation system by embracing the private sector. The State Road 91 project with its unique electronic tolled express lanes, scheduled to open by the end of 1995, may well determine the immediate future of other private-sector transportation endeavors in California.

CALIFORNIA ASSEMBLY BILL 680

Assembly Bill 680 authorized the California Department of Transportation to "exercise any power possessed by it with respect to the development and construction of state transportation projects to facilitate the development and construction of the privately constructed projects, and would require the agreement to provide for reimbursement for maintenance and police services." Section 1 (c) of the bill introduced the establishment of public/private partnerships for the construction of projects in California: "One important alternative is privately funded Build-Operate-Transfer (BOT) projects whereby private entities obtain exclusive development agreements to build, with private funds, all or a portion of public transportation projects for the citizens of California." The bill authorized Caltrans to enter into agreements with private entities for the construction of four transportation demonstration projects, including at least one in northern California and one in the southern part of the state.

The provisions of Assembly Bill 680 allowed Caltrans to perform certain environmental, developmental, and operational services for the private sector and receive payment for these services from the private developer's revenue stream once the project was up and running. Caltrans was authorized to enter into partnerships with developers and lending institutions for the purpose of leveraging loan funds. AB 680 also allowed the use of federal funds as seed money in order to develop these new tolled highways. Investors would be granted leases for up to 35 years to operate transportation facilities and recoup their investment and profits through toll collection and land development revenue. Caltrans would support these private initiatives by accelerating schedules and negotiating equitable leases and reasonable rates of return on investment with these developers.

Caltrans would offer assistance in the reduction of tort liability after these highways were completed by the private entities. According to Charles Stoll of Caltrans, the Build, Transfer, Operate (BTO) concept was preferred over a Build, Operate, Transfer (BOT) concept primarily because of the liability issue. Stoll said that the state recognized that it would be prohibitively expensive for a private developer or consortium to carry enough insurance to cover tort liability issues such as highway accidents and related property damage. And if the developer were required to do so these costs would somehow be reflected in their proposal. The BTO process would keep ownership of the project with the state of California upon completion of the construction portion and prior to its being opened for public use; therefore, the issue of tort liability would remain with the state.

PROJECT GENESIS

The process of selecting private developers and the demonstration projects was to be a two-stage affair. The first phase would be to pre-qualify prospective developers, and the second phase would be to review and select the projects from the detailed presentations submitted by the pre-qualified proposers.

Caltrans established the following schedule to initiate the formal process of selecting the four demonstration projects:

Issuance of requests for qualifications (RFQ)	November 1989
Period for response by private entities	45 days
Period for Caltrans to review/issue guidelines to private entities	30 days
Submission of developer's conceptual project proposals	120–210 days

Carl Williams, Assistant Director of the Caltrans Federal Fiscal Planning office, was designated by the Department to head the implementation of the AB 680 program. "Governments," says Williams, "in most cases, simply do not understand what inhibits private finance." His unique ability to speak the language of the entrepreneur as well as that of the bureaucrat makes him well qualified for the task. On November 15, 1989, Williams issued his department's request for qualification (RFQ). This document made it clear that the state was looking to attract experienced and qualified companies or consortiums capable of designing, financing, permitting, constructing, and operating toll revenue transportation facilities such as highways, bridges, tunnels, monorails, and light rail projects. The state had not identified any specific projects; however, a proposed project would not be selected unless a free transportation facility existed as an option for those who do not choose to use the toll facility.

Ten copies of the RFQ were to be submitted to Carl Williams' attention on or before January 16, 1990. The respondant's qualifications would be evaluated according to the following weighted criteria:

Experience of the principal organization and consortium members	30%
Record of financial strength to commit to a major project	30%
Ability to work cooperatively with a broad range of government agencies	20%
Individual qualifications of key project team personnel	10%
Organizational and management approach of the group	5%
Familiarity and experience with automated traffic operations	5%

Thirteen teams responded to the RFQ:

1. California Toll Road Company L.P. (CTRC)—composed of Parsons Municipal Services, Inc.; Compagnie Financière et Industrielle des Autoroutes (Cofiroute); the Banque Nationale de Paris; and Westpac Banking Corporation
2. Perini Corporation—with partners Daniel, Mann, Johnson & Mendenhall; Bank of America; Ferrovial/Europistas Concessionaria Espanola, S.A.; Bergstralh-Shaw-Newman, Inc.; and Science Applications International Corporation
3. Bouygues, the giant French construction conglomerate
4. Bechtel Corporation
5. California Transportation Ventures, Inc. (CTV)—consisting of Parsons Brinckerhoff Development Group; Fluor Daniel, Inc.; GIE Transroute; and Prudential-Bache Capital Funding
6. PrivaCal—a joint venture between Bechtel; D. J. Smith Associates, Inc.; William R. Gray and Company, Inc.; Kidder Peabody; and Wilbur Smith Associates
7. Enserch Development Corporation and subsidiaries—Ebasco services,

Ebasco Environmental, Ebasco Constructors, Ebasco Business Consulting Co., Ebasco Plant Services Inc.; VSL Corporation, and Earl & Wright

8. National Tollroad Authority Corporation—The Perot Group; Greiner Engineering; Putnam Hayes and Bartlett Inc.; Kiewit Pacific; Amtech Systems; First Boston Corporation; Wilbur Smith Associates; Traffic Consultants; and Nossaman, Gunther, Knox & Elliott
9. California Private Transportation Consortium (CPTC)—composed of Woodward-Clyde Consultants; CRSS Commercial Group; Citicorp; McDermott, Will & Emery; Vollmer Associates; Howard Needles Tammen & Bergendoff; Granite Construction; The John Meyer Company; and Putnam Hayes & Bartlett
10. Transportation Systems Associates—T. Y. Lin International; GTM International; Morrison Knudsen Engineers; and Kidder Peabody & Company
11. Alfred Hollander
12. Titan PRT Systems; Arvin Calspan Industries; Fishbach Corporation; Bank of America; Ferrovial/Europistas Concessionaria Espanola, S.A.; Bergstralh-Shaw-Newman, Inc.; and Science Applications International Corporation
13. Fletcher Challenge Ltd. of New Zealand and subsidiaries

The selection of prequalified groups would be accomplished by February 1990 so that a preproposal meeting could be scheduled for March 1, which according to previously established milestones would see the submission of the developer's conceptual proposals anywhere from July 1 to October 1, 1990.

In anticipation of the second stage of this two-stage process, on March 9, 1990, Caltrans issued the brochure *Guidelines for Conceptual Project Proposal.* At the same time Caltrans issued the RFQ statement for a financial consultant. The selected financial consultant would review and render an opinion as to the reasonableness and sufficiency of the financing plans presented by the groups or consortiums submitting proposals on the four demonstration toll road projects. After this selection process led to the appointment of Price Waterhouse as the financial consultant, Caltrans indicated to the proposers that the committee was now in place to review and evaluate all proposals and to select the best four. All other submissions would be given a priority rating and could move into the "acceptable" category if any of the first four selections dropped out or were disqualified for any reason.

The conceptual project proposal guidelines contained the following components.

A. Description of the Proposer Even though the initial RFQ had required proposers to submit a developer questionnaire, any changes to that initial team had to be indicated at the time of submission. The proposer's description had to provide:

Any updates to all previous questionnaires.

Evidence of ability to manager a project team.

A description of experience in major transportation facilities of the size and type being proposed.

B. Concept Report A report clearly describing the proposal in detail had to contain at the minimum:

1. The location and limits of the project with a suitable map at 1″ = 100′ scale.
2. A clear statement of the transportation services to be provided, including a description of any existing state owned facilities that would be supplemented.
3. A discussion of engineering concepts containing

   a geometric cross-section

   the estimated average daily traffic for the first year of operation

   the estimated average daily traffic for 20 years

   a two-way design hourly volume (DHV)

   the percentage of DHV in direction of heavier flow

   the truck increment as a percentage of DHV and design speed in mph

   the interchange locations and type

   the toll collection concept.
4. An identification and description of the alternatives—no-build alternates, design alternatives, and/or alignment alternatives.
5. A brief description of the right-of-way requirements, including the needs in acres and the width of the corridor. It must also describe any involvement or relocation of utilities, railroads, or residential or commercial space and whether relocation assistance is required. If applicable, the description should cover airspace usage, reserved utility corridors, transit, roadside rests, and commercial development. A statement about hazardous waste potential must also be appended.
6. A list of all agency permits required, including all anticipated local, city, county, regional, state, and federal permits.
7. Capital cost estimates, identified by major components and the basis by which the estimate was derived. Right-of-way costs and construction costs identified by major component must be listed.
8. The connection of or involvement of the proposed project with federal aid highways, and a statement as to whether National Environmental Policy Act documents or FHWA design approvals were required.

C. Preliminary Environmental Evaluation The proposer will describe how the proposal will further California's environmental policies, and discuss how the proposal will treat physical issues such as seismic exposure, erosion, air, noise, energy, solid waste, use of natural resources, fish, wildlife, vegetation, agriculture, and timber. The proposal must also consider and address housing, neighborhoods, schools, public facilities, heritage resources, recreation, park land, open spaces, aesthetics, and scenic resources. Potential mitigation measures must be incorporated into the proposal.

D. The Financial Plan A statement of opinion is required from the prequalified financial consultant along with a financial plan of sufficient detail to demonstrate a reasonable basis for project funding. If any deficiencies exist in the financial plan, the proposer will be given an opportunity to correct the deficiencies and resubmit the plan to the financial consultant.

E. A Schedule for Development A schedule is to be submitted listing important events and the dates on which these events will be met. The starting date in the schedule is to be the date when the proposer is notified by Caltrans that the proposal has been selected as one of the initial four. The end date will be the date when the project is to be opened to the public. A statement of how the schedule is to be managed and what strategy will be employed to deal with unanticipated delays must be included.

F. Documentation of Support The proposer must provide evidence of support for the project, which could take the form of resolutions passed by local government, state agencies, government officials, or other parties having an interest in the project.

G. State Services Required The proposer must list any optional, reimbursable services that the state will have to provide under separate contract such as highway maintenance and highway patrol services. Proposed dates when these services may be required to start and finish must also be included.

H. Support for State Civil Rights Objectives Proposers must include plans to support state civil rights objectives such as furtherance of Disadvantaged Business Enterprises (DBE) participation and the use of minority and women employees.

I. Filing Fee as Required Caltrans required a $50,000 check from each project sponsor. If a sponsor submitted more than one project, only one check was required. The $50,000 would be drawn down in $5,000 increments as each submission was reviewed. Unsuccessful bidders would have the unused portion of their filing fee returned once the successful candidates had been selected.

In the selection process, each portion of the proposal was weighted on a numerical scale:

Transportation service provided	20 points
Degree to which proposal encourages economic prosperity	10 points
Degree of local support for the project	15 points
Relative ease of proposal implementation	15 points

Experience/expertise of sponsors and support team	15 points
Degree to which proposal supports state's environmental quality and energy conservation goal	10 points
Degree to which non-toll revenues support proposal costs	5 points
Degree of technical innovation displayed in proposal	10 points
Degree of proposal's support for achieving the civl rights objectives of the state	10 points
Highest achievable score	*110 points*

PROJECT AND DEVELOPER SELECTIONS

By September 14, 1990, the four demonstration projects had been selected and with them the winning developer groups. Governor George Deukmejian approved Caltran's selections allowing the program to commence contract negotiations with the following four groups.

Project No. 1

The Consortium:	The National Toll Road Authority Corporation (the Perot Group)
The Project:	State Route 57 (SR 57), also known as the Santa Ana Viaduct Express
The Cost:	$700 million
The Completion Date:	Mid-1997

This project consists of construction of an 11.7 mile (19 km), four-lane limited access toll road which would extend Route 57 from its present termination point near the Anaheim Stadium, continuing on to the San Diego Freeway, and then turning slightly eastward to connect to Route 73 near the John Wayne Airport. Much of this "cars only" roadway would be elevated over a flood control channel being programmed by the U.S. Army Corps of Engineers. Construction of this highway would complete the missing portion of an Orange County north to south highway system.

Project No. 2

The Consortium:	California Toll Road Company L.P. (CTRC)
The Project:	Interstate 80 Route (I-80), also called the Mid-State Tollway
Cost Estimate:	$1.2 billion, all phases
Completion date:	January 1997

The project would entail construction of a five-lane toll road commencing at Interstate 680 (I-680) in the South San Francisco Bay area and traveling 85 miles (138 km) northward to State Road 4 (SR 4) near Antioch and then on to Interstate 80 (I-80). This highway would be built in two phases: The first phase would be the 40 mile (65 km) stretch to SR 4, and the second continuing the toll road from SR 4 to connect to I-80. This second phase would include two high-span bridges over the San Joaquin Delta and the Sacramento River. Construction of this highway would not only relieve congestion at the I-680 corridor, but would also provide an alternate route for commercial traffic around the Bay Area for freight moving to and from Silicon Valley.

Project No. 3

The Consortium: California Transportation Ventures Inc. (CTV)
The Project: State Road 125 (SR 125), also called the San Miguel Mountain Parkway
Cost Estimate: $400 million
Completion Date: December 1995

This highway would be a four-lane toll road with limited access, extending from State Route 905 (SR 905) near the Mexican border at Otay Mesa northward about 10 miles (16 km) to State Route 54 (SR 54) near Bonita. Expansion to 10 lanes would be possible if traffic warrants this (Figure 4.1). There is also a provision for future high-occupancy-vehicle (HOV) lanes and a fixed-rail system. The SR 125 project would fit into the federal government's NAFTA program and would increase commercial traffic to and from Mexico as well as generate tourist traffic across the border into the San Diego area.

Project No. 4

The Consortium: California Private Transportation Corporation (CPTC)
The Project: State Road 91 (SR 91), known as the SR 91 Median Improvement
Cost Estimate: $125 million
Completion Date: Late 1994

This project represents another new concept beyond that of public/private partnerships. The median strip within the Riverside Freeway (SR 91) would be converted to an all-electronic tolled four-lane express highway for car pools with two such lanes in each direction. The express lanes would be equipped with AVI systems which in turn would require the state to establish a proceedure for drivers to obtain the necessary transponders and deposit money on account. Of the total cost estimate, CPTC had set aside $65 million for construction and $80 million for the design, engineering studies, purchase, and installation of the AVI system and related toll equipment.

Determining Equitable Returns on Investment for The Projects

The state of California conducted a study to determine what might be a reasonable rate of return based upon the investment required and the risk associated with that investment. The risk factors, as identified by Caltrans, consisted of

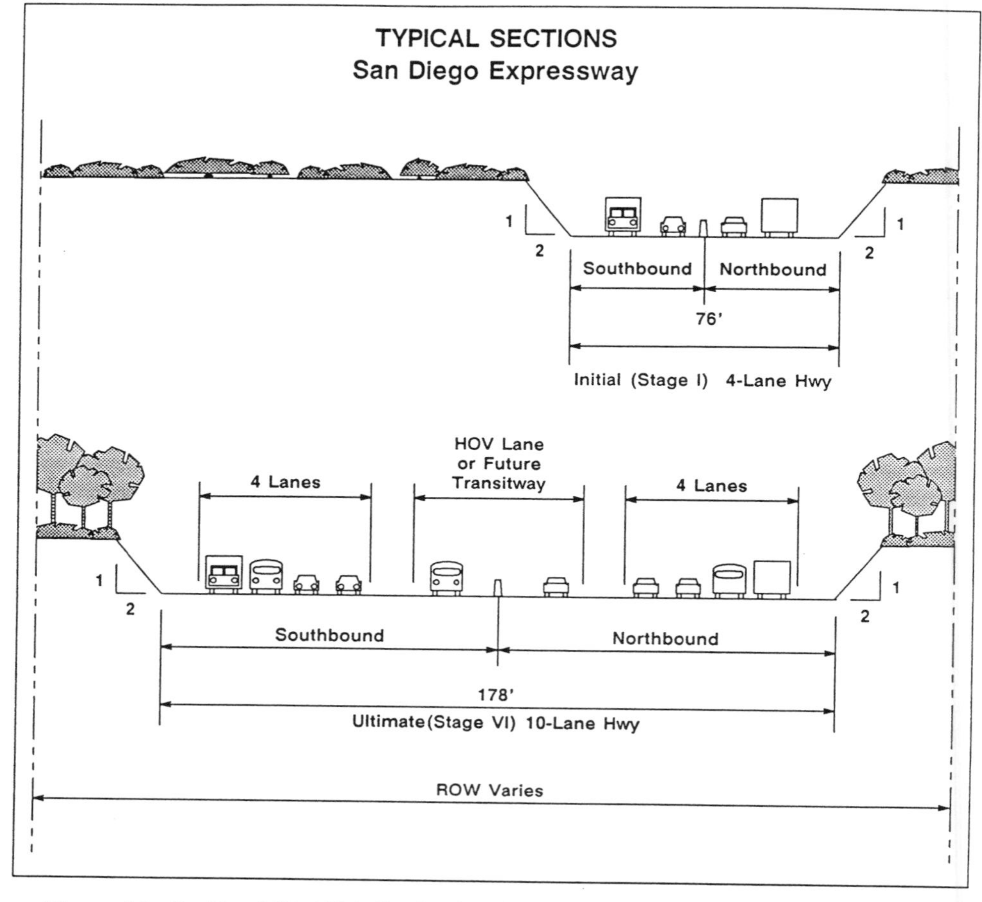

Figure 4.1. Profile of SR 125 indicating interim and ultimate stages. [Courtesy California Transportation Ventures, Inc.]

1. Adverse policy changes by future state administrations
2. Costly difficulties in obtaining financing
3. Lack of historical information from which to predict demand for the project's services
4. Potentially long delays before the revenue stream covers all costs including return on investment

The return on investment (ROI) philosophy would be reviewed with each sponsor along with other incentives to sweeten the pot. The annual return on investment was defined as the amount of revenue remaining at the end of each year after paying expenses that is available to investors (in the form of interest, principal, or dividends) measured against the total amount of capital invested in the project. For example, if $10 in revenue was received on a $100 investment in one year, the ROI would be calculated as 10 percent for that year.

Taking all of the risk factors into consideration, Caltrans arrived at the following ROIs for each project.

SR 91 An allowable rate of return for would be 17 percent. Of the four demonstration projects, State Road 91 was the least risky because the right-of-way had already been acquired and much of the design and permits for the project had been accomplished. Existing traffic counts made it somewhat easier to predict revenues for this project.

SR 125 The rate of return on investment for this project was pegged at 18.5 percent. It was slightly more risky than SR 91, but did not face the relatively high construction costs and business risks of the SR 57 and Mid-State projects.

SR 57 This was the second most costly project of the four and the rate of return was established at 20.25 percent because its construction costs and design risks were greater than those posed by the SR 91 and SR 125 projects.

The Mid-State Tollway the rate of return set for this project was 21.25 percent, this being the riskiest of the four demonstration projects regarding business risk, local policy uncertainty, and environmental mitigation risk.

In January 1991 each successful sponsor obtained a "handshake" agreement from Caltrans, but there was much work required by each developer group before the informal agreement would be converted into a real, viable, ready-to-go deal. The final report of the finance provisions of the AB 680 transportation project development franchise agreement was released on February 22, 1991 by Caltrans and contained a comparison of the project franchise agreements prepared by Price Waterhouse (see Table 4.1).

A CLOSER LOOK AT THE FOUR PROJECTS

When Caltrans announced the four demonstration projects in September 1994, the developers targeted one project completion in 1994, another in 1995, and two in 1997. As often happens, in their exuberance to project a positive image to their lenders and clients developers sometimes get so caught up in the optimism of the project that they overlook Murphy's Law: what can go wrong, often does go wrong. Construction time can be calculated with a great degree of certainty, but getting through the maze of financing requirements and approval procedures can be a nightmare, no matter how many times a developer has gone through the process. And these were no ordinary projects; the four developers were traversing new ground—public/private partnership transportation projects—and the milestone dates began to slip and slide forward.

TABLE 4.1 Caltrans Toll Road Privatization Project Financial Comparison

Subject	SR 91	SR 125	SR 57	Mid-State
A. RETURN ON INVESTMENT				
1. Base Return Rate	17%	18.5%	20.25%	21.25%
2. Base Return Method	NPV Method (Present value of project cash flows less accrued acceptance date capital costs.)	NPV Method (Present value of project cash flows less accrued base date capital costs.)	NPV Method (Present value of project cash flows less accrued acceptance date capital costs.)	Modified DRA Method (Base rate applied to capital base. Shortfalls accrued in suspense account for future realization.)
3. Incentive Return Criteria	Peak Hour Throughput	Mean Vehicle Occupancy	Mean Vehicle Occupancy Injury Accident Rate Operating Cost/Vehicle Number of Van Pools	Mean Vehicle Occupancy Injury Accident Rate Operating Cost/Vehicle
4. Incentive Return Method	Incorporated into NPV method Incentive return rate (base plus incentive percentage) used to calculate total NPV. Difference between base NPV and total NPV earned as incentive and split with Caltrans.	Deferred incentive account method Incentive percentage applied to capital base to calculate incentive earnings. Shortfalls accrued in suspense account for future realization. Incentive split with Caltrans.	Deferred incentive account method Incentive percentage applied to capital base to calculate incentive earnings. Shortfalls accrued in suspense account for future realization. Incentive split with Caltrans.	Deferred incentive account method Incentive percentage applied to capital base to calculate incentive earnings. Shortfalls accrued in suspense account for future realization. Incentive split with Caltrans.
5. Pre-Transfer Financing Costs	Capital costs include actual capitalized interest and a fixed development fee.	Hard capital costs (excluding capitalized interest and equity) escalated at the base rate on a quarterly basis to the base date.	Hard capital costs (excluding capitalized interest and equity) escalated at the base rate on a quarterly basis to the acceptance date.	Hard capital costs (excluding capitalized interest and equity) escalated at the base rate on a quarterly basis to the transfer date.
6. Phasing	One 35-year lease with the potential for locally negotiated extensions	One 35-year lease	One 35-year lease	Two overlapping 35-year leases allowed

B. RESERVE ACCOUNTS				
1. Limits on Debt Service Reserve	18 months, 24 months if required by lender, or 36 months if required by lender due to HOV arrangement.	18 months, or 24 months if required by lender.	18 months, or 24 months if required by lender.	18 months, or 24 months if required by lender.
2. Limits on Working Capital Reserve	180 days projected operating costs	180 days projected operating costs	180 days projected operating costs	180 days projected operating costs
3. Limits on Capital Improvement Fund	14% of capital base	15% of capital base	14% of capital base	20% of capital base
4. Limits on Maintenance Reserve	Contribution limited to 1% of capital base. Balance not limited.	Contribution limited to 1% of capital base. Balance not limited.	Contribution limited to 1% of capital base. Balance not limited.	Contribution limited to 1% of capital base. Balance not limited.
5. Trust Account	None	None	None	Limited to 150% of annual debt service requirements
C. INDEXING				
1. Pre-Transfer	No indexing	Increase in 6-month T-Bill yield from agreement date	Positive change in Index from agreement date	Increase in 3-month T-Bill yield from agreement date
2. At Transfer	Increase in 20-year T-Bill yield from agreement date.	Increase in 30-year T-Bill yield from agreement date.	Positive change in Index from agreement date.	Increase in Moody's BAA corporate bond average yield from agreement date.
3. Post-Transfer	20% of increase in 5-year T-Bond yield from Acceptance Date.	25% of increase in 3-year T-Bond yield from Base Date.	25% of positive change in Index from Acceptance Date.	One minus % of capital base financed with fixed rate debt, times increase in variable debt rate.
4. Index Basis	Based on monthly average of yield.	Based on monthly average of yield.	Based on yield on single dates.	Based on monthly average of yield.
5. Index Definition	Varies by index period above.	Varies by index period above.	Equal portions of 13-week T-Bill yield, 30-year T-Bond yield, and long-term corporate security yield.	Varies by index period above.

TABLE 4.1 *(Continued)*

Subject	SR 91	SR 125	SR 57	Mid-State
D. REPORTING				
1. Required Reports	Developer is required to submit to Caltrans detailed reports of financial performance similar to Exhibit J. [Sec. 3.6]	Developer is required to submit to Caltrans detailed reports of financial performance similar to Exhibit G. [Sections 10.2 and 12.9]	Developer is required to submit to Caltrans detailed reports of financial performance similar to Exhibit M. [Sec. 9.7(c)]	Developer is required to submit to Caltrans detailed reports of financial performance similar to Exhibit K. [Sec. 10.3(h)]
2. Caltrans Right to Audit	Caltrans may audit independent auditor, and, if problems are discovered, require a new audit by a new independent auditor. [Sec. 3.6(e)]	Caltrans may conduct an audit of reported matters for up to 3 years after matter is reported. [Sec. 12.9(c)]	Caltrans may conduct an audit of reported matters for up to 3 years after matter is reported. [Sec. 9.7(e)]	Caltrans may audit at reasonable times, and must provide advance notice. [Sec. 10.3(e)]
E. STANDARDS				
1. Related Party Definition	The term "Related Parties" shall mean any party that is a parent or subsidiary of CPTC; holds directly or through one or more intermediate entities a 10% or greater interest in CPTC or in which entity CPTC holds such interest; any party which otherwise effectively controls or is controlled by CPTC; members of the immediate families of principal owners of	The term "Related Parties" shall mean all Affiliated Entities, management of Developer, other parties with which Developer may deal which can significantly influence the management or policies of Developer, and members of the immediate families of principal owners of Developer and its management. [Sec. 2.68] The term "Affiliated Entity"	The term "Related Party" shall mean: (a) Any person or entity that is a general partner of Developer or that owns, directly or through one or more intermediate entities, a 10% or greater equity interest in Developer (a "Substantial Owner"); (b) Any entity in which Developer owns a 10% or greater interest; or (c) An entity effectively controlled by a Substantial	The term "Related Party" shall mean: (i) a party that directly or indirectly controls, is controlled by, or is under common control with Developer, (ii) an owner of record or known beneficial owner of more than 10% of the voting interests of Developer or such owner's immediate family, (iii) Developer's management, or the immediate family of Devel-

	CPTC and its management; or other parties with which CPTC may deal which can significantly influence the management or policies of the CPTC. [Sec. 2.90]	shall mean any Person which is a parent or subsidiary of Developer; which holds, directly or through one or more intermediate entities, a 10% or more interest in the voting securities, net income or liquidation value of Developer, or in which entity Developer holds such interest; or which is otherwise effectively controlled by or is under common control with Developer. [Sec. 2.5]	Owner as determined by Developer after a reasonable and diligent inquiry; or (d) Senior management of Developer and members of the immediate family of any individual Substantial Owner or of senior management of Developer. [Sec. 2.52]	oper's management, (iv) a trust for the benefit of employees, such as a pension or profit sharing trust, which is managed or is under the trusteeship of Developer's management, and, (v) any other party with which Developer deals if such party controls the management or operating policies of Developer (or if Developer controls the management or operating policies of such party) to the extent that one of the transacting parties might be prevented from pursuing its own separate interests. [Sec. 2.50]
2. Related Party Standard	With respect to transactions with Related Parties, CPTC agrees to conduct business on terms no less favorable to CPTC than would be incurred in the normal course of business with parties which are not Related Parties. [Sec. 13.2(h)] Independent auditor must	Items determined by Caltrans as a result of such audit to be materially at variance with normal commercial practices (including contractual relationships between Developer and Related Parties) and not resolved by negotiation shall be resolved in accordance with Article XIX of	With respect to Related Party Transactions, Developer agrees not to include in Costs or Operating Costs the portion of any amounts paid to a Related Party for services or materials that is materially at variance with amounts Developer would incur in normal commercial practices.	Developer further warrants and represents, with respect to transactions with a Related Party, the Developer shall conduct business at prices and on terms not in excess of those Developer would incur during the normal course of business with independent parties. [Sec. 16.2(f)]

TABLE 4.1 *(Continued)*

Subject	SR 91	SR 125	SR 57	Mid-State
	submit separate report on compliance with 13.2(h). [Sec. 3.6(d)]	this agreement. The amounts of such variances, if any, which are agreed by the parties or determined pursuant to Article XIX shall be taken into account and appropriate adjustments (including, where applicable, exclusions of expenditures from outflows) shall thereafter be made in the calculations required by Article X. [Sec. 12.9(c)]	For purposes of this Section 9.9, the portion of any amount paid by Developer to a Related Party for services or materials that will be considered as being "materially at variance with normal commercial practices" shall be that portion that is greater than 120% of the amount Developer would have paid to an independent third party contractor for the same services or materials under similar circumstances, taking into account any services rendered or materials furnished after the Start Date. Developer shall maintain an internal system for monitoring Developer's compliance with this Section 9.9(b). [Sec. 9.9]	Independent auditor must submit separate report on system for compliance with 16.2 (f). [Sec. 10.3(d)(6)]
3. Related Party Reporting	In the preparation of the aforesaid reports, CPTC	Annual reports shall include . . . a list of transactions	The statements and reports to be provided under Section	. . . Developer shall submit to Caltrans an annual fi-

	shall identify all Capital Costs, Operating Costs and all transactions with Related Parties which do not meet any of the following criteria: (i) Are expressly permitted by this Agreement; (ii) Are less than 110% of the amounts which Caltrans would have likely paid for comparable goods or services; (iii) Are consistent with generally available commercial list prices; (iv) Are justifiable by life cycle analysis, accelerated delivery or completion of goods or services; (v) Are consistent with industry practices, or (vi) Are on terms more favorable to CPTC than would be incurred in the normal course of business with independent parties. [Sec. 3.6(c)]	and contracts with Related Parties . . . [Sec. 12.9(b)]	9.7 above also shall identify any transaction between Developer and a Related party pursuant to which amounts are paid to a Related Party that are Costs or Operating Costs (a "Related Party Transaction"), and the reports to be furnished by the independent auditor under Section 9.7(c) shall include a report on Developer's internal monitoring system referenced in Section 9.9(b) below. [Sec. 9.9(a)]	nancial report . . . and shall include a description of any material related party or affiliate transactions and charges. [Sec. 10.3(d)]
4. "Good Faith" Standard	CPTC shall plan, design, construct and operate the Private Transportation Project in a safe, reasonable, and prudent manner,	This agreement shall be construed to impose an obligation of goof faith, honesty, reasonableness, and fair dealing upon each of the	Developer shall employ good business practices and appropriate management techniques in the development and operation of the	

TABLE 4.1 *(Continued)*

Subject	SR 91	SR 125	SR 57	Mid-State
	and shall use its best efforts to: (i) Employ good business practices and appropriate management techniques; and (ii) Conduct its commercial affairs in a manner consistent with good faith and fair dealing. [Sec. 13.2(g)]	parties, but shall not be construed to permit Caltrans to interfere with the day-to-day operations of Developer's business. [Sec. 21.8(c)]	Transportation Facility and shall conduct its commercial affairs in a manner consistent with good faith and fair dealing. [Sec. 11.6]	
5. Dispute Resolution	Any controversy arising with respect to reasonable return that has not been satisfactorily resolved by administrative determination, shall be resolved in arbitration. [Article 15]	Any unresolved controversy involving an amount not exceeding $500,000, with aggregate unresolved issues not exceeding $5 million shall be resolved in arbitration. [Article 19]	Any disputes seeking damages aggregating $100,000 or less may be resolved in arbitration at the option of the party instituting the arbitration. Disputes seeking damages in excess of $100,000 but less than $20,000,000 shall be heard and determined by a referee. [Article 16]	Any controversy arising with respect to reasonable return may be settled in arbitration. [Article 19]

Source: Price Waterhouse.

The Mid-State Tollway Project

The Mid-State Tollway sponsored by the California Toll Road Development Group is in a holding pattern of alternate route reviews. Its initial 85 mile (137 km), $1.2 billion program has been substantially reduced in scope. The Northern Reach section and the Solano County portion have been deleted, and just about 40 miles (65 km) remain to be built under the Caltrans demonstration program.

The State Route 57 Project

The Perot Group, developers of the Santa Ana Viaduct Express (the State Route 57 development) became one of many victims of the Orange County bankruptcy. One of the project's environmental obligations required a $24 million preliminary study prior to the granting of a right-of-way agreement over a flood control channel. The cost of this study was to have been split between the Perot Group and the Orange County Transportation Authority (OCTA). The right-of-way issue encompassed about 75 percent of the entire project's alignment, so it was of major concern to both parties. When the financial crisis erupted in Orange County and the government filed for protection from creditors under Chapter 9, the $12 million dollar payment from OCTA was put on hold. The Perot people must wait until this check can be written unless they wish to be responsible for the entire $24 million costs. SR 57 must wait until the financial, environmental, and right-of-way issues are resolved before the development process can proceed.

State Road 125

California Transportation Venture's August 1990 proposal to Caltrans for State Road 125 indicated that environmental studies were underway and that, with appropriate mitigation measures, approvals and permits of a highway facility in the corridor could be secured. CTV stated that public and private interest in the SR 125 Corridor was uniformly in favor of a toll road serving the region, and that the degree of local support was evident in actual commitments of donated rights-of-way by the principal landowners and developers. In reality, however, the project has been controversial from the start.

Citizen reaction against the program has been high, and many residents and neighborhood organizations have opposed to the various routings proposed by CTV. In response to the July 6, 1993 Draft Alternatives Report, for example, about 700 letters were received from residents and agencies:

- Residents and businesses along the Sweetwater Springs and Jamach alignments presented hundreds of objections to this route, ranging from the effect on air quality and noise levels to the impact on the newly created Little League field.
- Salt Creek residents were worried about displacement resulting from the construction of a substation, and they too were concerned about noise levels and air quality issues.

- Eastlake residents were also worried about noise and air quality levels and made a request for a highly landscaped facility.
- The Bonita-Sunnyside and Bonita Hills Estate mobile home residents continued to express their concern over loss of community character, noise, water quality, air quality, and loss of a Little League field, as well open space loss, and negative impact on the local golf course and horse trails.

More than a few favor the "No Project" alternative and feel that SR 125 South is not a justifiable project. Complaints about not being advised of public meetings and about the notice map being ambiguous have been heard in many quarters as well.

CTV's alternative routes had been evaluated in order to find the most environmentally desirable approach, and Environmental Impact Reports/Environmental Impact Statements (EIR/EIS) were created. The EIR/EIS provides detailed information about the effects certain highway routings will have on the environment from the standpoint of air and water quality, visual impact, biological and socioeconomic effects along with the engineering studies and traffic patterns that affect the project's cost effectiveness. On the basis of such studies, CTV could not route the highway east of the Sweetwater Reservoir, a largely undeveloped area; that route not only involved treating substantial biological factors, but the greater number of business and residential displacements that would take place would potentially affect the quality of the South Bay's supply of drinking water.

In September 1994, a representative of Caltrans who had knowledge of the SR 125 route, conceded that the environmental issues were somewhat unique. At the northern end the 10-mile (16 km) highway would pass through Bonita where $\frac{1}{2}$ acre zoning allows horses to be raised and considerable opposition to the project should be anticipated if the planned route went through that area.

As a result of public opposition, a number of alternative routings for SR 125 have been proposed by CTV. The Traffic Circulation Study Alternative Map of 1992 (Figure 4.2) provided the alternative routes for SR 125 that CTV was studying before deciding on the final alignment. CTV subsequently filed several draft alternative reports showing revised routings and sections deleted during the process (Figures 4.3 and 4.4). A final alternatives report was issued in August 1993 with alternative alignment segments substantially reduced from the previously submitted plan (Figure 4.5), and in January 1994 yet another alternative alignment segment plan was prepared which basically followed the route of the August 1993 submission. CTV's public relations newsletter, *The Community & SR 125 South,* attempted in the spring of 1995 to explain the environmental issues being faced by the developer, and invited participation and comments from concerned citizens.

The Citizens Advisory Committee (CAC)—which represents approximately 30 community organizations within the various proposed routings of SR 125 concerned about the impact the highway will have on natural habitat, homes, businesses, and recreational facilities—has had a significant impact on the alignment of SR 125. On February 9, 1995, Caltrans announced that the proposed alignment suggested by CAC would be included in the EIR/ESI. The acceptance of the CAC proposal to move the highway farther east, away from existing homes located west of Sweetwa-

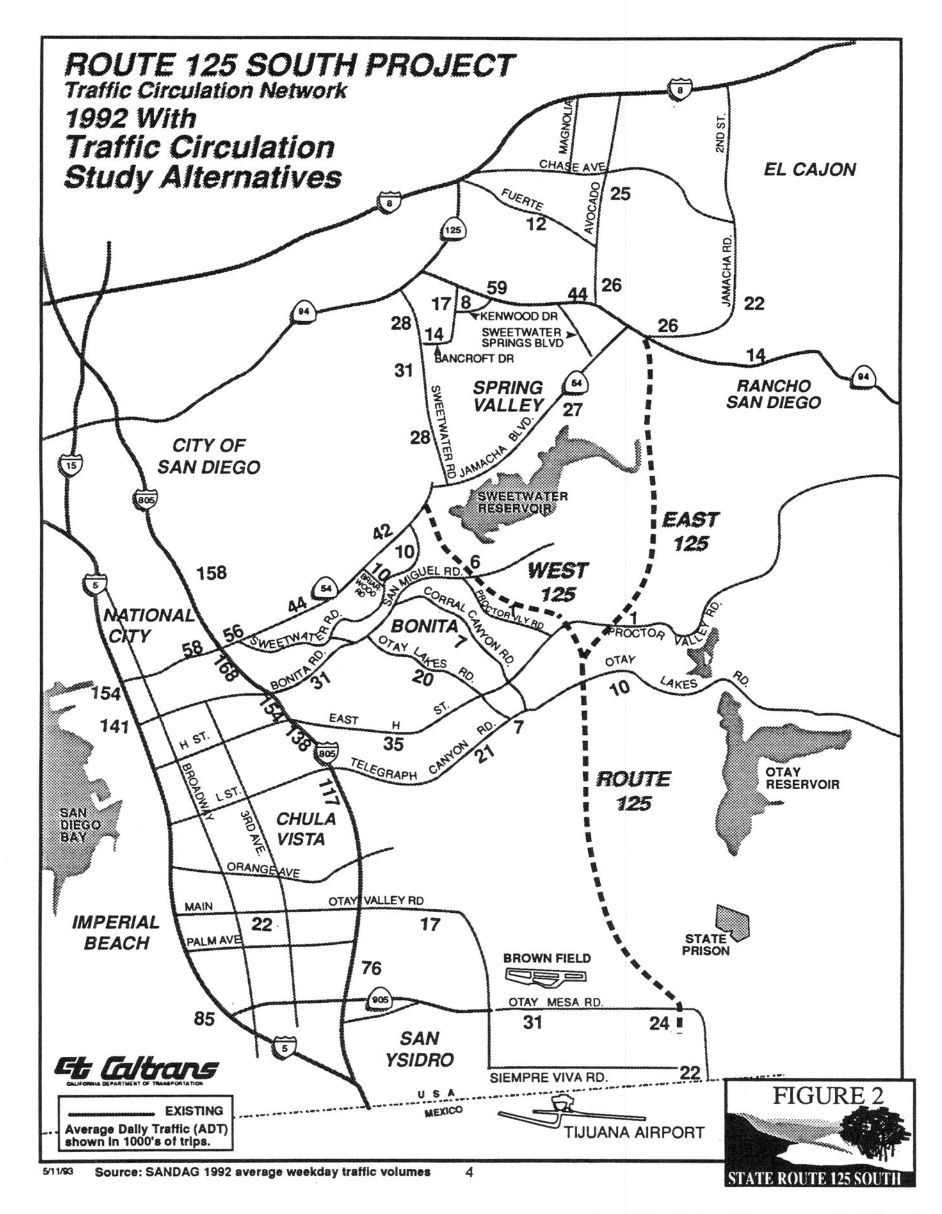

Figure 4.2. Traffic circulation study alternative map for SR 125. *[Courtesy California Transportation Ventures Inc.]*

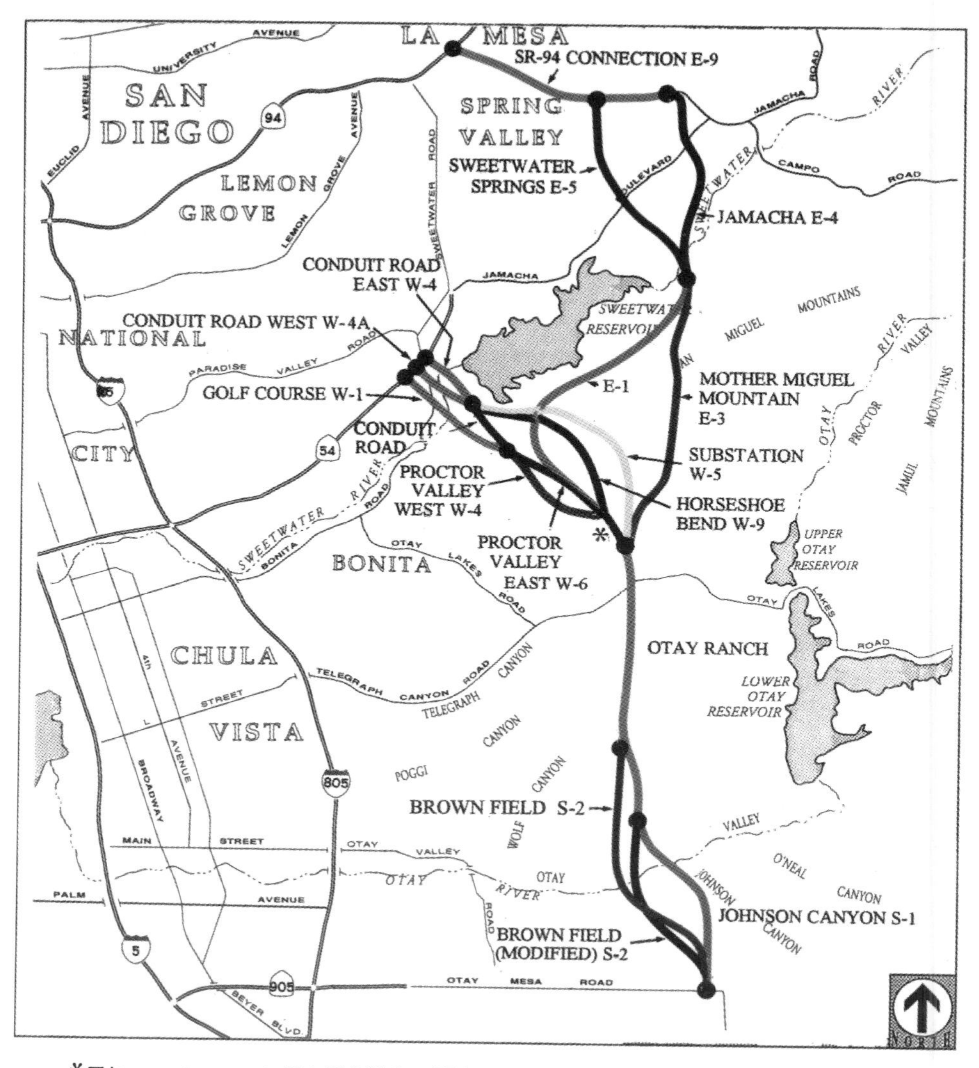

*This segment common to W-4, W-6, W-9, and E-1

Figure 4.3. One of the alternative alignment segments drafts for SR 125. *[Courtesy California Transportation Ventures, Inc.]*

ter Reservoir, has demonstrated the power of local citizen action groups. It also demonstrated that state agencies such as Caltrans do react positively to community concerns. Kent Olsen, President of CTV, indicated that this additional alignment alternative and its corresponding draft EIR/ESI would be released for the public hearings December 1995 or January 1996, which meant another 7 to 8 month project delay. Olsen expects financing to be completed in 1997, allowing construction to commence later that year.

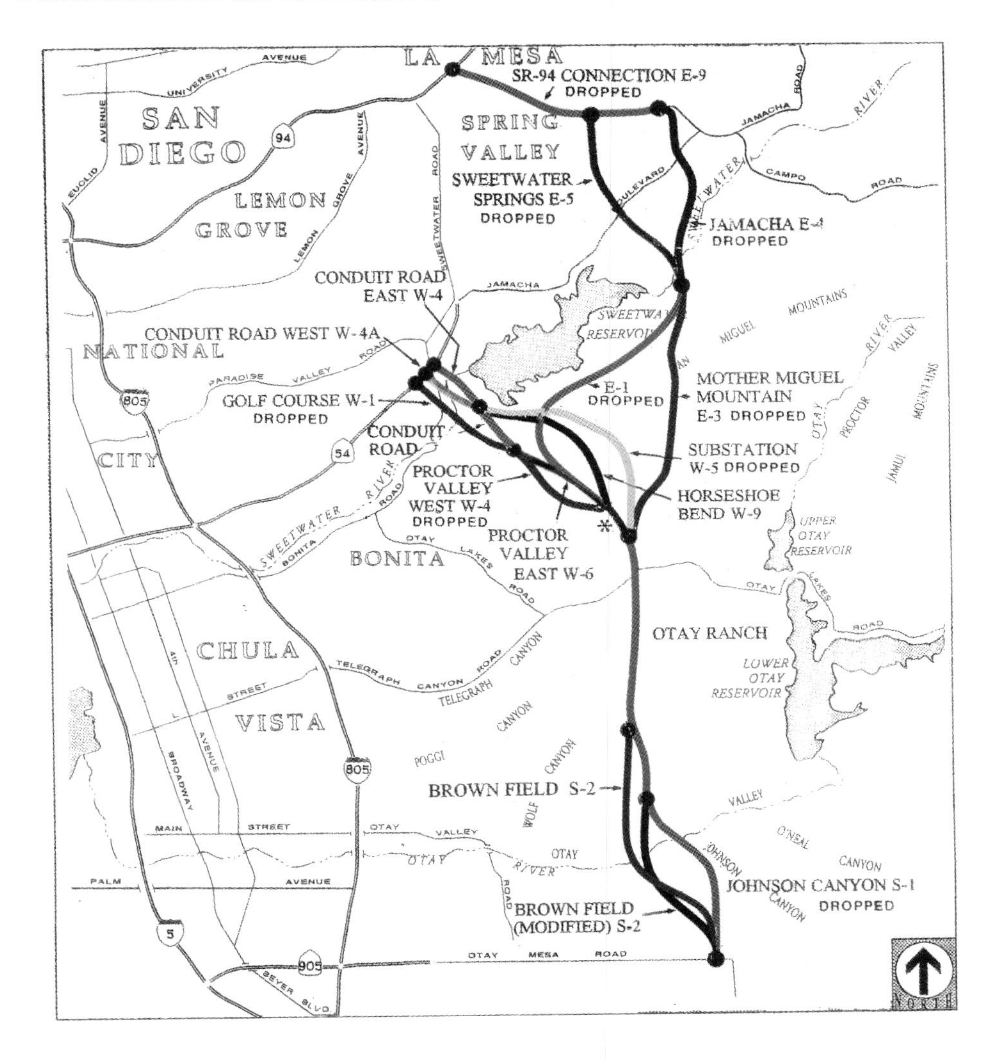

* This segment common to W-4, W-6, W-9, and E-1

Figure 4.4. Another alternative alignment segment draft for SR 125. *[Courtesy California Transportation Ventures, Inc.]*

Environmental controversies aside, CTV has done a credible public relations job. The developer has created an identity for the project with their San Miguel Mountain Parkway logo (Figure 4.6). CTV has made the public aware that along with improving traffic circulation, travel time, and road safety, the project will provide tangible economic benefits to the region. During the initial construction phase 500 direct local jobs will be created over a two-year period, and the operational end of the highway project will create 70 local jobs which will increase to 150 over the 35-year lifespan of the program. The highway should increase property

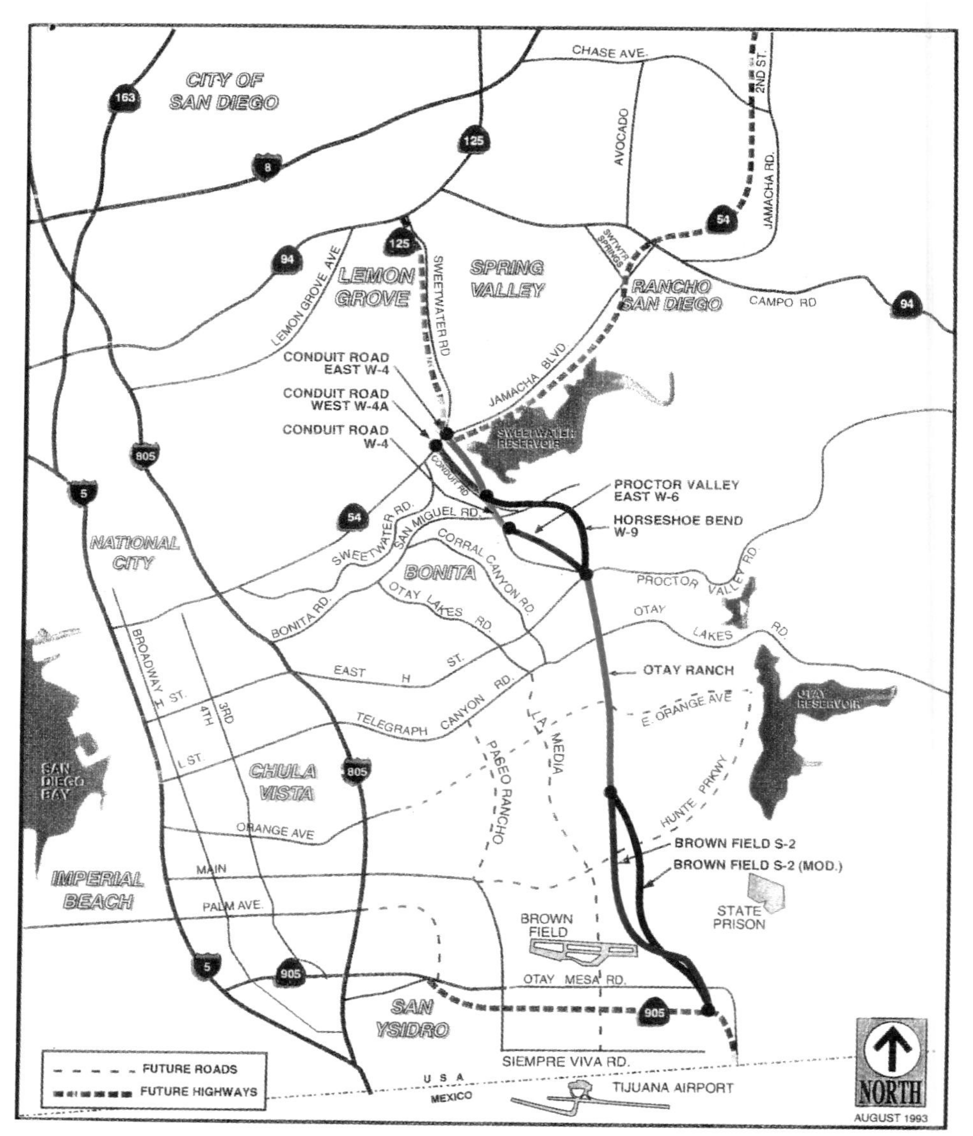

Figure 4.5. The final alternative segment report of August 1993 (similar to the January 1994 submission). *[Courtesy California Transportation Ventures, Inc.]*

Figure 4.6. The San Miguel Mountain Parkway logo. *[Courtesy California Transportation Ventures, Inc.]*

values as well—but to some residents' ways of thinking, many property values will decrease.

State Road 91

As of September 1995, according to Roy Nagy of Caltran's Directors Office, State Route 91 was moving right along. Although initially scheduled for a late-1993/early-1994 completion, project probably would open to the public in December 1995. SR 91 is the first new AVI toll road to open in the state of California. The project is shaping up as one of those deals where everything seemed to click: It is 100 percent financed and scheduled to be open to the public by the end of 1995, and the public response to the concept of tolled express lanes is very favorable.

The formidable team that worked hard and took risks to put the SR 91 project together is the reason why this first public/private AVI highway operational west of the Mississippi has fared so well. The general partners in the project are

Peter Kiewit's Sons, a 111-year-old Omaha, Nebraska–based construction dynamo

Granite Construction of Watsonville, California, one of the leading road builders in California and winner of a national quality control award

Cofiroute, the world's largest toll road developer, headquartered in France with annual turnover in the $500 million category

The State Road 91 business structure is shown in Figure 4.7.

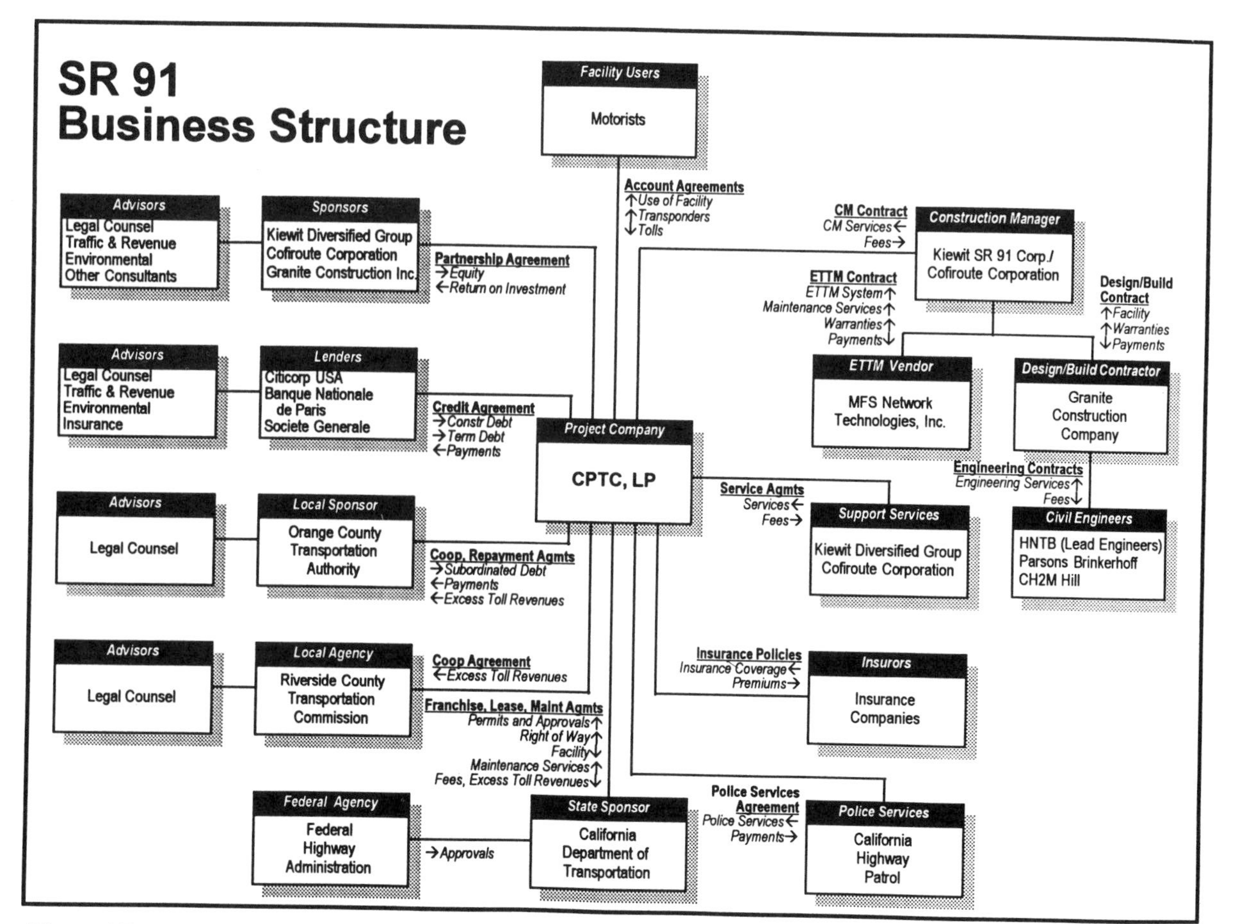

Figure 4.7. Business structure of SR 91. *[Courtesy California Private Transportation Company, L.P.]*

It's about time.

Think about all the time you spend on the 91. Now, think about life without all that traffic.

What would you do with the time you save? Spend time with your family? Or get more done? Or just relax, traveling without the stress of stop-and-go driving. That's a choice you don't have today on the freeway. But there's a new option — one that's fast, safe and reliable: The 91 Express Lanes.

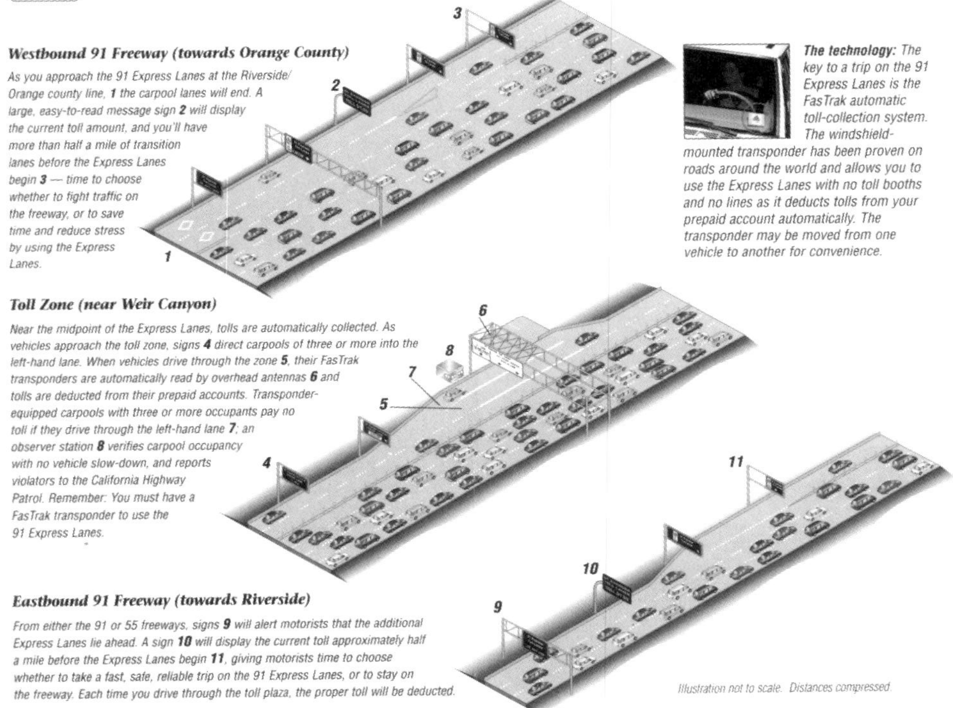

Figure 4.8. State road 91 FasTrak express lane operation. *[Courtesy California Private Transportation Company, L.P.]*

Kiewit had construction expertise, fiber optic technology experience via its MFS subsidiary, and cash. The large architectural-engineering design firm CRSS had been part of the original SR 91 franchise negotiations; when Kiewit bought out CRSS's interest in the project, team head Gerald Pfeffer came over to Kiewit as well, giving the company someone who was intimately involved in the project. With Kiewit's active pursuit of project financing another difficult hurdle was conquered.

Kiewit Diversified Group, Inc. and co-developer Cofiroute put up $19 million of sponsor equity. Granite Construction with a 25 percent stake in the project and a $56.8 million road construction contract also became an equity debt supplier. Kie-

New ideas for a new road

No doubt about it: The 91 Express Lanes are a new kind of driving experience for motorists. The new lanes use Value Pricing and automatic collection of tolls — two concepts new to Southern California drivers. But does that mean the Express Lanes are hard to use? Not at all.

FasTrak and Value Pricing

Here's how it works: As vehicles equipped with FasTrak transponders approach the 91 Express Lanes, a large, easy-to-read sign will indicate the current toll rate. Once you see the toll, you'll have more than half a mile to decide whether to merge into the Express Lanes or use the adjacent freeway. It's a concept that's fair: Only those who choose to use the new Express Lanes will pay for them.

But electronic toll collection is only part of the story. The key to a fast, reliable trip is Value Pricing — a system of varying tolls to reflect changing traffic conditions and to ensure a speedy commute. Using the sophisticated FasTrak system — an easy-to-install windshield-mounted device and roadside readers — tolls are deducted from your prepaid account automatically, so there are no toll booths to stop at. The system even guarantees that the Value Price you see posted at the entrance to the Express Lanes is the price you'll pay, every time. A tone will let you know that your transponder has been read and the proper toll has been deducted.

Sign up today

Customers can obtain FasTrak transponders by completing a FasTrak application, available by calling 1-800-600-9191.

Payment options include credit card, cash or check. With the credit card option, customers establish an account balance and a credit card imprint is taken as a transponder deposit. The 91 Express Lanes are authorized to replenish the account when the balance drops below an agreed amount.

With cash or checks, customers establish a prepaid account and place a deposit on the transponder. When the account balance drops below a set amount, the customer agrees to replenish the account by mail or in person.

Westbound 91 Freeway (towards Orange County)

*As you approach the 91 Express Lanes at the Riverside/Orange county line, **1** the carpool lanes will end. A large, easy-to-read message sign **2** will display the current toll amount, and you'll have more than half a mile of transition lanes before the Express Lanes begin **3** — time to choose whether to fight traffic on the freeway, or to save time and reduce stress by using the Express Lanes.*

Toll Zone (near Weir Canyon)

*Near the midpoint of the Express Lanes, tolls are automatically collected. As vehicles approach the toll zone, signs **4** direct carpools of three or more into the left-hand lane. When vehicles drive through the zone **5**, their FasTrak transponders are automatically read by overhead antennas **6** and tolls are deducted from their prepaid accounts. Transponder-equipped carpools with three or more occupants pay no toll if they drive through the left-hand lane **7**; an observer station **8** verifies carpool occupancy with no vehicle slow-down, and reports violators to the California Highway Patrol. Remember: You must have a FasTrak transponder to use the 91 Express Lanes.*

Eastbound 91 Freeway (towards Riverside)

*From either the 91 or 55 freeways, signs **9** will alert motorists that the Express Lanes lie ahead. A sign **10** will display the current toll approximately half a mile before the Express Lanes begin **11**, giving motorists time to choose whether to take a fast, safe, reliable trip on the 91 Express Lanes, or to stay on the freeway. Each time you drive through the toll plaza, the proper toll will be deducted.*

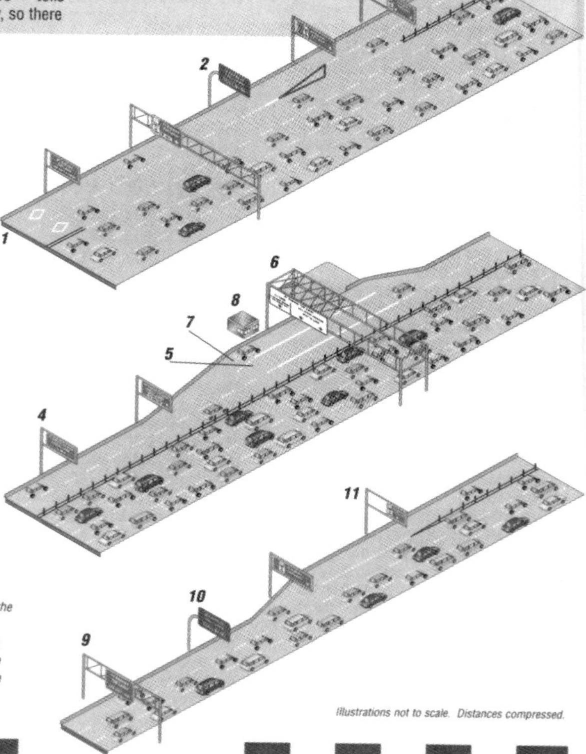

Figure 4.9. Entering the express lane AVI system. *[Courtesy California Private Transportation Company, L.P.]*

wit Diversified Group in addition took on the job of underwriting a $35 million institutional debt which they would sell as market conditions allowed. The Orange County Transportation Authority had a $7 million subordinated debt that represented seed money it had spent in the early stages of the project, and this would be repaid, including interest, by CPTC. Three banks—Citicorp, and French banks Banque National de Paris and Société Générale loaned the group $65 million on a 14-year variable rate basis to round out the financial arrangements. Without Peter Kiewit's faith, hard work, and financial risk taking, the SR 91 project may well have faltered and sputtered before breaking into the clear.

FasTrak—SR 91's Automatic Vehicle Identification System The SR 91 Express Lane concept will give riders a choice: Drive on the freeway portion of the highway and be subjected to the potential delays and frayed nerves of its peak hour congestion, or pay a reasonable toll and zip along in the express lanes. Toll rates for the express lanes average $1.75; they rise during peak hours and lower during off-peak hours. The operators of the roadway have coined the term "value pricing" for this variable rate structure. There are easy-to-read overhead signs to display the toll charges in effect at any time and this will assist motorists in making their decision as to which roadway to use. When traffic moves more freely on a highway, air pollution levels tend to drop and vehicle gas mileage increases—two other advantages that the public should not dismiss when analyzing the benefits to be derived from these new Express Lanes.

With the AVI system consisting of a windshield-mounted transponder and roadside antennas, there will be no need to stop at entrance and exit booths.

The illustration in Figures 4.8 and 4.9 indicate how the system will function and how it will allow motorists time in which to make their decision as to whether they will travel on the freeway or opt for the toll road and use the FasTrak Express Lanes.

The AVI suppliers and installers were testing their equipment and systems, and applications from drivers for vehicle transponders were already being received and filled in 1995. Still unresolved is how motorists will be able to make initial deposits to their FasTrak account and how this account can be replenished when it has been drawn down to the point where added funds are required. But these decisions will be made and announced to the public well in advance of the highway opening.

Once the SR 91 express lane project opens in late 1995 or early 1996 and the initial fanfare has died down, all eyes will be on the way drivers continue to react to its promise of hassle-free commuting at a price. This may be the watershed project for Caltrans and possibly other areas of the country where state governments are concerned about the viability of public/private transportation projects. If SR 91 proves to be successful, it will be a confidence-builder for Caltrans and the other three pioneers of toll road development in the state of California.

5

WASHINGTON STATE AND ARIZONA—WHAT WENT WRONG?

The public/private partnerships envisioned by Washington State and the state of Arizona somehow went awry. Other states and other developers may wish to clip a page from the process that saw good intentions and legislative efforts come to naught.

WASHINGTON STATE

In 1993 Washington State enacted legislation enabling it to enter into agreements with private transportation developers to build highways and ancillary structures valued at approximately $2.5 billion. Substitute House Bill 1006 passed the House of Representatives and the Senate in Olympia, Washington on April 20, 1993, with an effective date of July 1, 1993; its express purpose was

> providing a sound economic investment opportunity for the private sector. Such initiatives will provide the state with increased access to property development and project opportunities, financial and development expertise and will supplement state transportation revenues, allowing the state to use its limited resources for other needed projects.

Private entities would be authorized under negotiated agreements to impose user tolls or tolls within the project area so that a reasonable return on investment could be realized.

SHB 1006 was to provide for six demonstration projects to be selected by both public and private sectors and approved by the state transportation commission.

These agreements would allow for private ownership during construction; once the projects were completed, the state would lease them to the private entity to operate for periods of up to 50 years.

When the Washington State Department of Transportation (WSDOT) announced its intention to accept proposals from the private sector for public works projects, 14 groups responded with schemes of various sorts, with a total value of $5 billion. State Transportation Secretary Sidney Morrison selected the six winning projects in August 1994.

1. State Road 522 (SR 522) Corridor Improvements—a new four-lane highway, Phase 1 of which was valued at $155 million
2. State Road 520 (SR 520)—high occupancy vehicle (HOV) lanes and a new bridge across Lake Washington, a $440 million project
3. Puget Sound region congestion pricing, along with other phased highway construction estimated to cost $932.5 million
4. State Road 18 (SR 18)—a limited access four-lane highway valued at $476.4 million
5. King County park-and-ride lots with a value of $68 million
6. State Road 16 (SR 16) and Tacoma Narrows Bridge congestion relief, a $564 million venture

THE PROJECTS

State Road 522

The State Road 522 proposal was submitted by the SR 522 Community Highway Group, a nonprofit corporation formed by local residents, including Interwest Management Group, INCA Engineers, and Max J. Kuney Company. Their proposal was divided into three phases; the value had only been estimated for Phase 1.

Phase 1, estimated to cost $155 million, would consist of the design, construction, and operation of a 10.59 mile (17.16 km) grade separated, four-lane full access controlled urban tollway starting at the State Route 9 interchange and continuing to the State Road 2 interchange with two new interchanges to be added, one at Paradise Lake Road and the other at Fales Road.

Phase 2 would contain an additional toll plaza to facilitate a congestion pricing pilot program along with the modernization of about 2 miles (3.2 km) of SR 522 between I-405 and State Road 9. Also to be added during this phase was the installation of several park-and-ride lots.

Phase 3 was to consist of evaluating the practicality of extending improvements across State Road 2 to the northeast to create a bypass around the city of Monroe.

SR 520 Corridor Improvement Project

Sponsored by Washington Transportation Partners (WTP), a team composed of Parsons Brinckerhoff Privatization and Morrison-Knudsen, the State Road 520 proposal will improve SR 520 corridor in two phases. During Phase 1, the existing freeway ramps from SR 520 to the I-5 express lanes would be connected. The elevated SR 520 structures would be seismically upgraded, partial noise abatement instituted, and the unused ramps at the Arboretum removed. Phase 1 would also include complete environmental and design studies for the second phase. Phase 2 would create park land over selected portions of the freeway, along with the additions of a bus and carpool lane in each direction and a bicycle and pedestrian path across Lake Washington. Automatic vehicle identification (AVI) would be available. Phase 1 was estimated to take 2 years to complete, and Phase 2 was projected at 3 years to completion. The total environmental, design, engineering, and constuction costs for the entire SR 520 corridor improvement program were estimated at $440 million.

The Puget Sound Congestion Pricing Project

Congestion pricing, a new term that will appear in transportation-related articles with more frequency in the coming decade, refers to the practice of charging a higher toll during peak traffic periods. This concept can be implemented in one of two ways: Limit the use of a freeway during peak periods by charging a toll for its use at those times, or build a couple of extra lanes onto an existing highway and charge a much higher toll to use those lanes during peak periods or for single occupancy vehicles (SOVs). The Federal Highway Administration is carefully studying the concept of congestion pricing to determine if it represents an equitable way to relieve congestion on some of the country's highways. One of the questions raised is whether this method of pricing tolls places an unfair burden on those people who have inflexible work schedules. But on the other side of the coin, the additional funds raised by congestion pricing can be used to fund construction of other highway projects or bridges or even mass transit alternatives to relieve highway traffic flow problems. Washington State thought the idea was a good one and WSDOT Economic Development Director Jerry Ellis was quoted by ENR Magazine in December 1994 as saying that congestion pricing in Seattle could raise more than $100 million per year to be used for transportation enhancement.

This third demonstration project, proposed by United Infrastructure Washington, Inc., a permanent joint venture between Bechtel and Peter Kiewit, included provisions for congestion pricing in the Puget Sound region.This project, to be implemented in four phases, was estimated to cost $325 million for Phase 1 and slightly more than $900 million for Phase 2. Costs for Phases 3 and 4 would be contingent upon certain milestone events in previous phases.

United Infrastructure, an advocate of congestion pricing, was of the opinion that properly administered variable fees—levied to encourage drivers to use the highway during slack periods and seek other means of transportation at peak periods—should create traffic flows that would permit traffic to move at speeds of 45 mph or higher during the peak periods. This off-peak pricing concept could be applied in much the same manner as telephone companies and electric power suppliers had used to shift usage from peak to off-peak periods.

To accomplish their goal, United Infrastructure proposed to proceed with Phase 1 and convert underutilized HOV lanes to fare lanes where buses and carpools would ride free, and low occupancy vehicles would pay a toll. During Phase 2, additional portions of the planned HOV network would be financed, constructed, and operated by United Infrastructure. The new lanes would operate as fare lanes with free access to buses and cars with more than three occupants. Phase 3 would see the gradual conversion of freeway lanes to fare lanes, one at a time. In Phase 4, the remaining lanes of the limited access system would be converted to fare lanes. According to United Infrastructure, these four phases would restore the free flow of traffic on the area's freeways to speeds of 45 mph or better during peak travel periods. Air pollution would be lowered as a result of this smoother flow of traffic, and energy savings would accrue since motorists would use less fuel.

SR 18 Improvements

The two-phase State Route 18 project was proposed by the National Transportation Authority (NTA), a joint venture of the Perot Group of Dallas, Texas and Greiner Engineering, Inc. The project would provide the design, financing, and construction of a 12 mile (19.4 km) section of the highway starting at the SE 312th Way and ending north of the Issaquah-Hobart Road interchange. Phase 2 would provide an opportunity for WSDOT and NTA to complete the SR 18 improvements from their termination point at the Issaquah-Hobart Road interchange to Interstate 90. Phase 1 was estimated to cost $256 million and Phase 2 approximately $220.4 million.

METRO/King County and WSDOT Park-and-Ride Capacity Enhancement Program

ABAM Engineering and Perini Corporation, operating as Perini-ABAM, proposed to design, develop, finance, and construct single-level parking decks over 23 existing park-and-ride lots located along the METRO in King County, effectively creating an additional 7,000 parking spaces. Although there would be no additional charge to park on these elevated decks, the proposal anticipated that some patrons would be willing to pay to park in secured, guarded areas. Perini-ABAM also planned to build revenue-producing amenities such as dry cleaning drop-offs, latte shops, and newstands at each location. Perini-ABAM believed secured parking areas and amenities would encourage more people to use the METRO.

The developer would lease the park-and-ride facilities to WSDOT; as WSDOT would receive additional income from increased passenger usage of the METRO line, the idea of leasing the parking decks from Perini-ABAM would be attractive. Perini-ABAM estimated that an initial daily parking fee of $2 would be sufficient to cover increased operating costs and offset WSDOT lease payments. At the end of the proposed lease period, ownership of these Perini-ABAM constructed facilities would revert to WSDOT providing WSDOT with another source of revenue. The proposed cost for the 23 planned park-and-ride structures was estimated to be $68 million.

The Tacoma Narrows Project

The Tacoma Narrows Proposal, submitted by United Infrastructure, offered four alternatives to reduce congestion on the existing bridge and along State Road 16.

Alternative 1. Build a new suspension bridge parallel to the existing one.

Alternative 2. Build a new cable-stayed bridge parallel to the existing one.

Alternative 3. Construct a lower deck on the existing bridge and create two HOV lanes on that deck. This alternative would also involve phased construction of the remaining HOV lanes on SR 16 from Interstate 5 to Gig Harbor.

Alternative 4. Outfit the existing bridge with a movable barrier for reversable lane operations and toll it in both directions to provide HOV fare lanes on SR 16 from Interstate 5 to Gig Harbor. HOVs would travel free but SOVs could buy excess fare lane capacity.

Double-decking the bridge would cost $564 million; the Transportation Demand Management option would cost $216 million. The proposals presented by United Infrastructure would reduce congestion and increase safety on the bridge, SR 16, and adjacent surface streets. The proposer would obtain revenue and repay the investment from the toll charges.

PROJECTS IN JEOPARDY

Due in part to changing political tides and a growing public mood of NTFIMBY—No Toll Facility In My Back Yard—all six of the proposed public/private partnership projects went awry. In August 1994 the State Transportation Commission dropped the SR 18 Corridor project and only five projects were to be considered, but it was not long thereafter that citizen opposition started to mount protesting the imposition of tolls on these proposed new highways.

In January 1995, some 300 citizens calling themselves Citizens Against Unfair Gouging of Highway Tolls (CAUGHT) demonstrated on the steps of the state legislature in Olympia against these tolled road projects. They felt that big developers and state officials meeting behind closed doors were plotting to extract more dollars out of their pockets. Ironically, most of the resentment was directed against three highway systems that the state badly needed but could not afford to build at the time.

The composition of the legislative body had changed from November 1994;

Democrats, the original backers of these public/private partnerships, were now in the minority and the Republicans took up the public outcries. Governor Mike Lowry, although pro-privatization, sniffed the political winds and said he would sign a bill to repeal SHB 1006 if it were presented to him.

Quite naturally developers and project proposers began to get very nervous. Some had spent considerable amounts of money preparing their proposals, and having had conditional approval several anticipated completing the design and starting construction. If these projects were delayed for months or even years, who would reimburse the developers? Consequently, Parsons Brinckerhoff and Morrison-Knudsen announced that they might abandon efforts to construct the State Road 520 project.

House Bill 1317

On January 28, 1995, the House Committee on Transportation submitted a Substitute House Bill 1317 to the legislature. While continuing to endorse a public/private sector transportation program, HB 1317 shifted a great deal of the responsibility for project approval to the voters. Also, HB 1317 would ban the congestion pricing mechanism and place a two-year moratorium on any new public/private partnership projects. Any such projects already under negotiation would be subjected to increased public and legislative scrutiny. This bill would limit the collection of tolls to the point when the project's cost had been satisfied and it would eliminate the collection of tolls beyond the economic life of the project.

Section 1 of HB 1317 is seemingly encouraging to public/private ventures:

> The ability of the state to provide an efficient transportation system will be enhanced by a public-private sector program providing for private entities to undertake all or portion of the study, planning, design, development, financing, operation and maintenance of transportation systems and facility projects.
>
> The secretary of transportation should be permitted and encouraged to test the feasibility of building privately funded transportation systems and facilities or segments thereof through the use of innovative agreements with the private sector.
>
> The department of transportation should be encouraged to take advantage of new opportunities provided by federal legislation under section 1012 of the Intermodal Surface Transportation Efficiency Act of 1991 (ISTEA). That section establishes a new program authorizing federal participation in construction or improvement of publicly or privately owned toll roads, bridges, and tunnels, and allows states to leverage available federal funds as a means for attracting private sector capital.

However, a later section of HB 1317 bill alerted the proposed developers that public involvement in the decision-making process would be mandated.

> The department shall not enter into an agreement with a private entity for a project selected prior to September 1, 1994 unless that agreement includes a process that

> provides for public involvement in decision making with respect to the development of the project.

Further, the bill authorizes the department to establish "incentive" rates of return beyond the negotiated maximum rate of return on investment:

> The incentive rates of return shall be designed to provide financial benefits to the affected public jurisdictions and the private entity, given the attainment of various safety, performance, or transportation demand management goals. The incentive rates of return shall be negotitated in the agreement.

A new section to be added to existing chapter 47.46 RCW spells out in detail how the public involvement would be implemented, and placing a substantial risk factor on any consortium submitting a proposal for a public/private partnership transportation project.

> In carrying out the public involvement process required in RCW 47.46.040, the private entity shall proactively seek public participation through a process appropriate to the characteristics of such project, or in the case of a project developed in phases or segments, such phase or segment, that assesses overall public support among: Users of such projects, phase or segment; residents of communities in the vicinity of such project, phase or segment; and residents of communities impacted by such project, phase or segment.

This public involvement process would provide opportunities for users and residents to comment upon key issues regarding each project including

- alternative sizes and scopes
- design
- environmental assessment
- right of way and access plans
- traffic impacts
- tolling or user fee strategies and tolling or user fee ranges
- project cost
- construction impacts
- facility operation
- any other salient characteristics

To facilitate public involvement the private entity seeking project approval is required to establish a committee of individuals who represent cities and counties in the vicinity of the area impacted by the project. This committee, known as the "local involvement committee," is to be composed of (1) various elected officials from each city and county within the project, (2) two representatives from an organization formed in support of the project, and (3) two representatives from an organization formed to oppose the project. The private entity must also conduct a comprehensive inventory of public positions of users and residents of communities within the vicinity of the project. This survey must be conducted by an independent professional or accountant jointly selected by the private entity and WSDOT in consultation with the local involvement committee.

And if that weren't enough to scare away any private entity, the proposed bill additionally requires that WSDOT and the developer provide the legislative transportation committee and the local involvement committee with progress reports on the status of the public involvement process and the inventory of public positions. Within 15 days of submission of this project information, the local involvement committee must submit a report to the department and the legislative transportation committee supporting or opposing the results of the inventory. Within 45 days of the submission of the local involvement committee's report, the legislative transportation committee must conduct a public hearing on the results of both reports. The tenor of the changes proposed by the legislature in HB 1317 placed all six proposed demonstration projects authorized under Substitute House Bill 1006 in mortal danger.

Engrossed Substitute Senate Bill 6044: The Highway 522 Project Killer

On March 6, 1995, the Senate Committee on Transportation introduced another section to Chapter 47.05 that temporarily killed the Highway 522 project. The Washington State legislature renewed its opinion that to maintain the high quality of life, the residents of the state must have an efficient transportation system and the public/private initiative would provide a sound economic investment for the private sector; however,

> the legislature finds that in the case of Highway 522, selected under this chapter, public support has not been demonstrated and therefore the secretary shall not proceed.
>
> Among the demonstrations of non-support for inclusion of Highway 522:
>
> (1) Over sixteen thousand citizens have signed petitions in opposition to the toll project.
> (2) The majority of city council members in Monroe, Duvall and Index have made public statements opposing the toll project, and that the Woodinville chamber of commerce has officially opposed the toll project.
> (3) No city council or chamber of commerce in the areas has favored the toll project.
> (4) Of the five hundred individuals who attended the public information hearings on the toll proposal, four hundred fifty eight signed a petition requesting that the proposal be rejected.
> (5) Businesses in Monroe, Woodinville, Duvall, Snohomish, Sultan, Startup, Gold Bar, Index, Skyomish, and Stevens Pass are extremely dependent on Highway 522 for commerce, that due to the rural nature of these areas no alternative for commerce exists, and that a toll on Highway 522 would severely inhibit their ability to stay in business.

(6) In an informal poll of residents who currently use Highway 522 to shop, eighty one and one-half percent of the respondents claimed that they would be unlikely to continue shopping at these stores if a toll were imposed.

Washington State may have to look far and wide for a private intiative proposal that could withstand the withering public scrutiny and the lengthy review process that most likely occur when a developer presents their plans for such a project.

According to Jerry Ellis, serving in her role as Director of the Transportation Economic Partnership of WSDOT, HB 1317 does not end public/private partnership programs but merely amends the original Public Private Initiatives in Transportation Act. The amendments to the original bill provide for a very thorough public involvement process including an advisory vote for those projects where known opposition exists. As defined in the legislation, "opposition" is defined as being a petition signed by 5,000 citizens. Ellis said that thus far three projects have been qualified for this advisory vote process: Highway 522, State Road 520, and the State Road 16 Tacoma Narrows project. The congestion pricing proposal is on hold, and the park-and-ride proposal is proceeding to final negotiations with an advisory vote expected in the spring of 1997.

Was it a case of too much too soon? Was it a case of one political party making short term hay possibly to the long term detriment of the public and the public transportation systems in the state? Or was it a citizenry unwilling to pay for transportation improvements?

The Puget Sound Light Rail System On March 14, 1995, the state Senate voted to kill the second public/private initiative, a $6.7 billion light rail project, which would have been the largest public works project in Washington's history. A poll conducted in August 1994 had indicated that 70 percent of the residents questioned supported a proposed light rail system linking the major population centers of Puget Sound. In February 1995 anti-tax sentiment had scuttled a $240 million light rail system for Clark County near Portland, Oregon, but Citizens for Sound Transit, an organization backing the Seattle rail plan, predicted that Washington's light rail vote would be positive as Seattle's traffic congestion was far more severe than Oregon's. Everyone misread the public: The light rail system was defeated by a 53 percent majority. Even more puzzling was the organized oppositions claim that the area's transportation needs could be best filled by highway expansion, car pooling, and the use of HOV lanes. The citizens of Washington State appear to want relief from traffic congestion but do not want to have their taxes increased nor do they want to pay user fees for these new roadway improvements.

WHAT WENT WRONG

As the public/private partnership process gathers steam in other parts of the country, the nation will be looking the Washington State example closely. Politicians and public alike will be asking what the alternate solutions to traffic congestion problems are, where the money to cure them should come from, and whether improvements should be financed with statewide tax increases or user fees. Many Americans travel the interstates and intrastates toll free and, when the idea of resurrecting the toll system was proposed in Washington State, both politician and private developer misjudged the public's objection to tolling. As one resident put it, "I'll walk before I'll pay $1,200 a year to drive into Seattle to work." One argument espoused by critics to these plans was that the public had a right to decide if public transportation improvements in their communities could be financed by tolls. However, if protecting or improving a structure that had been paid for by taxes levied statewide, shouldn't all taxpayers have a say in these matters?

In Washington citizens felt isolated from the process that would create a $2.5 billion highway program employing a new system known as a public/private partnership. Clearly, the public needed to be educated to support user fee methods of infrastructure financing, but the method by which the proposals were prepared and evaluated by WSDOT was not conducive to gaining public support. When private developers presented their proposals to WSDOT, these proposals contained a great deal of proprietary information that, if made public, could divulge confidential corporate information and thereby lessen the proposer's competitive edge for the project at hand and for future projects. WSDOT had not devised a system to review the proposals under public scrutiny while still preserving the proprietary information in these proposals. So the public, always wary of government officials and developers, felt that deals were being made behind their backs.

Already several other states have taken notice of the Washington controversy. Minnesota is seeking private developers to submit proposals on Build, Operate, Transfer highway projects for some of its $5 billion unfunded transportation projects. Unlike WSDOT, however, MinnDot will require developers to submit their proposals to local communities after they submit to the state but before entering into negotiations with the government. Developer's proprietary information will not be available for competitor's scrutiny until after the submission date. South Carolina's DOT has devised a unique system to review proposals whereby each proposal is identified by letter only and is evaluated that way. Once the successful bidder has been selected, the identify of the proposer or consortium is divulged; the identities of the unsuccessful bidders are not revealed.

The lesson of Washington State has not been lost on other states or on other developers who realize that there is new depth to the risk they are expected to take when bidding on these public/private partnership projects.

ARIZONA AND THE VUE 2000 EXERCISE

The State of Arizona has enjoyed rapid growth over the last decade. The population of the city of Phoenix grew by 24.5 percent; to the south, Tucson experienced a 23 percent increase during the same period. With these increases in population and the state's growing tourist trade, more and more demands are being placed on the existing infrastructure systems. Not only is Arizona faced with a billion dollar estimate for highway repairs over the next 30 to 35 years but billions more will be required to finance future capital costs.

These were some of the compelling reasons for the state legislature's decision in 1991 to pass a law that would enable Arizona to embark on a highway privatization program similar to that being instituted in other parts of the country. Arizona's privatization law, commonly refered to as "Article 2," is Article 2, Chapter 26 of Title 28—Transportation Project Privatization of the Arizona Revised Statutes [sections 28–3061 to 28–0374].

Shortly after this law was enacted, 10 proposals were submitted to ADOT by various groups of developers.

Proposal 1 Submitted by HDR Engineering, Ebasco Infrastructure, JHK & Associates, and Hughes Aircraft, this proposal encompassed much of ADOT's Proposition 300 highway system. The plan would build much of the Proposition 300 valley urban highways (see Figure 5.1), and carried a price tag of $3.75 billion.

Proposal 2 Submitted by HDR, JHK & Associates, and the Fort Mojave Indian Tribe, this proposal was designed to connect California's I-40 highway north of Needles to Arizona Route 95.

Proposals 3, 4, and 5 These three proposals were submitted by Private Roadbuilders of Arizona, a group headed by the Spanish construction firm Dragados y Construcciones SA and a local Phoenix company, Pulice Construction. All of three proposals were slightly different variations on the construction of 15.7 miles (25.4 km) of highway through Scottsdale known as the Pima Highway.

Proposals 6, 7, and 8 Three proposals were submitted by three consortiums that addressed the expansion of the partially completed Squaw Peak Parkway running north through Phoenix. One proposal, submitted by National Transportation Authority, a group led by the Perot Group and Greiner, was the most costly of the three with a price tag of $410 million for a six-lane, 12.2 mile (19.76 km) stretch of new highway. Another less costly proposal came from the Arizona Transportation Corporation, a group that included Andrade Gutierrez Construction Inc. from Brazil, ICA Construction Group from Mexico, and the Sundt Corporation from Tucson. The third consortium that submitted a proposal was composed of Parsons Brinckerhoff Privatization, Kiewit Western, and several other firms.

Proposal 9 Perini Corporation of Framingham, Massachusetts led a team that included Daniel Mann Johnson & Mendenhall, Vollmer Associates, and Westinghouse Electric. This group proposed a $70 million toll bridge to connect two roads in Scottsdale.

Proposal 10 Arizona Tollroad, Inc., a group of 15 firms including Kiewit Construction Group and AT&T proposed to build a toll road parallel to the existing freeways along I-10 and the Superstition Freeway at a cost of $180 million.

Three proposers were shortlisted in June 1992. However, there was strong opposition to the Dragados y Construcciones project voiced by an Indian tribe whose land was adjacent to the path of that proposed roadway. Objections to a proposed toll road south of Phoenix through Gila Bend were raised by a local group, and the opposition was able to scuttle that project as well. The only proposal that survived scrutiny was that submitted by the HDR team. Their proposal, Valley Urban Expressways for the 21st Century, became known as VUE 2000.

The Proposition 300 Highway System

The Proposition 300 highway system was a comprehensive highway plan for Maricopa County that was endorsed by the voters in 1985. Only a shortfall in funding prevented the Arizona Department of Transportation from implementing the plan. The total value of the 231 mile (374 km) Proposition 300 highway system was estimated to be $3.8 billion and ADOT funding lacked $2.8 billion.New state highway construction in Arizona is funded from sales tax revenues under the Regional Area Road Fund (RARF) and a portion of the State Highway User Revenue Fund (HURF). RARF funding will expire in the year 2005 and if construction of the entire Proposition 300 highway system had started, only 33 percent would have been put into place by the time state funds were depleted. Without the nearly $3 billion in additional funding required to complete this Maricopa County highway system, the traffic congestion already in evidence in the late 1980s would build to the point where travel time would be seriously impeded and both driver tempers and air pollution would rise to new levels. Phoenix, Scottsdale, Tempe, and adjacent areas would begin to experience many of the urban ills that their newly arrived citizens thought they had left behind.

In November 1994 voters rejected a $5 billion transportation-related sales tax increase leaving ADOT high and dry as far as implementing any plans for a highway system around Phoenix. Although a $\frac{1}{2}$ cent increase in the sales tax in 1985 funneled some funds into the Arizona Department of Transportation (ADOT), inflation and increased costs of land acquisition resulted in a shortage of funds for these repairs and construction. Without a private initiative the state would probably have to resort to raising the necessary funds by issuing bonds. This, in turn, would require new taxes on gasoline to pay for this new project and to provide funding for the maintenance of the expanded highway system. The state appeared to be caught between a rock and a hard place: the need to implement the Proposition 300 Highway System, versus insufficient revenue and the reluctance to consider new taxes. Thus, the appeal of a public/private partnership was enticing.

THE VALLEY URBAN EXPRESSWAY FOR THE 21ST CENTURY—VUE 2000

The VUE 2000 project proposed by the HDR team excluded three Proposition 300 highway systems in their entiriety—Sky Harbor, East Papago, and Hohokam—and

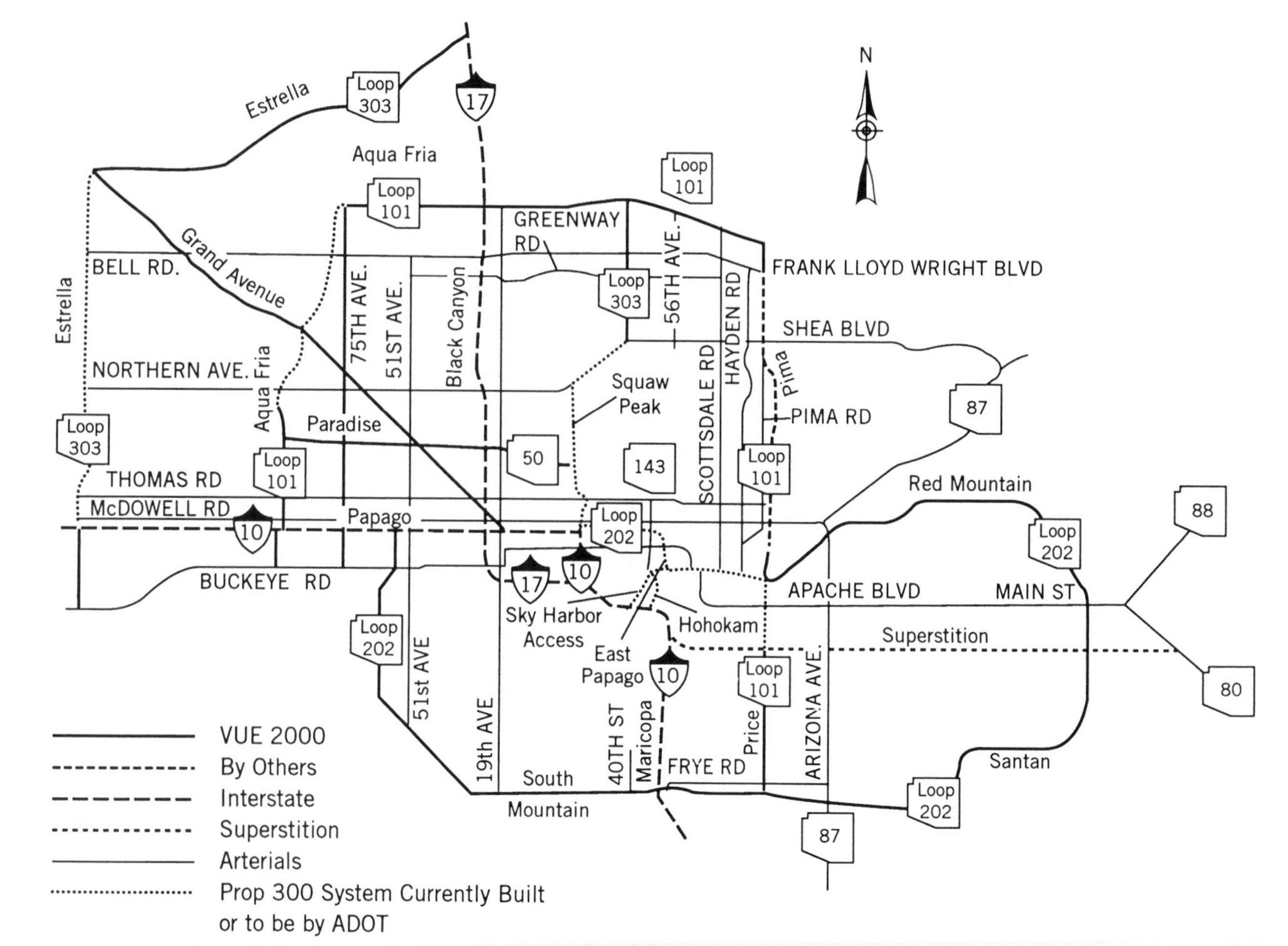

Figure 5.1. Completion of Proposition 300 highway system per VUE 2000 proposal. *[Courtesy HDR Project Services Corporation of Phoenix.]*

TABLE 5.1 The Extent of VUE 2000 Highway System

Corridor	Description	Length in Miles
Agua Fria	75th Avenue to I-17	6.0
Agua Fria	I-10 to about Glendale Avenue	6.2
Estrella (interim facility)	Grand Avenue to I-17	18.0
Grand	17th Avenue to Beardsley	23.0
Paradise	Squaw Peak to Agua Fria	12.5
Pima *	I-17 to Frank Lloyd Wright	17.5*
Price	North of Guadalupe Road to Frye road	6.0
Red Mountain	Entire	19.0
Santan	Entire	24.3
South Mountain	Entire	21.0
Squaw Peak	Shea to Pima Corridor	6.0
Total VUE 2000 Mileage (Rounded)		160.0

Source: HDR Project Services Corporation of Phoenix.

*The Excluded Pima Segment (Red Mountain Interchange to Frank Lloyd Wright Boulevard) would add approximately 15 miles.

excluded portions of four other corridors, but the balance of the Proposition 300 system was left intact. The total VUE 2000 scheme would encompass the routing and mileage depicted in Table 5.1. The project was scheduled for completion by the year 2001 and project financing would create low-cost, tax exempt bond proceeds to be used to supplement the remaining tax-based resources.

The VUE 2000 Team

HDR, a Nebraska-based engineering firm with offices throughout the United States, is a subsidiary of the French construction giant Bouygues, a multibillion dollar firm with interests in road building, offshore platforms and pipelines, telecommunications and television production, food production, and private management of public utilities. The other members of the VUE 2000 team were equally impressive:

Motorola, Inc.—a world leader in technology and communications equipment.

Goldman, Sachs and Company—one of the nation's largest privately held investment banking firms.

Science Application International Corp.—a leading provider of diversified research, engineering hardware, and software, with experience in transportation systems.

Dillon, Read and Company, Inc.—investment bankers.

Snell & Wilmer—one of the largest law firms in the western United States.

JHK & Associates—nationally recognized engineering consulting firm specializing in systems engineering, transportation planning, and traffic engineering.

Hawkins, Delafield & Wood—a general practice law firm specializing in publicly supported real estate development and public contracting.

Ametch Systems Corporation—a leading provider of automatic equipment identification for transportation and intelligent vehicle highway systems.

Apogee Research Inc.—specialists in public works economics and finance.

Coopers & Lybrand—a global accounting and financial consulting firm.

Lachel and Associates—a transportation tunnel engineering firm.

The Organizational Vehicle—A Nonprofit Corporation

The VUE 2000 group planned create a nonprofit corporation to develop, finance, design, build, operate, and maintain the proposed toll road system. The not-for-profit enterprise would allow the team to have access to low-cost, tax exempt bonds which would make the project financially viable and also save the government billions of dollars over the 30-year debt servicing period when compared to the cost of funds obtained by other private sector means.

A financial model was created to test the debt serving capacity over the life of the project extending from 1992 to 2029, based upon obtaining low-cost tax exempt long-term bonds. Assuming an average coupon rate of 8 percent, debt services projected to the year 2005 would be $441 million. Offsetting a portion of this sum would be the earnings of the debt service reserve fund (DSRF). Based upon earning 6 percent per annum, the DSRF would yield $32.8 million per year. Appendix 5.1 at the end of this chapter contains the base case financial analysis.

The $1 billion estimated RARF and HURF funds that the state anticipated collecting by the year 2005 would be leveraged by VUE 2000 to provide the total $3.8 billion needed for the complete urban highway development plan. Therefore, VUE 2000 could be looked upon as a private initiative augmenting government efforts and funding to construct the complete highway system as quickly as possible—a superior method than building the system in sections as funding became available over a much more extended period of time.

Revenue for debt service and developer profits would come from user-fees. Arizona's privatization law permits private-sector toll road operators to set their toll rates; therefore, an anticipated revenue stream can be published to make the yield on bond issues attractive. Toll rates would be "market oriented," that is, the private-sector developer would have to establish a toll charge schedule that would attract users as well as make the bond issue interesting to potential investors. A study of nine urban toll road systems throughout the country found toll rates to be in the $0.063 to $0.143 per mile range; VUE 2000 developers therefore considered capping the tolls at $0.15 per mile for noncommercial automobiles. Also according to their study, the HDR group anticipated that 96 percent of all traffic would be generated by passenger cars and two-axle vehicles. Only 4 percent of projected revenue would come from trucks with more than two axles. Assuming a toll rate for trucks at twice the rate for passenger cars, a $0.15 toll rate would have generated an estimated $538 million per year by 2005.

Base Case Debt Service Coverage Ratio (DSCR). Excess Revenue projected to the year 2005 produced an excess of $77 million by that year based upon the following assumptions:

Toll revenue	$538 million
Less: O&M Costs	($53 million)
Revenue Net of O&M	$485 million
Plus: Debt Service Reserve Fund	$33 million
Equals: Revenue Available for Debt Service	$518 million
Less: Debt Service	($441 million)
Excess Revenue	$77 million

Although it appeared unlikely that toll rates could be less than $0.10 per mile, VUE 2000 consultants indicated this was possible under the following conditions:

1. If long-term interest rates were closer to 6.5 percent rather than the 8 percent projected
2. If the traffic capture is nearer to 80 percent rather than the 60 percent used in the financial pro forma
3. If the project could be built for $3.5 billion instead of $3.8 billion
4. If ADOT elected to make the most cost-effective allocation of RARF and HURF funds

In approximately 30 years, when construction costs have been recouped via repayment of debt service and when all other debts incurred to develop the expressway have been satisfied, all private sector interests in the facility would end and responsibility for future operation, management, and maintenance would revert to the state of Arizona. VUE 2000's proposal was therefore basically a modified BOT project.

Value Engineering—A Private-Sector Advantage

Private developers point to another advantage when proposing a Build, Operate, Transfer project and that is the ability of their team to establish meaningful value engineering alternatives to reduce overall project costs. The VUE 2000 team had several team members who assembled value engineering suggestions which probably would have not been brought to the attention of ADOT if this highway project had proceeded to bid in the conventional manner. But the HDR group came up with some worthwhile suggestions that, if accepted, would yield a minimum $47.6 million in savings.

1. Use enhanced prestressed concrete girders to save both construction dollars and time. If all bridges could use prestressed girders, a savings of $16 million could be effected.

2. Limit the width of the overcrossing structures and defer widening for a later date. Although the amount of savings would be dependent on the amount of reduction deemed acceptable for traffic considerations, savings in the magnitude of $25 million could be realized.
3. Defer or eliminate painting of the structures at a savings of about $1.6 million.
4. Standardize design for similar street bridges, thereby limiting design costs and construction contingencies. This could result in a $5 million saving.

HDR developed some other potential costs savings that could not be quantified due to the limited nature of the study concducted by the VUE 2000 group.

1. Reduce the width of undercrossing structures to that required by current traffic projections.
2. Do not build structures at every mile crossing on remote areas of the MAG system.
3. Standardize column sizes and shapes.
4. Take advantage of ultimate strength and save as much as one-third of the reinforcing bars in bridge decks.

Furthermore, according to HDR, a study of proposed drainage plans could produce even more savings and when they included seven design changes, more that $25 million in other savings could be realized.

The VUE 2000 team also included 13 pages of additional potential deferral or cost reductions considerations ranging from postponement of landscaping (worth $119 million) to delaying installation of an entire section of freeway (saving $110.79 million) A comprehensive list of value engineering suggestions such as this would have been nonexistant during the course of public bidding in the conventional manner.

Seeking Project Enhancements—Another Private-Sector Initiative

The VUE 2000 proposal also investigated possible enhancements that could add to the value of the highway project and yield either additional state revenue or more tax dollars.

Park-and-ride facilities could be built to encourage ride-sharing, thereby reducing traffic on other local highways near the VUE 2000 system. The Park-and-ride facilities could become part of the electronic tolling facilities planned for VUE 2000 where the user fees would be based on time parked rather than distance traveled on the toll road.

As another suggestion, the developers of VUE 2000 indicated that some portion of the planned surface roadway could be placed underground via a cut-and-cover tunnel or as a bored tunnel.Placing a section or sections of roadway underground could reduce not only rights-of-way costs but also pollution, which could be cleaned from the tunnel air before it was expelled to the atmosphere. The need to relocate overhead utilities might not be necessary, and the displacement of some surface facilities, such as neighborhood businesses which might otherwise have had to relocate, could be prevented. The VUE 2000 group speculated that in the future these tunnels could be widened to accomodate a light rail system or other forms of mass transit. And if park-and-ride facilities were built, mass transit users would find parking areas convenient.

The group also suggested that the electronic billing that would be in place for toll road users could be modified to accomodate certain specific programs:

- Toll charges for employees commuting to work could be billed directly to their employers to create a meaningful employee benefit.
- Parents could make arrangements to be billed directly for their childrens tolls on weekends or during hours when their children attend music lessons, participate in a team sport, and so forth.
- Incentives for local residents to use mass transit could be arranged by discounting the toll charges paid by those carriers, which should result in reduced fares for the users.

Economic Advantages and Development

Any major construction project will inject money into the local economy as workers buy food, the companies working on the site purchase supplies from local merchants, and so forth. But the proposers of VUE 2000 estimated that nearly $400 million per year in project-related expenses would flow directly into local and statewide construction and engineering firms as well as to material and equipment suppliers. The activity created by the construction work would boost the sales of local retail shops, restaurants, real estate offices, banks, and utility companies. Overall economic activity was projected to rise $7 billion or $900 million per year. Construction costs were estimated to be as much as $3.3 billion in current dollars and according to statistics obtained from the U.S. Bureau of Economic Analysis, for each $1 million in construction spent, 30.7 man work-years in all industries would be generated. Shipping costs to and from the area could conceivably drop because truckers would be able to deliver more economically, further enhancing the area's reputation as an attractive place in which to locate a business.

It was anticipated that VUE 2000 would generate 7,200 jobs over an 8 to 10 year period, or effectively 74,000 man work-years. Each work-year would generate $3,000 in state and local taxes, for a total of $222 million flowing into government coffers. For comparison purposes, the entire state's personal income tax receipts in 1992 totalled $1.3 billion. From a tax standpoint, a "user fee" is more politically saleable than a tax increase. Users of the VUE 2000 facility would be the prime source of revenue to be used for repayment of the facility whereas the levying of a statewide sales tax would result in many citizens paying for a highway they would probably never use. And lastly, a sales tax is regressive; lower income residents pay a greater percentage of their personal income than do higher income residents.

But just as important, this new highway system would contribute to the quality of life in the area. A study conducted by the developer revealed that commuting time would be decreased by an average of 10 minutes per day and this gain would equate to a 2 percent increase in productivity. These figures were based upon an eight-hour average work day containing 48 to 10 minute segments and assuming an average wage of $11 per hour. (Of course time spent commuting does not necessarily detract from one's workday productivity, it merely lessens the amount of time available for nonwork activities.)

The VUE 2000 project would have provided a much needed highway system for which no tax increase would have been required. The ensuing construction and post-construction activities would have had a positive economic effect on the immediate area, and potentially on the entire state.

The Fate of VUE 2000

Proposition 300 was a key item on the November 1994 state election ballot and it was defeated. ADOT officials and private developers attributed its defeat primarily to a failure of public relations.

In previous years, ADOT had promised voters that it could build the Proposition 3000 system with a $\frac{1}{2}$ cent increase in the sales tax. Circumstances beyond ADOT's control prevented the Department from doing so, and it lost a great deal of credibility with the voters. ADOT did not realize the political implications of the proposal; it didn't coordinate the formulation and review process with other state and local agencies, but instead handed over control of the project negotiations to the Attorney General's Office.

The original VUE 2000 proposal had recognized the need to gain public support for the project. Public opinion polls conducted by ADOT and a local newspaper indicated little support for new taxes but strong support for user fees to finance the completion of the Proposition 300 highway system. The VUE 2000 team was ready, willing and able to conduct community forums, institute a direct mail campaign and commence a radio and TV public relations program. But the public relations end of the VUE 2000 project was sorely lacking, and neither the developers or ADOT made any concerted effort to obtain local support.

Opponents of the project, on the other hand, were very vocal. The Sierra Club donated $100,000 to defeat Proposition 300. Former Senator Barry Goldwater sent out a letter in August 1994 stating "With the list of prominent people who oppose Proposition 300, and the number of people who have spoken to me about it, I would be an idiot if I didn't vote against it."

The MetroRoad Project

The VUE 2000 project is dead and will not be resurrected but the hope for a public/private transportation project in Arizona lives on. In December 1994 Governor Fife Symington announced a Regional Freeway Plan, and a map issued in January 1995 contained proposed the new routes. One section, known as Loop 202, extended

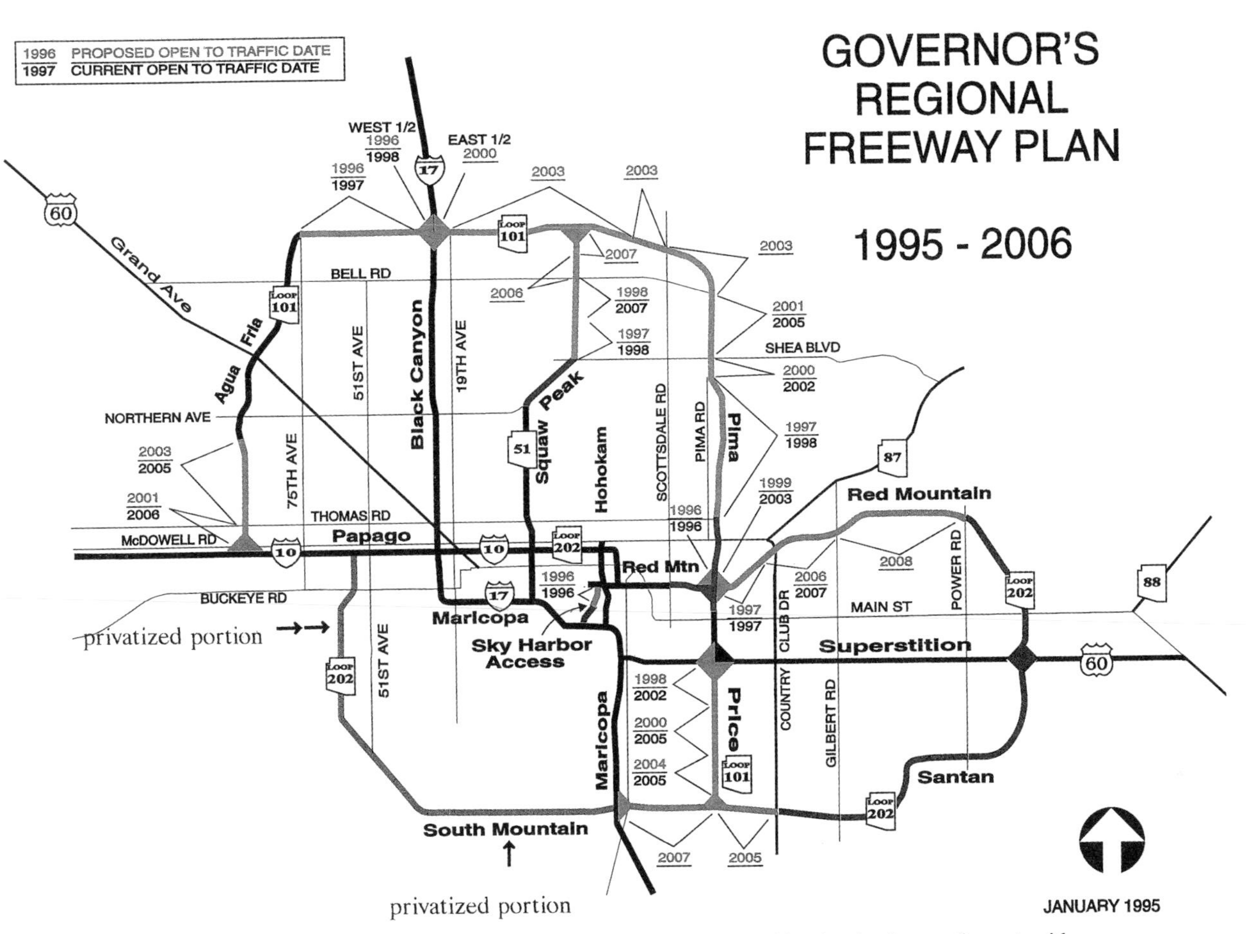

Figure 5.2. Governor Symington's Regional Freeway plan with privatization section set-aside.

TABLE 5.2 MetroRoad Selection Criteria Matrix

ADOT Criteria	Metroroad Highlights
1. Team Composition and Qualitications	• A firm and abiding commitment to help meet Arizona's transportation needs. • Comprehensive experience with ADOT and Regional Freeway System. • Depth and range of experience to handle all aspects of project management. • Team expertise in finance, infrastructure privatization, toll systems, and complex transportation projects. • Commitment to partnering, public outreach, and community involvement.
2. Capability and Capacity	• Team firms have a worldwide total of over 116,000 employees and $16.3 billion annual revenues. • Tax-exampt financing structure assures financing capability as big as the market. • Procurement plan assures all resources needed to design, build, and operate facility.
3. Application of Privatization Law	• Project organized to meet all requirements. • Innovative yet conforming strategy for nonprofit structure saves community hundreds of millions of dollars.
4. Commitment to Schedule	• Internal methods and procedures based on experience with accelerated delivery and understanding of ADOT processes. • Corporate culture, commitment, and leadership to achieve goals.
5. Contributions to State's Needs	• Provides approximately $1 billion of new transportation funding. • Completes all of State Route Loop 202 on an accelerated schedule. • Relieves congestion on 35 miles of ADOT highways by adding express lanes. • Enhances access and mobility for over a million people. • Enhances the life and capacity of existing South and East Valley streets and highways. • Provides congestion relief equivalent to adding 120 miles of 6-lane arterial streets. • Improves air quality for large region. • Supports regional and State economic development.

Source: HRD Project Services.

from Highway 10 to Route 17, and was noted as "Potential Privatization" (Figure 5.2). Under the existing Arizona Transportation Project Privatization Law, amended in the spring of 1995, ADOT sought competitive proposals, and the competition was closed in September of that year. The three proposals submitted were under review and evaluation, and it was anticipated that a conditional project approval would be granted before January 1, 1996.

HDR, persistent in its efforts to crack the Arizona privatization market submitted an unsolicited proposal called The MetroRoad Project to ADOT in June 1995. HRD's MetroRoad Project is designed to provide $1 billion of new financing to complete State Route Loop 202. The project would add an express lane in each direction on the Superstition Freeway and on the Price/Pima Freeway, and accelerate completion of the toll-free Price Expressway. Loop 202 and the express lane would become toll facilities to provide the necessary revenue stream, and HDR anticipated that this toll road portion could become toll-free by or before 2015.

Since the voters were emphatic about not wanting to pay for new highway construction by having their taxes raised, HDR would provide the financing and construction of these highways by creating a toll road. The Metropolitan Roadway Corporation (MetroCorp) was formed by the HDR Group as a nonprofit corporation under Arizona Law and would be therefore be permitted to issue tax-exempt bonds. The MetroRoad program would create a Loop 202 toll road and present the 1 million residents of South and East Valley with an alternative highway system of new and enlarged highways to ease congestion and avoid the loss of quality of life that inevitably follows traffic jams and overcrowded highways.

The parent company of HDR, French construction giant Bouygues, would head a strong team of financial managers, toll system technology and operations companies, environmental consultants, and legal advisors. Their selection criteria mix comparing ADOT's requirements with the MetroRoad plan of attack (Table 5.2) addresses the issues that just might make HDR the prime mover of the state of Arizona's venture into the public/private transportation market.

It is projected that Phoenix will grow from its present 2 million population to 6.5 million by the year 2020. As with Washington State, the citizens will have little excuse to complain about traffic congestion if they continue to turn a deaf ear to economically and environmentally sound projects proposed by the private sector.

APPENDIX 5.1

Base Case Financing Analysis for VUE 2000

Valley Urban Expressways for the 21st Century
VUE 2000: An Arizona Privatization Project
Submitted to: Arizona Department of Transportation

Financing Analysis: Summary
Project Overview
Sources and Uses of Funds—Construction Period
Assumptions and Parameters
Debt Service Cash Flows
Operating Cash Flows
Construction fund
Cumulative Net Revenues
Revenue and Cost Inputs

Sources and Uses of Funds—Construction Period

Source of Funds:		
Par amount of debt	$5,039,396,832	91.68%
Federal funds	0	0.00%
Other sources	0	0.00%
Construction fund earnings	207,675,374	3.78%
DSRF earnings (until segments are self-sufficient)	249,448,745	4.54%
Total Funds	$5,496,520,951	100.00%
Uses of Funds:		
Debt funded construction draws	$3,307,094,795	60.17%
Other funded construction draws	0	0.00%
Upfront letter of credit costs	0	0.00%
Annual letter of credit costs	0	0.00%
Bonded isuance costs (1.5% of par)	75,590,952	1.38%
Debt service reserve fund (gross)	503,939,683	9.17%
Shortfall (cumulative until each segment is self-sufficient)	1,609,895,520	29.29%
Contingency	$ (0)	−0.00%
Total Funds	$5,496,520,951	100.00%

* Before DSRF earnings.

Year Ending	Gross Toll Revenues	Interest Earnings*	Projected Revenues Avail. for O&M and Debt	O&M	Debt Service	Coverage Ratios	
						Gross†	Net‡
12/31/92	0	0	0	0	0	NM	NM
12/31/93	0	0	0	0	0	NM	NM
12/31/94	0	5,424,801	5,424,801	0	66,766,788	0.08	0.08
12/31/95	0	11,130,122	11,130,122	0	136,986,115	0.08	0.08
12/31/96	3,468,430	16,851,864	20,320,294	3,080,959	207,407,553	0.10	0.08
12/31/97	23,166,718	22,177,413	45,344,131	9,612,591	272,952,781	0.17	0.13
12/31/98	59,763,891	27,421,818	87,185,709	16,661,824	337,499,300	0.26	0.21
12/31/99	122,076,335	32,756,079	154,832,414	24,259,616	403,151,747	0.38	0.32
12/31/2000	229,259,472	32,756,079	262,015,552	32,442,058	403,151,747	0.65	0.57
12/31/2001	428,576,218	32,756,079	461,332,297	41,237,460	403,151,747	1.14	1.04
12/31/2002	454,096,502	32,756,079	486,852,581	46,785,773	403,151,747	1.21	1.09
12/31/2003	480,762,414	32,756,079	513,518,494	48,657,204	403,151,747	1.27	1.15
12/31/2004	508,660,652	32,756,079	541,416,732	50,608,930	421,389,269	1.28	1.16
12/31/2005	537,758,792	32,756,079	570,514,872	52,633,287	440,569,862	1.29	1.18
12/13/2006	568,144,847	32,756,079	600,900,927	54,738,619	459,805,661	1.31	1.19
12/31/2007	599,870,419	32,756,079	632,626,499	56,928,163	477,709,511	1.32	1.21
12/31/2008	633,038,843	32,756,079	665,794,922	59,211,652	495,340,561	1.34	1.22
12/31/2009	667,606,353	32,756,079	700,362,432	61,580,118	513,273,699	1.36	1.24
12/31/2010	703,678,433	32,756,079	736,434,512	64,043,323	513,273,699	1.43	1.31
12/31/2011	735,660,617	32,756,079	768,416,697	66,605,056	513,272,699	1.43	1.37
12/31/2012	769,158,679	32,756,079	801,914,758	69,276,702	513,273,699	1.56	1.43
12/31/2013	804,116,941	32,756,079	836,873,020	72,047,770	513,273,699	1.63	1.49
12/31/2014	840,664,056	32,756,079	873,420,135	74,929,680	513,273,699	1.70	1.56
12/31/2015	878,872,237	32,756,079	911,628,316	77,926,868	513,273,699	1.78	1.62
12/31/2016	918,891,392	32,756,079	951,647,471	81,052,651	513,272,699	1.85	1.70

12/31/2017	960,655,006	32,756,079	993,411,085	84,294,757	513,272,699	1.94	1.77
12/31/2018	1,004,316,776	32,756,079	1,037,072,855	87,666,548	513,273,699	2.02	1.85
12/31/2019	1,049,962,973	32,756,079	1,082,719,052	91,173,210	513,273,699	2.11	1.93
12/31/2020	1,097,772,687	32,756,079	1,130,528,767	94,830,327	513,273,699	2.20	2.02
12/31/2021	1,147,666,456	32,756,079	1,180,422,535	98,623,540	513,273,699	2.30	2.11
12/31/2022	1,199,827,896	32,756,079	1,232,583,976	102,568,482	513,273,699	2.40	2.20
12/31/2023	1,254,360,074	116,214,564	1,370,574,638	106,671,221	513,273,699	2.67	2.46
12/31/2024	1,311,476,943	115,104,438	1,426,582,380	110,949,992	428,269,389	3.33	3.07
12/31/2025	1,371,083,570	109,652,755	1,480,736,325	115,387,991	338,869,468	4.37	4.03
12/31/2026	1,433,399,318	97,835,750	1,531,235,069	120,003,511	249,212,231	6.14	5.66
12/31/2027	1,498,547,317	91,261,815	1,589,809,132	124,803,651	165,763,153	9.59	8.84
12/31/2028	1,566,783,170	87,399,819	1,654,182,989	129,809,745	83,585,583	19.79	18.24
12/31/2029	1,637,993,465	0	1,637,993,465	135,002,135	0	NM	NM
Total	27,501,137,893	1,486,622,066	28,987,759,959	2,466,105,414	14,296,991,436		

*Limited to interest earnings available for debt repayment; therefore, excludes earnings in construction fund. Includes transfer from debt service reserve funds in year each bond issue is fully retired.

†Total revenues/Debt service.

‡(Total revenues—O&M expenses)/Debt service.

Cumulative Net Revenues

Year Ending	Net Revenues	Cumulative Net Revenues
12/31/92	0	0
12/31/93	0	0
12/31/94	0	0
12/31/95	0	0
12/31/96	0	0
12/31/97	0	0
12/31/98	0	0
12/31/99	0	0
12/31/2000	4,060,761	4,060,761
12/31/2001	27,412,334	31,473,095
12/31/2002	47,545,211	79,018,306
12/31/2003	69,276,111	148,294,417
12/31/2004	98,516,673	246,811,090
12/31/2005	77,311,723	324,122,813
12/31/2006	86,356,647	410,479,460
12/31/2007	97,988,824	508,468,284
12/31/2008	111,242,709	619,710,993
12/31/2009	125,508,616	745,219,608
12/31/2010	159,117,491	904,337,099
12/31/2011	188,537,943	1,092,875,042
12/31/2012	219,364,358	1,312,239,400
12/31/2013	251,551,552	1,563,790,952
12/31/2014	285,216,756	1,849,007,708
12/31/2015	320,427,750	2,169,435,458
12/31/2016	357,321,121	2,526,756,579
12/31/2017	395,842,629	2,922,599,208
12/31/2018	436,132,609	3,358,731,817
12/31/2019	478,272,144	3,837,003,961
12/31/2020	522,424,741	4,359,428,702
12/31/2021	568,525,296	4,927,953,998
12/31/2022	616,741,795	5,544,695,794
12/31/2023	750,629,718	6,295,325,512
12/31/2024	887,363,000	7,182,688,512
12/31/2025	1,026,478,865	8,209,167,378
12/31/2026	1,162,019,327	9,371,186,705
12/31/2027	1,299,242,328	10,670,429,033
12/31/2028	1,440,787,661	12,111,216,694
12/31/2029	1,502,991,330	13,614,208,025
Total	13,614,208,025	

*Estimated at 5% of inflated original cost.

Assumptions and Parameters

Summary			
Design, Right of Way, and Construction:			
Start date	12/31/93	Projected Funds:	
Completion date	12/31/2001	Construction fund:	
First full year operations (segment 1)	12/31/97	Reinvestment rate	4.00%
Unescalated cost	2,710,000,000	Debt service reserve fund:	
Escalation rate (beginning in 1995)	4.00%	Sized at % of par	10.00%
Debt Information:		Reinvestment rate	6.50%
First issue date	12/31/93	Letter of Credit:	
First interest payment	12/31/94	Letter of credit fee (annual)	N/A
First principal payment	12/31/2004	Letter of credit fee (upfront)	N/A
Last principal payment	12/31/2028	Size of Letter of Credit	0
Interest rate	8.00%	Other Sources of Funds:	
Project costs and revenues:		Federal funding contribution	
Estimated operations cost (1992)	$18,000,000	(% of exc. construction)	0.00%
Estimated maintenance (1992)	$13,600,000	Other (% of exc. construction)	0.00%
Projected O & M escalation rate	4.00%	Miscellaneous:	
MAG-based traffic through 2010		Costs of issuance (% of par)	1.50%
Toll revenues (current 2005 $)	537,758,792		
Projected annual toll rate increase	3.00%		
Toll rate (1992 $)	$0.15		
Annual increase in traffic (after 2010)	1.50%		

Construction Fund

	Construction Fund									
Year Ending	Beginning Balance	Par Value Bonds Issued	Less: Bond Issuance Costs	Less: Debt Service Fund Req.	Other Source of Funds	Less: Shortfall*	Upfront/ Annual Letter of Credit Fees	Escalated Const. Draws	Interest Earnings	Ending Balance
12/31/92	0	0	0	0	0	0	0	0	0	0
12/31/93	0	834,584,845	12,518,773	83,458,485	0	0	0	0	0	738,607,588
12/31/94	738,607,588	877,741,596	13,166,124	87,774,160	0	61,341,986	0	159,720,625	20,701,799	1,315,048,089
12/31/95	1,315,048,089	880,267,976	13,204,020	88,026,798	0	125,855,993	0	332,218,900	34,278,928	1,670,289,281
12/31/96	1,670,289,281	819,315,348	12,289,730	81,931,535	0	190,168,218	0	518,261,484	38,474,383	1,725,428,045
12/31/97	1,725,428,045	806,831,491	12,102,472	80,683,149	0	237,221,241	0	538,991,943	37,968,594	1,701,229,325
12/31/98	1,701,229,325	820,655,577	12,309,834	82,065,558	0	266,975,415	0	560,551,621	34,948,092	1,634,930,566
12/31/99	1,634,930,566	0	0	0	0	272,578,948	0	582,973,686	31,175,117	810,553,049
12/31/2000	810,553,049	0	0	0	0	177,639,013	0	404,195,089	9,148,758	237,867,704
12/31/2001	237,867,704	0	0	0	0	10,469,243	0	210,181,446	688,681	17,905,695
12/31/2002	17,905,695	0	0	0	0	10,630,149	0	0	291,022	7,566,568
12/31/2003	7,566,568	0	0	0	0	7,566,568	0	0	0	0
12/31/2004	0	0	0	0	0	0	0	0	0	0
12/31/2005	0	0	0	0	0	0	0	0	0	0
12/31/2006	0	0	0	0	0	0	0	0	0	0
12/31/2007	0	0	0	0	0	0	0	0	0	0
12/31/2008	0	0	0	0	0	0	0	0	0	0
12/31/2009	0	0	0	0	0	0	0	0	0	0
12/31/2010	0	0	0	0	0	0	0	0	0	0
12/31/2011	0	0	0	0	0	0	0	0	0	0
12/31/2012	0	0	0	0	0	0	0	0	0	0
12/31/2013	0	0	0	0	0	0	0	0	0	0
12/31/2014	0	0	0	0	0	0	0	0	0	0
12/31/2015	0	0	0	0	0	0	0	0	0	0
12/31/2016	0	0	0	0	0	0	0	0	0	0
12/31/2017	0	0	0	0	0	0	0	0	0	0

12/31/2018	0	0	0	0	0	0	0	0	0	0
12/31/2019	0	0	0	0	0	0	0	0	0	0
12/31/2020	0	0	0	0	0	0	0	0	0	0
12/31/2021	0	0	0	0	0	0	0	0	0	0
12/31/2022	0	0	0	0	0	0	0	0	0	0
12/31/2023	0	0	0	0	0	0	0	0	0	0
12/31/2024	0	0	0	0	0	0	0	0	0	0
12/31/2025	0	0	0	0	0	0	0	0	0	0
12/31/2026	0	0	0	0	0	0	0	0	0	0
12/31/2027	0	0	0	0	0	0	0	0	0	0
12/31/2028	0	0	0	0	0	0	0	0	0	0
12/31/2029	0	0	0	0	0	0	0	0	0	0
	5,039,396,832	75,590,952	503,939,683	0	1,360,446,776	0	3,307,094,795	207,675,374		

*Represents funds required to meet shortfall in cash flow during construction and ramp-up period.

Revenue and Cost Inputs

	Revenues	O & M	Construction Costs			
Year Ending	Gross Toll Revenues*	Escalated O & M	Annual Escalation Rate	Compound Escalation Rate	Construction Costs	Escalated
12/31/92	0	0		1.0000	0	0
12/31/93	0	0	2.50%	1.0250	0	0
12/31/94	0	0	3.50%	1.0609	150,555,556	159,720,625
12/31/95	0	0	4.00%	1.1033	301,111,111	332,218,900
12/31/96	3,468,430	3,080,959	4.00%	1.1474	451,666,667	518,261,484
12/31/97	23,166,718	9,612,591	4.00%	1.1933	451,666,667	538,991,943
12/31/98	59,763,891	16,661,824	4.00%	1.2411	451,666,667	560,551,621
12/31/99	122,076,335	24,259,616	4.00%	1.2907	451,666,667	582,973,686
12/31/2000	229,259,472	32,442,058	4.00%	1.3423	301,111,111	404,195,089
12/31/2001	428,576,218	41,237,460	4.00%	1.3960	150,555,556	210,181,446
12/31/2002	454,096,502	46,785,773	4.00%	1.4519	0	0
12/31/2003	480,762,414	48,657,204	4.00%	1.5100	0	0
12/31/2004	508,660,652	50,608,930	4.00%	1.5704	0	0
12/31/2005	537,758,792	52,633,287	4.00%	1.6332	0	0
12/31/2006	568,144,847	54,738,619	4.00%	1.6985	0	0
12/31/2007	599,870,419	56,928,163	4.00%	1.7664	0	0
12/31/2008	633,038,843	59,211,652	4.00%	1.8371	0	0
12/31/2009	667,606,353	61,580,118	4.00%	1.9106	0	0
12/31/2010	703,678,433	64,043,323	4.00%	1.9870	0	0
12/31/2011	735,660,617	66,605,056	4.00%	2.0665	0	0
12/31/2012	769,158,679	69,276,702	4.00%	2.1491	0	0
12/31/2013	804,116,941	72,047,770	4.00%	2.2351	0	0
12/31/2014	840,664,056	74,929,680	4.00%	2.3245	0	0
12/31/2015	878,872,237	77,926,868	4.00%	2.4175	0	0
12/31/2016	918,891,392	81,052,651	4.00%	2.5142	0	0

12/31/2017	960,655,006	84,294,757	4.00%	2.6148	0	0
12/31/2018	1,004,316,776	87,666,548	4.00%	2.7193	0	0
12/31/2019	1,049,962,973	91,173,210	4.00%	2.8281	0	0
12/31/2020	1,097,772,687	94,830,327	4.00%	2.9412	0	0
12/31/2021	1,147,666,456	98,623,540	4.00%	3.0589	0	0
12/31/2022	1,199,827,896	102,568,482	4.00%	3.1812	0	0
12/31/2023	1,254,360,074	106,671,221	4.00%	3.3085	0	0
12/31/2024	1,311,476,943	110,949,992	4.00%	3.4408	0	0
12/31/2025	1,371,083,570	115,387,991	4.00%	3.5785	0	0
12/31/2026	1,433,399,318	120,003,511	4.00%	3.7216	0	0
12/31/2027	1,498,547,317	124,803,651	4.00%	3.8705	0	0
12/31/2028	1,566,783,170	129,809,745	4.00%	4.0253	0	0
12/31/2029	1,637,993,465	135,002,135	4.00%	4.1863	0	0
Total	27,501,137,893	2,466,105,414			2,710,000,000	3,307,094,795

* Annual rate increase at 3%. Assumes MAG-based forecast of VMT through 2010 and 1.5% annual increase in demand thereafter.

Debt Service Cash Flows

	Debt Service				
Year Ending	Payments Applied to Principal	Average Interest Rate	Interest Expense	Debt Service	Principal Outstanding
12/31/92	0	N/A	0	0	0
12/31/93	0	N/A	0	0	834,584,845
12/31/94	0	8.00%	66,766,788	66,766,788	1,712,326,441
12/31/95	0	8.00%	136,986,115	136,986,115	2,592,594,417
12/31/96	0	8.00%	207,407,553	207,407,553	3,411,909,765
12/31/97	0	8.00%	272,952,781	272,952,781	4,218,741,256
12/31/98	0	8.00%	337,499,300	337,499,300	5,039,396,832
12/31/99	0	8.00%	403,151,747	403,151,747	5,039,396,832
12/31/2000	0	8.00%	403,151,747	403,151,747	5,039,396,832
12/31/2001	0	8.00%	403,151,747	403,151,747	5,039,396,832
12/31/2002	0	8.00%	403,151,747	403,151,747	5,039,396,832
12/31/2003	0	8.00%	403,151,747	403,151,747	5,039,396,832
12/31/2004	18,237,522	8.00%	403,151,747	421,389,269	5,021,159,310
12/31/2005	38,877,117	8.00%	401,692,745	440,569,862	4,982,282,193
12/31/2006	61,223,086	8.00%	398,582,575	459,805,661	4,921,059,108
12/31/2007	84,024,783	8.00%	393,684,729	477,709,511	4,837,034,325
12/31/2008	108,377,815	8.00%	386,962,746	495,340,561	4,728,656,510
12/31/2009	134,981,178	8.00%	378,292,521	513,273,699	4,593,675,332
12/31/2010	145,779,672	8.00%	367,494,027	513,273,699	4,447,895,660
12/31/2011	157,442,046	8.00%	355,831,653	513,273,699	4,290,453,614
12/31/2012	170,037,409	8.00%	343,236,289	513,273,699	4,120,416,205
12/31/2013	183,640,402	8.00%	329,633,296	513,272,699	3,936,775,802
12/31/2014	198,331,634	8.00%	314,942,064	513,273,699	3,738,444,168
12/31/2015	214,198,165	8.00%	299,075,533	513,272,699	3,524,246,003

12/31/2016	231,334,018	8.00%	281,939,680	513,273,699	3,292,911,985
12/31/2017	249,840,740	8.00%	263,432,959	513,273,699	3,043,071,245
12/31/2018	269,827,999	8.00%	243,445,700	513,273,699	2,773,243,246
12/31/2019	291,414,239	8.00%	221,859,460	513,273,699	2,481,829,007
12/31/2020	314,727,378	8.00%	198,546,321	513,273,699	2,167,101,629
12/31/2021	339,905,568	8.00%	173,368,130	513,273,699	1,827,196,061
12/31/2022	367,098,014	8.00%	146,175,685	513,273,699	1,460,098,048
12/31/2023	396,465,855	8.00%	116,807,844	513,273,699	1,063,632,193
12/31/2024	343,178,813	8.00%	85,090,575	428,269,389	720,453,380
12/31/2025	281,233,198	8.00%	57,636,270	338,869,468	439,220,182
12/31/2026	214,074,616	8.00%	35,137,615	249,212,231	225,145,566
12/31/2027	147,751,507	8.00%	18,011,645	165,763,153	77,394,058
12/31/2028	77,394,058	8.00%	6,191,525	83,585,583	0
12/31/2029	0	N/A	0	0	0
Total	5,039,396,832		9,257,594,604	14,296,991,436	

Operating Cash Flows

Year Ending	Revenues*			Costs*				Net Revenues	Coverage Ratios	
	Toll Revenues	DSRF Earnings	Total Revenues	O & M Expenses	Debt Service	Letter of Credit Fees	Total Costs		Gross §	Net‖
12/31/92	0	0	0	0	0	0	0	0	NM	NM
12/31/93	0	0	0	0	0	0	0	0	NM	NM
12/31/94	0	5,424,801	5,424,801	0	66,766,788	0	66,766,788	(61,341,986)†	0.08	0.08
12/31/95	0	11,130,122	11,130,122	0	136,986,115	0	136,986,115	(125,855,993)†	0.08	0.08
12/31/96	3,468,430	16,851,864	20,320,294	3,080,959	207,407,553	0	210,488,512	(190,168,218)†	0.10	0.08
12/31/97	23,166,718	22,177,413	45,344,131	9,612,591	272,952,781	0	282,565,372	(237,221,241)†	0.17	0.13
12/31/98	59,763,891	27,421,818	87,185,709	16,661,824	337,499,300	0	354,161,125	(266,975,415)†	0.26	0.21
12/31/99	122,076,335	32,756,079	154,832,414	24,259,616	403,151,747	0	427,411,362	(272,578,948)†	0.38	0.32
12/31/2000	229,259,472	32,756,079	262,015,552	32,442,058	403,151,747	0	435,593,804	(173,578,253)†	0.65	0.57
12/31/2001	428,576,218	32,756,079	461,332,297	41,237,460	403,151,747	0	444,389,207	16,943,090	1.14	1.04
12/31/2002	454,096,502	32,756,079	486,852,581	46,785,773	403,151,747	0	449,937,520	36,916,062	1.21	1.09
12/31/2003	480,762,414	32,756,079	513,518,494	48,657,204	403,151,747	0	451,808,950	61,709,543	1.27	1.15
12/31/2004	537,758,792	32,756,079	570,514,872	50,608,930	421,389,269	0	471,998,199	98,516,673	1.35	1.23
12/31/2005	537,758,792	32,756,079	570,514,872	52,633,287	440,569,862	0	493,203,149	77,311,723	1.29	1.18
12/31/2006	568,144,847	32,756,079	600,900,927	54,738,619	459,805,661	0	514,544,280	86,356,647	1.31	1.19
12/31/2007	599,870,419	32,756,079	632,626,499	56,928,163	477,709,511	0	534,637,675	97,988,824	1.32	1.21
12/31/2008	633,038,843	32,756,079	665,794,922	59,211,652	495,340,561	0	554,552,214	111,242,709	1.34	1.22
12/31/2009	667,606,353	32,756,079	700,362,432	61,580,118	513,273,699	0	574,853,817	125,508,616	1.36	1.24
12/31/2010	703,678,433	32,756,079	736,434,512	64,043,323	513,273,699	0	577,317,021	159,117,491	1.43	1.31
12/31/2011	735,660,617	32,756,079	768,416,697	66,605,056	513,273,699	0	579,878,754	188,537,943	1.50	1.37
12/31/2012	769,158,679	32,756,079	801,914,758	69,276,702	513,273,699	0	582,550,400	219,364,358	1.56	1.43
12/31/2013	804,116,941	32,756,079	836,873,020	72,047,770	513,273,699	0	585,321,468	251,551,552	1.63	1.49
12/31/2014	840,664,056	32,756,079	873,420,135	74,929,680	513,273,699	0	588,203,379	285,216,756	1.70	1.56
12/31/2015	878,872,237	32,756,079	911,628,316	77,926,868	513,273,699	0	591,200,566	320,427,750	1.78	1.62
12/31/2016	918,891,392	32,756,079	951,647,471	81,052,651	513,273,699	0	594,326,350	357,321,121	1.85	1.70
12/31/2017	960,655,006	32,756,079	993,411,085	84,294,757	513,273,699	0	597,568,456	395,842,629	1.94	1.77
12/31/2018	1,004,316,776	32,756,079	1,037,072,855	87,666,548	513,273,699	0	600,940,246	436,132,609	2.02	1.85

12/31/2019	1,049,962,973	32,756,079	1,082,719,052	91,173,210	513,272,699	0	604,446,908	478,272,144	2.11	1.93
12/31/2020	1,097,772,687	32,756,079	1,130,528,767	94,830,327	513,273,699	0	608,104,026	522,424,741	2.20	2.02
12/31/2021	1,147,666,456	32,756,079	1,180,422,535	98,623,540	513,273,699	0	611,897,239	568,525,296	2.30	2.11
12/31/2022	1,199,827,896	32,756,079	1,232,583,976	102,568,482	513,273,699	0	615,842,181	616,741,795	2.40	2.20
12/31/2023	1,254,360,074	116,214,564	1,370,574,638	106,671,221	513,273,699	0	619,944,920	750,629,718	2.67	2.46
12/31/2024	1,311,476,943	115,105,438	1,426,582,380	110,949,992	428,269,389	0	539,219,380	887,363,000	3.33	3.07
12/31/2025	1,371,083,570	109,652,755	1,480,736,325	115,387,991	338,869,468	0	454,257,459	1,026,478,865	4.27	4.03
12/31/2026	1,433,399,318	97,835,750	1,531,235,069	120,003,511	249,212,231	0	369,215,741	1,162,019,327	6.14	5.66
12/31/2027	1,498,547,317	91,261,815	1,589,809,132	124,803,651	165,763,153	0	290,566,804	1,299,242,328	9.59	8.84
12/31/2028	1,566,783,170	87,399,819	1,654,182,989	129,809,745	83,585,583	0	213,395,328	1,440,787,661	19.79	18.24
12/31/2029	1,637,993,465	0	1,637,993,465	135,002,135	0	0	135,002,135	1,502,991,330	NM	NM
	27,530,236,033	1,486,622,066	29,016,858,099	2,466,105,414	14,296,991,436	0	16,763,096,850	13,592,592,515‡		

*Excludes construction fund earnings/costs.

†Financed through construction fund.

‡Sum of positive revenues in each segment only.

§Total revenues/(Debt service + Letter of credit fees)

||(Total revenues − O&M expenses)/(Debt service + Letter of credit fees)

6

SOUTH CAROLINA'S PARTNERSHIPS FOR PROGRESS

Over the past decade, the number of miles driven in the state of South Carolina increased by 50 percent, registered drivers increased 18 percent, and registered vehicles increased 26 percent—all signs of a robust economy. Americans and foreign visitors to this country spent more than $416 billion on travel and tourism in the United States in 1994, and states are eager to provide more services to capture their fair share of this lucrative market. The Myrtle Beach area has become a major tourist attraction with more than 12 million travelers visiting the area in 1994 and predictions of even more in 1995. A new roadway system would be a step in that direction. Larry Duke, Director of Finance and Administration for SCDOT, in an August 1995 interview for this book, said that South Carolina had always been a "donor" state; it has contributed more money to the federal highway fund than it has received. According to Duke, a state transportation authority today should not build a project on a pay-as-you-go plan, augmenting funding from the federal government with funds from state general funds, sales tax proceeds, or bonding issues. "You just can't count on these sources of funds year after year," he said.

Construction of a road project nowadays might have to be phased over 20 years because of the way funding is forthcoming, Duke observed. If all funding were in place at the start of construction, this 20-year cycle could probably be reduced to a four- or five-year period. Extended construction cycles also encounter more complex environmental problems over the years, thereby adding to their cost. But increased traffic and projected expenditures created a need for a road program that would cost the state about $1 billion for five key targeted projects: the Conway ByPass system, the Cross Island Connector, the Southern Connector in Greenville, the Bobby Jones Expressway in North Augusta, and the Grace Bridge replacement in Charleston. South Carolina could not put all of its financial eggs in one basket;

because other highway work would capture part of the available highway money, a new initiative was needed.

With other demands on the state budget, SCDOT was seeking alternative solutions in order to build a better road system, and the Conway ByPass RFP represented the first attempt to find another way to build and finance these projects. Since this was the first such public/private sector partnership project to be put into place in the state, the qualifications and experience of the proposer were deemed to be of major importance. Duke said, "We've given that weight because basically this is our first project, and we really don't want to fumble the ball on the first hand-off of the game."

On September 1, 1994, the South Carolina Department of Transportation (SCDOT) issued a Request for Proposal (RFP) for the Conway ByPass Project, the state's first attempt to solicit private-sector participation in a design, build, finance, operate highway project. Ninety individuals representing top-flight engineering, construction, and development firms requested copies of the RFP. From the number of responses to the Conway ByPass RFP received by SCDOT, it appeared that interest in this type of public/private highway venture was running high. The successful response to the Conway ByPass project triggered RFPs for two more SCDOT design, build, finance, operate maintain projects: the Southern Connector and the Sea Islands Expressway.

THE REQUEST FOR PROPOSAL PROCESS

SCDOT, in its first request for a public/private sector highway project, clearly stated the proposal requirements.

1. A nonnegotiable, nonrefundable fee of $20,000 was to accompany each proposal. It had been determined that this was the cost to the state to hire expert consultants to review each proposal properly.
2. The respondent's proposal was to describe the project in sufficient detail that SCDOT could determine its scope and intent.
3. All assumptions used in the preparation of the proposal were to be included.
4. A list of critical factors necessary for the project's success were to be included.
5. An estimate of costs with breakdowns for design, right-of-way acquisition, construction, maintenance, and operations were required with the submission.
6. A proposed schedule was required listing the anticipated start of the project and the projected completion date.
7. The respondent was to clearly state the assumptions relating to ownership of the facility, the legal liabilities, law enforcement considerations, and facility operational details.

8. A development and operation plan was to be prepared to include the anticipated funding schedule and the proposed sources of funds, and, upon notification of selection, when equity and debt financing would be finalized.
9. A schedule of project revenue, project cost, and return on investment was required.
10. Assumptions about proposed toll rates, user fees, risks, operations, and usage of the facility were to be included in the proposal.
11. A statement about the assumption of liability during the design and construction phases and assurances for timely completion of the project were to be submitted.
12. A statement that all required permits are the responsibility of the respondent was to be enclosed.
13. The proposed project must meet or exceed SCDOT design standards, and respondents were to acknowledge this requirement.
14. The proposal must include an executive statement using cross referencing to explain the project. This document was not to exceed 40 pages, excluding appendixes.

SCDOT indicated in its RFP that the evaluation of respondent's proposals would be performed by a voting committee of four SCDOT employees, one South Carolina treasury office employee, and a nonvoting group of experts from the field of financial management, the bond counsel, and the Federal Highway Administration. The evaluation process to be used by the committee ranks each acceptable proposal on the basis of

The proposer's qualifications and experience	25 points maximum
Project cost and financing arrangements	50 points maximum
Project completion time	25 points maximum

The RFP expanded on the elements in the proposal that would be considered by the committee during the weighted evaluation and ranking process:

A. The proposer's qualification and experience
   - Experience with similar projects.
   - Demonstration of an ability to perform the work.
   - Leadership structure.
   - Project manager's experience.
   - Management approach.
   - Financial condition of the entity.

B. Cost and proposed financial arrangements
   - Investments—define amounts and private placement objectives schedules for design, right-of-way acquisition, construction, operation, and maintenance.

Rights-of-way issues—how they are to be acquired, and provisions for relocation and/or displacement of residential or commercial occupants.

C. Completion time

Implementation of a phase-by-phase schedule to include proposed date of completion.

Plans for partial use of the project prior to full completion.

THE CONWAY BYPASS PROJECT

The issuance of the Conway ByPass RFP signaled that the state of South Carolina desired to seek qualified, experienced respondents who could design, finance, construct, equip—and, according to the request, potentially operate—a transportation facility. Specifically, the request dealt with the creation of a link-up with U.S. 17 and other major roads that lead to the Grand Strand/Myrtle Beach area. Figure 6.1 is a map of the Myrtle Beach, South Carolina, area with its existing network of interstate and state highways. The Conway ByPass envisioned by SCDOT would be a limited access four- and six-lane highway covering 28.5 miles (46 km) from U.S. 501 midway between Aynor and Conway, northeast to U.S. 17, which is slightly north of Myrtle Beach. This new route would provide improved east to west highway capacity between the Grand Strand and interior portions of the state. Four lanes would be built between U.S. 501 and U.S. 701, and six lanes constructed for the remaining sections of highway.

The following milestones were included in the RFP:

Advertise RFP	September 1, 1994
Preproposal meeting	September 21, 1994
Deadline for proposers to submit written questions	October 17, 1994
Deadline for SCDOT to respond to these questions	October 29, 1994
Submittal of proposals	November 15, 1994
Presentation of proposals	November 15, 1994 to December 9, 1994
Notification of rankings	January 13, 1995
Begin development agreement negotiations	January 16, 1995

In the meantime, SCDOT began to work its way through the Conway ByPass proposal review, ranking, and contract award process.

The Conway ByPass Preproposal Meeting

Right on schedule, the Conway ByPass preproposal meeting took place at the Embassy Suites in Columbia, South Carolina, on September 21, 1994. High-level managers and directors from SCDOT and other state agencies were present along with a representative of the Federal Highway Administration to review the project with

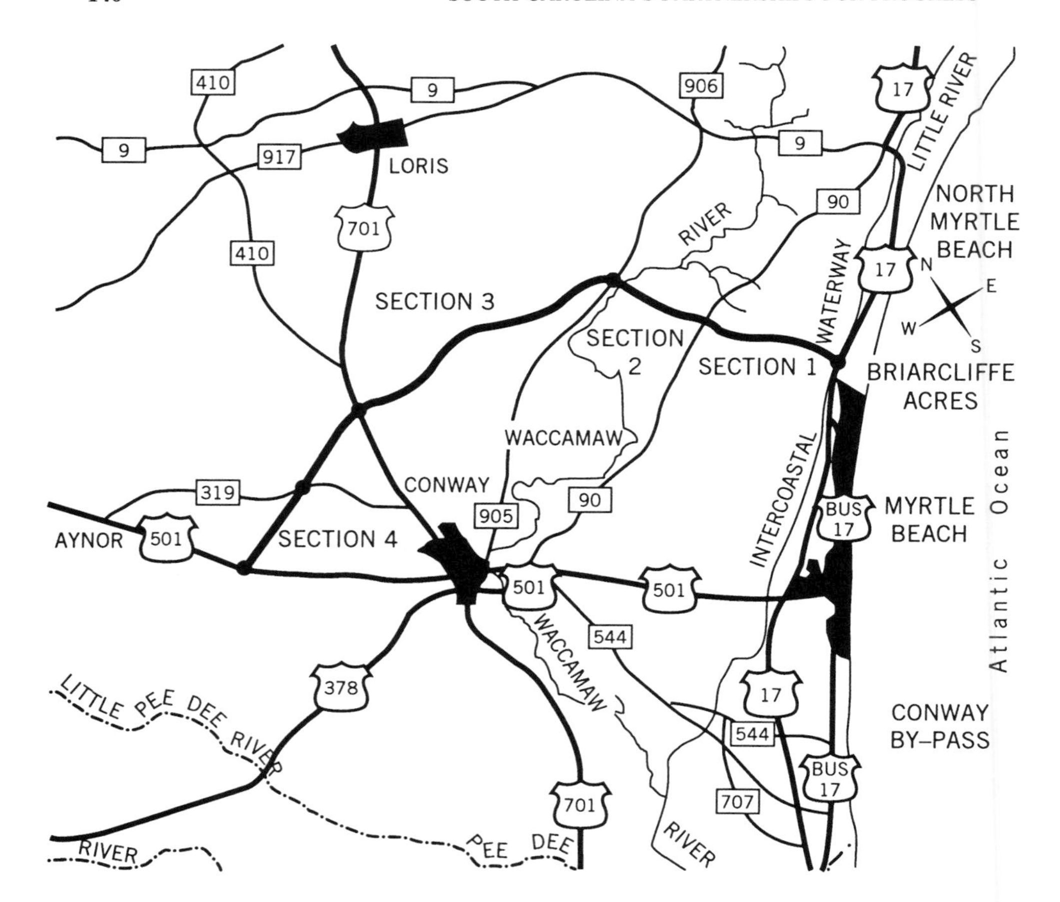

Figure 6.1. Conway ByPass and Myrtle Beach area. *[Source: South Carolina DOT]*

executives from the engineering, design, construction, and developer communities. Representatives from Wilbur Smith Associates, Interwest Management Group, Hewitt Construction Group, Bechtel, Sloan Construction, the American Concrete Pavement Association, Fluor Daniel Construction, Great Britain's Balfour Beatty Construction, and the National Transportation Authority were all set to raise a number of questions.

The meeting kicked off with Deputy Director for Finance Larry Duke and Chairman of the Proposal Review Committee Elizabeth Mabry presenting an overview of the project. The meeting proceeded to clarify the project requirements.

The subject of permitting came up quickly, and William Dubose, Director of Preconstruction for SCDOT, spent considerable time reviewing the state's position in this matter. He said, "We've got the Corps of Engineer's 404 permit, the Coast Guard permit, and Budget and Control Board permits. There are a couple of conditions in the Corps permit that we have to meet before we can let any additional contracts that would impact wetlands." The state's obligations are to acquire the wetlands permit and a Declaration of Protected Covenants, a form of protective deed that must be recorded in Horry County where the project would be located. The state also would prepare a site mitigation plan for two archeological sites located within the project boundaries and this should be completed by the end of 1995. Although permits had been obtained for some of the toll facility structures, additional permits might be required of the developer. If a proposer went beyond the limits of the existing permits then it would be the proposer's responsibility to acquire additional permits.

Duke's statement about the state's goal in the issuance of the RFP, "to select qualified and experienced proposers who can complete the design; secure the right-of-way; provide financing support, construction, and possibly operate the Conway ByPass," brought up the question of eminent domain. One attendee asked if the state would use its right-of-way power to obtain the remaining portions of the rights-of-way. Mabry responded that the state's power of eminent domain would be available if the state owned the project. Because this could be interpreted to mean that this project could either be a BTO project or a true BOT project depending upon how and when title would pass, Mabry discussed the state's three options:

> The proposer could include design and construction and transfer the project to the Department of Transportation and then lease it back and operate it, or transfer the project directly to SCDOT to operate and then the Department could opt to operate it themselves or contract the operation to a private concern. A third option would also be considered, one by which the proposer would own the project and operate it. [This would make it a true BOT project.]

The state was not setting any hard-and-fast rules, but it would obviously select the proposal that best benefited the people of South Carolina.

One of the builders present at the meeting asked about the requirement for concrete paving and whether consideration would be given to alternate paving materials with an equivalent life cycle. The state's answer was that its cost benefit analysis pointed to concrete paving, but the proposers could submit bituminous concrete as an alternate and this would be considered along with other aspects of their proposals during the evaluation process.

When it was announced that the project cost had been estimated to be $535 million, including costs for design, construction, and right-of-way acquisition, one attendee asked if a Bill of Quantities (a British term somewhat equivalent to a detailed quantity take-off) was available to support the estimated cost. The answer was that some exact quantities were known, others are not. The matter of project financing was a topic requiring clarification. Duke said that at that time, there was no allotment of either federal, state, or local funds to support the project. During the proposal evaluation and selection process, the state would actually commit to whatever funds were required by the winning proposal. The state was prepared to arrange for financing of any missing portion not included in the selected proposal. In response to a question about whether the state would guarantee toll revenue or compensate franchise holders if competing facilities were constructed in the future, Duke said simply, "I don't think we can guarantee the toll revenues."

With respect to tort liability, the state's position was that if the proposer was planning to maintain the project, the proposer would be responsible for all tort liability and risks. If the state were to own the project and be responsible for maintenance, then the state would assume liability for tort action. Therefore, proposers should be very clear in their proposals as to which party would maintain the highway.

SCDOT's Unique Selection Process

When all proposals had been received and reviewed by SCDOT, a short list was compiled. There were four short-listed groups: United Infrastructure—UIC, South Carolina Transportation Authority—SCTP, National Transportation Authority—NTA, and Fluor Daniel (see Table 6.1). The state, cognizant of the fact that these proposals which contained proprietary information might have to be made public, wished to protect each company's sensitive material. So upon receipt of each proposal for evaluation, the proposer's name was replaced by a letter designation to mask their identities during the proposal review. The four short-listed firms became known as Proposers A, B, C, and D. When the successful proposal had been selected, the name of the group would be announced, but the names of the three unsuccessful proposers would not be revealed individually.

A "value" chart prepared by SCDOT displayed each of the four short-listed groups' scope of work, total project cost, right-of-way acquisition process, maintenance and law enforcements provisions, and toll collection policies (Table 6.2).

The state would have to consider initial costs and future costs in its evaluation. The value chart included any additional future costs that would be incurred if a particular proposal was accepted. For instance, Proposers A and B planned to construct concrete roadways, but Proposers C and D were going to use bituminous concrete and asphalt, which would require overlays at some future date. Proposer C anticipated a future cost of $60 million for this work which would take place in 2010, 2020, and 2030; Proposer D placed a value of $90 million on the overlay costs.

Another evaluation chart prepared by SCDOT listed the proposer's source of revenue and how much funding would be required from revenue bonds, toll collections, state obligation bonds, investment earnings, and the like. A third chart compared the cash outlays that would be required of SCDOT for each proposal under consideration. And a fourth chart was a risk analysis chart in which components of each proposal's financial plan dependency were rated in one of four categories as high, medium, medium-high, and very high.

An overview/conclusion tabulation was then prepared for each proposal as reflected in Table 6.3. The committee recommended that SCDOT should begin negotiations with Proposer B. Only then was Proposer B indentified as Fluor Daniel, and SCDOT began these negotiations in July 1995 anticipating that a contract award could be made by the end of the year.

The contract between SCDOT and Fluor Daniel was dependent upon passage of pending legislation that would increase the sales tax by one penny, and that legisla-

TABLE 6.1. Composition of Conway ByPass Proposal Consortiums

United Infrastructure Group (UIC)	
Bechtel—Lead Management	Joint Venture
Kiewit—Advisory and Legal	Joint Venture
DeLeuw, Cather & Co.—Design	
Law Engineering—Geotechnical	
Gilbert Southern—Construction	
Metric Constructors—Construction	
J. A. Jones—Construction	
Rea Construction—Construction	
Becon Construction—Construction	
Goldman, Sachs & Co.—Financial	
Ogletree, Deakins, Nash, Smoak, Stewart, et al.—Legal	
Vollmer—Traffic and Revenue	
MFS Network Technologies—Toll Systems	
South Carolina Transportation Authority Group (SCTP)	
Morrison Knudsen—Design and Construction	
Sverdrup—Design and Construction Management	
Wilbur Smith Assoc.—Traffic, Revenue Studies And Design	
PAN Inc.—Right of Way	
Civil Engineering Consulting Services—Civil Engineering Services	
McKim & Creed—Field Surveys	
CZR Inc.—Environmental	
Lehman Brothers—Investment Banker	
JC Bradford & Co.—Investment Banker	
Sinkler & Boyd—Legal	
Thompson, Henry, Gangi, Bultan and Stevens PA—Legal	
The Wordsmith Inc.—Community Relations	
National Transportation Authority Group (NTA)	
Perot Group—Finance and Management	Joint Venture
Greiner—Design	Joint Venture
Parsons Brinckerhoff—Design	
LPA Group—Design and Financing	
Moreland Altobelli—Right of Way	
Law Engineering—Engineering and Testing	
Ashmore Bros.—Construction	
Eagle Construction—Construction	
Republic Construction—Construction	
Ballenger—Construction	
Granite Construction—Construction	
CS First Boston—Finance	
Traffic Consultants Inc.—Finance and Management	
Nelson, Mullins, Riley—Legal	
Fred Allen Assoc.—Legislative	

TABLE 6.1. *(Continued)*

AMTECH Systems Corp.—Toll Collection System
URS Consultants—Traffic and Revenue
Roy Jorgensen Assoc.—Maintenance

The Fluor Daniel Group
Fluor Daniel—Prime Contractor/Program Manager
Davis & Floyd—Civil/Survey/Environmental
Figg Engineering—Bridge Design Engineer
Daniel, Mann, Johnson & Mendenhall—Lead Engineer/Highway Engineer
Law Engineering—Geotechnical Engineering
A. O. Hardee & Sons—Construction
Cherokee Inc.—Construction
Willis Construction—Construction
Ballenger—Construction
APAC Carolina—Construction
Traylor Bros.—Construction
Recchi America—Construction
NationsBank—Finance
Smith Barney—Finance
Haynsworth Law Firm—Legal
Vollmer Associates—Traffic and Toll Revenue
Grupo ICA—Toll Systems and Operations
Transroute International—Toll Systems and Design

Source: South Carolina DOT.

tion was a November 1995 referendum. A lump sum contract for the design and construction of the Conway Bypass project between SCDOT and Fluor Daniel was subsequently signed by both parties. The $489,583,000 contract makes for interesting reading, and a copy of a similar contract, absent the volumes of documents mentioned as exhibits, is Appendix 6.1 at the end of this chapter.

TWO MORE PARTNERSHIPS FOR PROGRESS INVITATIONS

On July 1, 1995, SCDOT launched two more of its Partnerships for Progress requests for proposal. One of them was for the Sea Islands Expressway, a highway linking several islands together in the southwestern portion of Charleston County, and the other one for a Southern Connector around the city of Greenville.

These RFPs, like that for the Conway ByPass, were issued under the authorization of Section 57–3–200 of the 1976 South Carolina Code (as amended) which contained the following provisions:

> From the funds appropriated to the Department of Transportation and from other sources which may be available to the Department, the Department of Transportation

TABLE 6.2 SCDOT Value Chart to Compare Developer Proposals

	Scope	Time	Cost $ Millions		Additional Cost	Right-of-Way	Maintenance & Law Enforcement	Toll Collection
			Nontoll	Toll				
SCDOT	4–6 lanes Concrete Full Row			$495.00	N/A			
A	4–6 lanes Concrete Build, Transfer, Operate	51 months		$509.00	N/A	SCDOT pays for settlements and verdicts. "A" pays some administrative costs.	SCDOT maintains SC law enforcement	"A" collects tolls
B	4–6 lanes Concrete Build, Transfer	36 months	$465.00	$485.00	N/A	"B" pays all costs including verdicts.	SCDOT maintains SC law enforcement	SCDOT collects tolls
C	4 lanes with footprint Asphalt U.S. Route 501 toll booths Build, Transfer, Operate	60 months		$499.71	Overlay in 2010 Overlay in 2020 Overlay in 2030 ($10 million/ per overlay) 6 Lanes in 2020 ($50 million)	"C" pays most costs.	SCDOT maintains SC law enforcement	"C" collects tolls
D	4 Lanes with no footprint Asphalt Build, Transfer, Operate	39 months		$471.00	Overlay in 2010 Overlay in 2020 Overlay in 2030 ($10 million/ per overlay) 6 Lanes in 2020 ($80 million)	"D" pays all costs including verdicts.	"D" routine maintenance (SCODT pays) SC law enforcement	"D" collects tolls

Source: South Caroline DOT.

TABLE 6.3 Overview/Conclusions of Proposer Submissions

Proposer A

A. Construction members of the team are strong—overall good team.
B. Third shortest completion time duration (51 months).
C. Proposes staged opening with Segment 1 opening in mid-1998.
D. Proposes to build SCDOT's original project concept.
E. Proposes highest state general obligation bond issuance ($269.6 million).
F. Proposes two forms of revenue: gas tax and tolls, although they have not investigated the risks or public acceptance of gas tax and tolls.
G. Expects SCDOT to obtain its own letter of credit to support debt.
H. Counts on nonexistent $15 million in federal money.
I. Chain of command is a layered, convoluted management structure.
J. Does not appear to have given much thought about leadership structure or quality control.
K. Provided no detail on plan for building the project.
L. Uses SCDOT gross estimate but does not back out (deduct) what SCDOT already spent.
M. Proposes the highest design/build cost ($509 million).
N. Rather than supply complete details, asks SCDOT to rely on the Proposer's reputation.

Proposer B [Fluor Daniel Group]

A. Least risky of all proposals.
B. Simple financing arrangement (less chance of things going wrong).
C. Lowest finance costs ($6 million).
D. Lowest net cash outlay by SCDOT ($95 million).
E. Creates a means, or generates funds, to build other projects in Horry County.
F. Not having a toll results in (1) no slowdown or traffic, (2) the deterrent to use the Bypass goes away, (3) no toll operation costs, (4) reduces total initial design/build costs, and (5) avoids negative impact on tourism.
G. Good overall proposal team.
H. Has a well-thought-out plan on means and methods of performing work. Appears that each team member understands its role.
I. In the chain of command, one person and firm make decisions.
J. Lowest design/build cost ($465 million).
K. Low rehabilitation cost ($2.4 million).
L. Lowest total project cost ($471 million).
M. Proposes to build SCDOT's original project concept.
N. Bypass would be opened in the shortest time duration (36 months).

Proposer C

A. Good team members.
B. Medium/High financing risk.
C. Very high cash outlay required by SCDOT ($223.7 million).
D. Major decisions, including financing and major maintenance, made by a board—proposes that SCDOT indemnify the Board for liability.
E. Proposes early tolling of Route 501—early tolling of 501 risky and financing plan depends on early tolls coming in.
F. Has not put together a construction team.

TABLE 6.3 *(Continued)*

G. Next to highest design/build cost (499.7 million) but does not propose to build SCDOT's original project concept. Proposes four lanes with asphalt surface using a downgraded pavement design. Does, however, propose to build the "footprint" to add more lanes later.
H. Longest time duration to completion and no staged or intermediate openings (60 months or 5 years).

Proposer D

A. Well-thought-out and creative toll structure.
B. Good team members.
C. Has given thought on how to build the project.
D. Second lowest design/build cost ($471 million), but does not build SCDOT's original project concept. Proposes to build four lanes with asphalt surface and no "footprint" for adding future lanes.
E. Second shortest time duration for completion of project—built in phases (39 months).
F. Opened Segment 1 the earliest of all proposers.
G. Highest risks in terms of financing.
H. Very high cash outlay by SCDOT ($256.4 million).
I. Did not furnish complete information on financial condition.
J. Decisions would be made by a management committee—committee members spread throughout the country.

Source: South Carolina DOT.

> may expend such funds as it deems necessary to enter into partnership agreements with political subdivisions including authorized transportation authorities and private entities to finance, by tolls and other financing methods, the cost of acquiring, constructing, equipping, maintaining and operating highways, roads, streets and bridges in this State.

THE SEA ISLANDS EXPRESSWAY PROJECT

In the southwestern portion of Charleston County lie Johns Island, Seabrook Island, Wadmalaw Island, and Kiawah Island (Figure 6.2). These islands are separated from the mainland and the city of Charleston by three rivers, the North Edisto, the Wadmalaw, and the Stono. The Limehouse Bridge on Main Road spans the Intercoastal Waterway and connects Johns Island with the mainland. The Maybank Highway has a bridge across the Stono River connecting James Island to the city of Charleston. Both bridges are moveable and allow large boats to pass down the Intercoastal.

Johns Island and Wadmalaw Island are primarily agricultural; Kiawah and Seabrook Islands are residential and resort oriented. Both Kiawah and Seabrook Islands are deemed prime real estate, and their populations are growing quickly. Between 1990 and 2015 it is anticipated that their populations will increase by 300 percent,

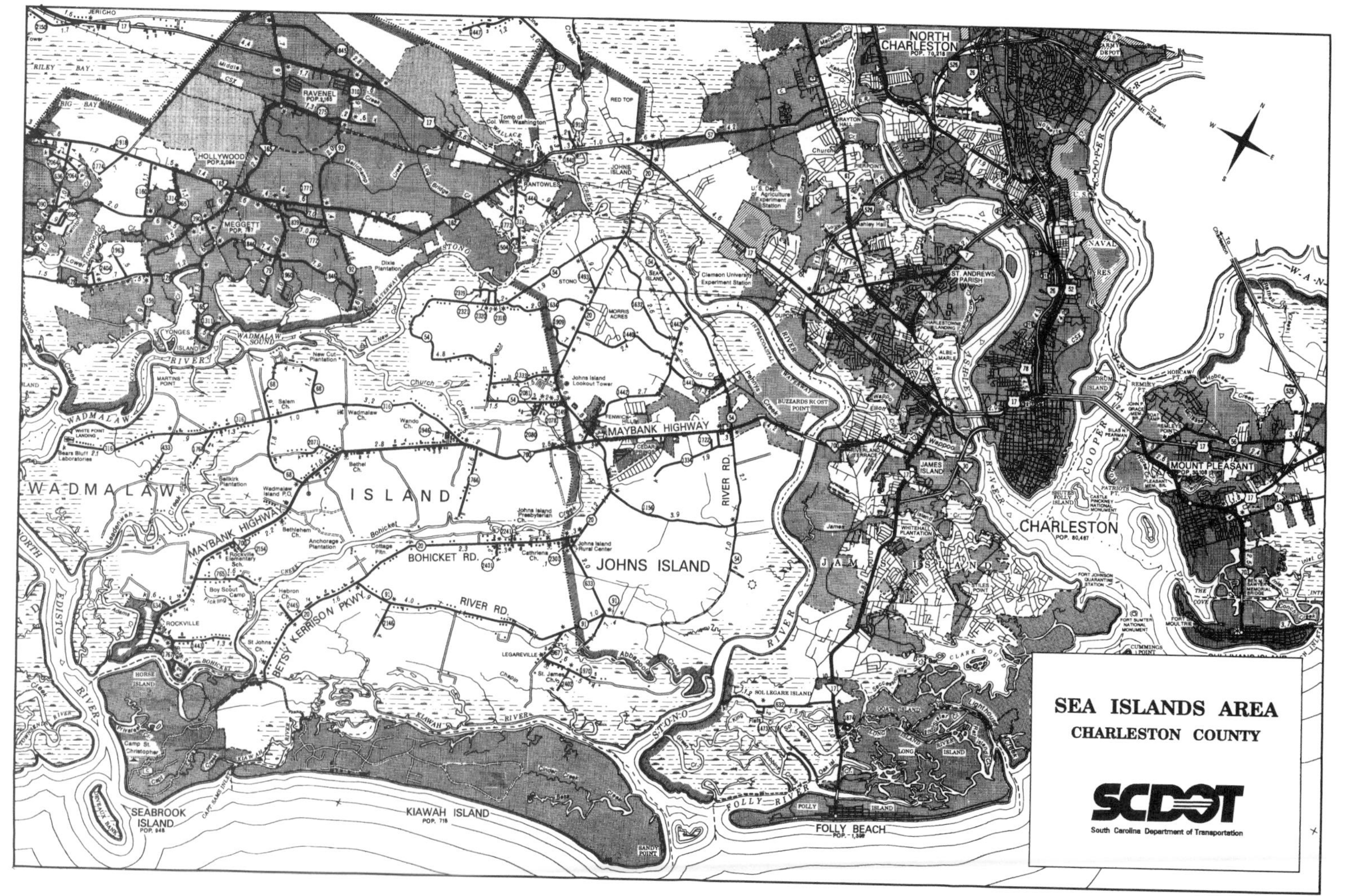

Figure 6.2. Map of the area of the Sea Islands Express project. *[Source: South Carolina DOT]*

and dwelling units are expected to double in the next 25 years to slightly more than 6,000 units. Population on Johns Island is also expected to soar from its present 8,775 residents to more than 15,000 by the year 2015.

Because of the anticipated growth on these islands, a nonprofit group was formed to investigate ways to provide safe and efficient methods of routing traffic for residents and visitors alike. A feasibility study conducted by HNTB Corporation in Atlanta provided the background information for SCDOT to develop a plan to interest private investors in building a toll road to service this area.

The RFP issued by SCDOT on July 1, 1995, contained the following milestones:

Preproposal meeting	August 4, 1995.
Deadline for proposers to submit written questions	January 26, 1996
Deadline for SCDOT to respond to written questions	February 12, 1996
Submittal of proposals	March 8, 1996
Presentation of proposals	April 8, 1996 to April 12, 1996
Notification of rankings	May 24, 1996
Begin development agreement negotiations commence on May 27, 1996	May 27, 1996

Each proposal had to be accompanied by a nonnegotiable, nonrefundable proposal review fee of $5,000 which the state estimated to be the cost of review by a group of expert consultants.

The Sea Island Review Process

The proposal review committee would be composed of four SCDOT employees and one employee from South Carolina's Office of the Treasury, and experts would be available to analyze various technical aspects of each proposal. This was similar to the way in which the Conway ByPass proposals had been reviewed. Nonvoting members would include experts in the field of financial management, the environment, and engineering—also similar to the makeup of the group for the Conway ByPass project. The Sea Island Expressway project review and evaluation process contained four critical categories, versus the three for the Conway ByPass review, and the weighting of each category was somewhat different because it included the environmental issue.

Qualifications and experience	20 points maximum
Cost effectiveness of the proposal	40 points maximum
Impact to development and the environment	20 points maximum
Financing proposal for the project	20 points maximum

The impact to development and the environment portion of the RFP was rather detailed.

1. Impact to the environment

   Is the project planned in a manner to minimize adverse impact to the environment?

   Does the proposal include a detailed plan to mitigate any impacts to the environment?

   Does the proposal identify all permits required to construct the project?

   Does the proposal clearly describe a plan for complying with applicable environmental laws and for obtaining environmental permits?

2. Impact to current development

   Is the project planned in a manner to minimize adverse impacts to existing development, such as displacement of existing businesses, residences, farms, and other developments?

3. Rights-of-way

   Does the proposal present a satisfactory plan by phase (appraisals, rights-of-way plan, negotiations, document preparation, recording, etc.) for the acquisition of rights-of-way, including the relocation of any displaced resident or business?

   Does the proposal clearly state who will be responsible for implementing each phase of right-of-way acquisition?

   Does the proposal state whether or not the SCDOT's power of eminent domain is expected to be used, and if so, does the proposal identify who will provide legal representation?

4. Impact to future development

   Does the proposal clearly explain how the project will benefit or enhance future development and tourism?

THE SOUTHERN CONNECTOR PROJECT

Concurrent with the Sea Islands Expressway proposal issuance, the Southern Connector RFP was also launched on July 1, 1995. This proposal would seek private development of a major roadway around the perimeter of Greenville, a project that had been under consideration by SCDOT since 1970.

In 1968 the Greenville Area Transportation Study (GRATS) established the need for a beltway around the southern portion of the city. Existing Interstate 85 had been improved to create an I-85 to I-185 interchange; from this interchange the Southern Connector would extend in a generally southern direction parallel to the Saluda River, and then turn southeasterly to cross the CSX and Southern Railroad lines and State 20, before proceeding east to cross State Road 85 and end at Route 276 southeast of Greenville. GRATS had endorsed a "preferred" alignment consisting of eight segments (Figure 6.3), and this plan was included in the Southern Connector RFP. However, SCDOT was quick to add that proposals were not restricted to the preferred alignment, allowing greater flexibility in proposal develop-

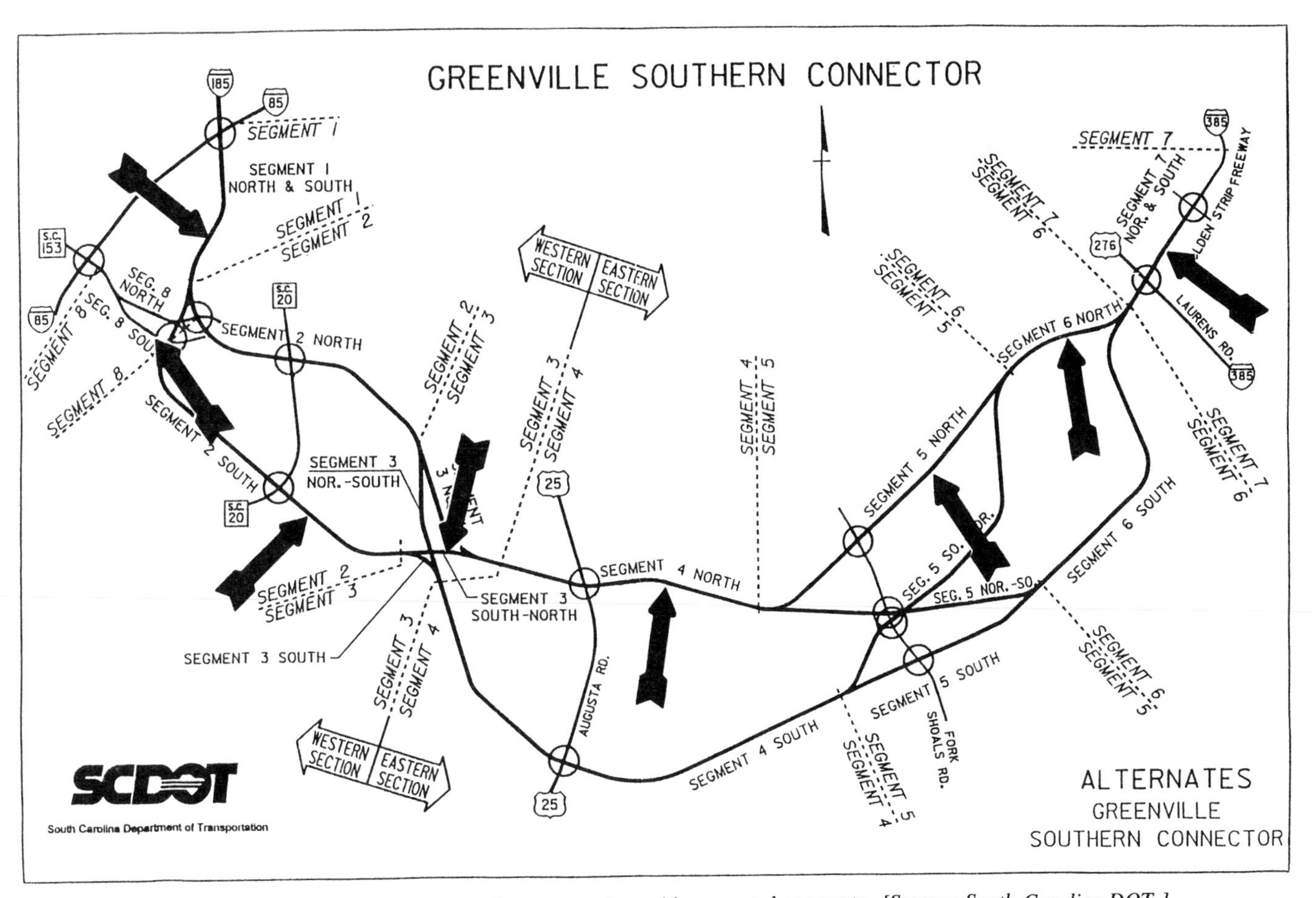

Figure 6.3. The Southern Connector alignment project with suggested segments. *[Source: South Carolina DOT.]*

ment and maximizing potential cost reduction or revenue enhancement opportunities.

The maximum rate of return on investment was to be established in a development agreement after the winning proposal had been selected, but all terms and conditions pertaining to user fees and toll charges were to be included in each proposal. SCDOT stated that it would entertain proposals that included loans or loan agreements with SCDOT or other governmental entities; however, the RFP made it clear that no federal, state, or other government funds had been earmarked or allocated for this project. Each proposal, therefore, had to develop its own revenue stream and financing structure to support the project, bearing in mind that one of the important evaluation criterion was cost effectiveness of the project.

The evaluation process for the Southern Connector project listed three weighted categories:

Qualifications and experience	30 points maximum
Cost and financing proposed for the project	35 points maximum
Proposer's ability to meet project goals	35 points maximum

The time line for this project was listed by SCDOT in their RFP:

Advertise for the RFP	June 28, 1995
Preproposal meeting	August 4, 1995
Deadline for proposers to submit written questions	December 1, 1995
Deadline for SCDOT to answer written questions	December 15, 1995
Submittal of proposals	January 5, 1996
Presentation of proposals	January 29, 1996 to February 2, 1996
Notification of rankings	March 8, 1996
Begin development agreement negotiations	March 11, 1996

Although SCDOT's milestone schedule on its first Partnerships for Progress program has slipped somewhat, it seems to have picked up the pace now. The thoroughness of its approach to this public/private investment program will undoubtedly reap big rewards by the end of this decade.

APPENDIX 6.1

LUMP SUM CONTRACT FOR THE DESIGN AND CONSTRUCTION OF THE BYPASS BETWEEN THE DEPARTMENT OF TRANSPORTATION AND THE CONSORTIUM

This contract is reprinted in its entirety except that the name of the succesful bidder has been deleted and the term "consortium " has been substituted.

TABLE OF CONTENTS

Notwithstanding the foregoing, the parties agree that the Consortium is authorized to modify the scope of the Bypass in any or all of the following ways:

A. *U.S. 007 Interchange:* Exhibit A depicts the Bypass as crossing over U.S. 007. The Consortium may design the Bypass so that U.S. 007 crosses over the Bypass instead. Otherwise, the basic configuration of the interchange is to remain the same;

B. *Big River Bridge:* Exhibit A calls for the bridge over the Big River to have a minimum vertical clearance of thirty-five (35) feet above the ordinary high-water elevation. U.S. Coast Guard and State Y Budget and Control Board permits were obtained on the basis of this thirty-five foot minimum clearance. If the Consortium is able to obtain necessary modifications of relevant permits, this clearance height may be reduced in accordance with any such modification;

C. *Long Road Crossover:* Exhibit A calls for Long Road to cross over the Bypass. The Consortium may design the Bypass so that the Bypass crosses over Long Road, provided that a minimum fifteen (15) foot vertical clearance is maintained at all points;

D. *Acme Industries' Access Road:* Exhibit A depicts Acme Industries' access road as crossing over the Bypass. The Consortium may design the Bypass so that the Bypass crosses over Acme Industries' access road, provided that a minimum fifteen (15) foot vertical clearance is maintained at all points. Any modification to Acme Industries' access road must also comply with the relevant agreement between DOT and Acme Industries, Inc. (PRE-173); and/or

E. *Elimination of Undulations:* The Consortiums may review and adjust the vertical geometry for Segment Numbers 1, 2, and 3 as depicted in Exhibit A to minimize undulating gradelines.

Such modifications are reflected in the Contract Price and no such modification shall be an occasion for an adjustment in the Contract Price, notwithstanding the fact that such modifications may vary from the right-of-way configuration shown on Exhibit A. Nor shall the Consortium's failure for any reason to make such modifications, including but not limited to the Consortium's own decision or a decision denying necessary permits, be an occasion for an adjustment in the Contract Price. Design and implementation of any such modifications shall be performed in accordance with the requirements of Exhibit B and Exhibit C.

WHEREAS, the People of County X, State Y in particular, and the People of the State of Y in general, stand to benefit from the construction of a bypass around the city of City Z; and

WHEREAS, the State Y Department of Transportation, as a servant of the People of the State of Y, wishes to see this strategic project completed; and

WHEREAS, limitations imposed by traditional methods of financing, designing, and constructing highways would mean that the Bypass could be completed only after an unacceptable delay, if at all; and

WHEREAS, the State Y Department of Transportation, working with the People of County X, the federal government, and other agencies of the State Y has devised an innovative plan to allow the commencement and completion of the Bypass in a timely and cost-effective manner; and

WHEREAS, after a competitive process, the Consortium has been selected to participate in this venture by designing and building the Bypass; and

WHEREAS, the State Y Department of Transportation wishes to avail itself of and rely on the Consortium's expertise and proven track record in designing and constructing such complex projects, on time and under budget; and

WHEREAS, the Consortium wishes to provide that expertise and to participate in this venture for the good of the People of County Z and the State of Y;

NOW THEREFORE, this Agreement is executed and made, effective as of the Effective Date as defined herein, between the STATE Y DEPARTMENT OF TRANSPORTATION ("DOT") and the Consortium ("Consortium"). In consideration of the covenants hereinafter set forth, the parties hereto mutually agree as follows:

I. LOCATION AND SCOPE OF THE BYPASS.

As used in this Agreement, the "Bypass" shall refer to a four-lane, fully controlled access highway between U.S. 005 and S.R. 009, and a six-lane, fully controlled access highway between S.R. 009 and U.S. 001, each with a 70 miles per hour design speed and interchanges at major intersecting roadways as specified, suitable for use by the traveling public. The location and scope of the Bypass are more particularly described in Exhibit A hereto, which is incorporated herein by reference.

II. THE CONSORTIUM'S DESIGN, PROCUREMENT, AND CONSTRUCTION RESPONSIBILITIES.

The Consortium shall perform all design, engineering, right of way acquisition, procurement, and construction services (the "Services") set forth in this Agreement, or otherwise required for the Bypass, as that term is defined in Article I of this Agreement and in Exhibit A hereto. It shall be the responsibility of the Consortium to furnish all personnel and furnish or procure all necessary equipment for execution of the design, engineering, right of way acquisition, procurement, and construction of the Bypass.

III. DESIGN AND REVIEW.

1. The Consortium shall design the Bypass in accordance with Exhibit B and this Agreement.
2. All final design plans issued for construction by the Consortium for the Bypass shall bear the seal of a State Y licensed professional engineer.
3. The Consortium shall in timely fashion provide to DOT, in mutually agreed upon formats, all plans and related documents for the Bypass, in order to allow DOT to review those plans when issued for construction, but prior to commencement of construction. DOT shall have the right, but not the obligation, to review and comment upon those plans. This review and comment is

fully discretionary, however, and no review or comment by DOT, nor any failure to review or comment, shall operate to absolve the Consortium of its responsibility to design and build the Bypass in accordance with this Agreement, or to shift any responsibility to DOT.

IV. CONSTRUCTION SPECIFICATIONS.

The Consortium shall construct the Bypass in accordance with Exhibit C.

V. PROJECT MANAGEMENT.

The Consortium shall be responsible for ensuring that the Bypass is constructed in conformance with this Agreement, including the Exhibits hereto.

The Consortium shall provide project management services sufficient to supervise the activities of its subcontractors. The Consortium shall provide a sufficient number of persons on site to provide for the construction management of the Bypass. The Consortium shall develop and be responsible for the implementation of construction inspection procedures that will provide control of the work, the materials and installation of materials in accordance with Exhibit I to this Agreement.

DOT may provide representatives assigned to the Bypass to monitor the construction and provide necessary coordination between DOT and the Consortium. All costs for salary and equipment to maintain DOT employees shall be provided by DOT at no expense to the Consortium. DOT representatives shall have full and complete access to the Bypass, the work in progress, the "Daily Diaries" (substantially in the form of DOT's form 647, as revised 7/95), and to other technical documents and project records associated with quality control, materials verification, materials installation, and testing. DOT representatives shall receive notice of and have the opportunity to participate in any meetings which may be held concerning the Bypass or the relationship between the Consortium and their consultants and subcontractors when such meetings are associated with technical matters, progress, or quality of the Bypass. As used in this paragraph, "notice" shall require actual notice to DOT's Agent, as defined in Article XXIV, to the extent reasonably practicable under the circumstances.

VI. RIGHT-OF-WAY ACQUISITION.

The Consortium shall be responsible for acquisition of all rights of way necessary for construction of Segments 2-4 of the Bypass, as those Segments are defined in Exhibit A (hereinafter "Rights of Way"). This responsibility shall be carried out as follows:

1. Title in fee simple (except that DOT may in its sole discretion direct the Consortium to acquire a right of way easement, in lieu of fee simple title,

with respect to any portion of the Rights-of-Way) shall be conveyed to "The State Y Department of Transportation" by general warranty deed substantially in the form attached hereto as Exhibit D, free of all liens and encumbrances except current property taxes. The Consortium shall prepare, obtain execution of, and record documents conveying Rights-of-Way to DOT. The Consortium shall handle all negotiations with property owners for the acquisition of the Rights-of-Way and shall provide copies of Rights-of-Way acquisition documents to DOT prior to Final Completion.

2. Because these acquisitions are being made on behalf of the State of Y, the State's power of eminent domain shall be available where the Consortium reasonably determines that condemnation proceedings are the most appropriate means of acquiring a portion of the Right-of-Way. It shall be the responsibility of the Consortium to prepare in the name of DOT, file on behalf of DOT, serve appropriate condemnation documents, and prosecute condemnation proceedings to final judgment in compliance with the procedures required by State Y Eminent Domain Procedure Act.
3. The Consortium shall exercise reasonable care in determining whether there is reason to believe that property acquired for the Rights-of-Way may contain concealed or hidden wastes or other materials or hazards requiring remediation or treatment, and shall take all prudent and necessary steps to investigate such possibilities where they are reasonably determined to exist. The Consortium shall timely notify DOT of any concerns regarding such materials on property to be acquired for the Rights-of-Way, to allow DOT to consider whether and in what form it wishes title to be acquired.
4. In conjunction with Rights-of-Way acquisition and condemnation activities, the Consortium shall use only appraisers from the list of appraisers approved by DOT.
5. The Consortium shall comply with all requirements of the Uniform Relocation and Real Property Acquisition Policies Act of 1970, as amended (the "Uniform Act").
6. It is understood by the parties that the Contract Price includes Six Million Dollars ($6,000,000.00) for the acquisition of Rights-of-Way. In the event that the Cost of Acquisition of the Rights-of-Way shall exceed Six Million Dollars ($6,000,000.00), and only in that event, such cost in excess of Six Million Dollars ($6,000,000.00) shall constitute an Eligible Additional Payment, as that term is defined in Article XIV below, which shall allow use of the Owner's Contingency Fund to fund such excess amount, in accordance with the procedure set forth in Article XIV below. In the event that the Cost of Acquisition of the Rights-of-Way is less than Six Million Dollars ($6,000,000), the difference between Six Million Dollars and the Cost of Acquisition shall constitute a Qualified Cost Reduction, as that term is defined in Article XIV below, and that differences shall be handled in accordance with Article XIV below. As used herein, the Cost of Acquisition of Rights-of-Way shall consist of: (i) direct payments for ownership or other property rights; (ii) direct payments for eligible relocation expenses as pro-

vided for under the Uniform Act; (iii) the expense of appraisers, land agents, title searches, right-of-way agents and surveyors; (iv) attorneys' fees, expert witness fees, filing fees, and deposition costs and expenses arising from condemnation proceedings; and (v) expenses incurred to cure or avoid damages proximately related to the acquisition of the Rights-of-Way. The Cost of Acquisition shall not include any overhead allocations, profit, or other mark-up by the Consortium.

7. The Consortium shall, during construction of the Bypass and for a period of three years after Final Completion, maintain and make available for inspection or audit by DOT, at DOT's reasonable request, all documents and records relating to the Cost of Acquisition of Rights-of-Way, and all documents and records necessary to determine compliance with the Uniform Act or other laws relating to acquisition of Rights-of-Way.
8. The Consortium shall provide to DOT a complete set of Right-of-Way plans indicating the date and manner of acquisition of each tract, along with all deeds and all Notices of Condemnation, annotated with the date and amount of verdict or settlement, recorded with the office of the Register of Mesne Conveyances for County X.

VII. UTILITIES.

1. The Consortium shall have the responsibility of coordinating construction on the Bypass with all public utilities that may be affected thereby. This responsibility shall include reimbursement to utility companies for the costs attributable to the Relocation (as the term "Relocation" is defined in 23 C.F.R. § 645.105 (i)) of utility facilities as may be necessary for the Final Completion of the Bypass.
2. The Consortium shall develop a procedure for the relocation of utilities in substantial conformity with 23 C.F.R. § 645. This procedure shall include obtaining a written agreement with each affected utility, such agreement to be substantially in the form of DOT Form 3088-A (Revised 7/1/93).
3. The Consortium shall have the responsibility for identifying the location of all utilities affected by the construction of the Bypass and shall establish whether the State or the utility has prior rights to each location. In the event that utilities have prior rights that require relocation, the Consortium shall be responsible for the cost of these relocations. Insofar as may be practicable, the Consortium shall design the Bypass to avoid conflicts with existing utilities. For those utilities requiring relocation, the Consortium shall conform to the DOT "Policy for Accommodating Utilities" dated March, 1968.
4. The Consortium shall provide all affected utilities with a copy of that portion of the Bypass plans sufficient to allow the utility to design for the relocations or the installations that may be required.
5. The Consortium shall be responsible for coordinating the work of its subcontractors and the various utilities. The resolution of any conflicts between utili-

ties and the construction of the Bypass shall be the responsibility of the Consortium. No additional compensation will be allowed for any delays, inconveniences or damage sustained by the Consortium or its subcontractors due to interference from utilities or the operation of relocating utilities, except as relief may be available under Article XV, Force Majeure.

6. At the time that the Consortium notifies DOT pursuant to Article XVIII that the Consortium deems the Bypass to have reached Final Completion, the Consortium shall certify to DOT that all utilities have been identified and that those utilities with prior rights, or other claims related to relocation or coordination with the Bypass, have been relocated or their claims otherwise satisfied by the Consortium.
7. The Consortium shall accurately show the final location of all utilities, including those requiring relocation, on the as-built drawings for the Bypass.
8. The Costs of Relocation of utilities shall constitute Eligible Additional Payments, as that term is defined in Article XIV below, which shall allow use of the Owner's Contingency Fund to fund such Costs of Relocation in accordance with the procedure set forth in Article XIV below. As used herein, the term "Cost of Relocation" shall be interpreted in general conformity with the relevant provisions of 23 C.F.R. § 645. The Cost of Relocation shall not include any upgrade, betterment, or other improvement of relocated facilities of the utility, nor shall it include overhead allocations, profit, or other mark-up by the Consortium.
9. The Consortium shall, during construction of the Bypass and for a period of three years after Final Completion, maintain and make available for inspection or audit by DOT, at DOT's reasonable request, all documents and records relating to the Cost of Relocation of utilities.

VIII. ENVIRONMENTAL COMPLIANCE.

1. In connection with the Services for the Bypass, the consortium shall comply with all applicable federal, state, and local environmental and wetland laws and regulations. This responsibility shall include, but not be limited to:
   a) Compliance with those restrictions and agreements specifically agreed to or entered into by DOT in obtaining permits for the Bypass;
   b) Compliance with those stipulations and conditions under which DOT received approval of the Draft and Final Environmental Impact Statements and permits necessary for construction in wetlands;
   c) Compliance with applicable laws and regulations relating to potential or actual hazardous materials that may be encountered in the course of carrying out this Agreement;
   d) Carrying out all necessary social, economic, and environmental studies required by regulatory authorities in the course of construction; and
   e) Updating or extension of approved permits.

2. The Consortium shall conduct one (or more, if appropriate) preconstruction conference(s) to discuss environmental and permitting issues, which conference shall include all subcontractors, and, to the extent feasible, representatives from the U.S. Army Corps of Engineers, the S.C. Department of Health and Environmental Control's Office of Ocean and Coastal Resource Management, the Federal Highway Administration, Fluor Daniel, and DOT.

IX. TRAFFIC CONTROL AND RESTRICTIONS.

The Consortium shall be responsible for establishing and maintaining traffic control adequate to protect the public and maintain jobsite safety, and to minimize inconvenience to the public. The Consortium shall comply with the seasonal and temporal restrictions set forth in Exhibit E hereto. The latest edition of the State Y Manual on Uniform Traffic Control Devices and Exhibit E shall provide minimum standards and guidance for this traffic control. The Consortium shall submit for review to DOT (and not DOT's Assignee) a traffic control plan for work affecting any existing highway prior to affecting traffic upon any such highway.

X. APPLICABLE LAW.

It shall be the responsibility of the Consortium to determine and comply with all applicable federal, state, and local law in connection with the services set forth in this Agreement. This obligation shall include but not be limited to compliance with those federal laws and regulations designated as applicable to the Bypass in Exhibit F hereto (including FHWA Form 1273 which shall be incorporated as required in all subcontracts), as well as the obligation to procure all permits and licenses (provided, however, that with respect to any permit or license that must be obtained in the name of DOT, the Consortium shall perform all functions within its power to obtain the permit, and DOT shall fully cooperate in this effort and perform any functions that must be performed by DOT), pay all charges, fees, and taxes, and give all notices necessary and incident to the due and lawful prosecution of the Bypass.

XI. WARRANTY.

1. The Consortium warrants that it will perform the Services in accordance with the standards of care and diligence normally practiced by recognized engineering and construction firms in performing services and obligations of a similar nature. The Consortium shall properly perform, at the written request of DOT made at any time within the three (3) year period after Final Completion of the Bypass as defined in Article XVIII, all steps necessary to satisfy the foregoing guarantee and correct any element of the Bypass or the Services

that does not reflect such standards of care and diligence. The cost of such corrective services shall be for the Consortium's account. With respect to any element of the Bypass or the Services on which corrective work is required pursuant to this warranty (including corrective work performed during an extension of the warranty which has been made pursuant to this provision), the period of warranty with respect to that element shall be the longer of one (1) year from completion of such corrective work or the remainder of the original three-year warranty period provided herein. The Consortium's makes no representations, covenants, warranties or guarantees, express or implied, other than those expressly set forth herein. The Consortium shall not be responsible for damages attributable to a failure by DOT to provide timely notification of potentially damaged or defective work of which DOT had actual knowledge.

2. The Consortium shall take all steps necessary to transfer to DOT any manufacturer's or other third-party's warranties of any materials or other services used in the construction of the Bypass.
3. Neither the Consortium nor its subcontractors shall be liable to DOT or its Assignee for defective work or latent defects after the expiration of this warranty, other than as provided for under the Indemnity provision, Article XXI, of this Agreement, or to the extent of any right of indemnity that DOT or its Assignee may have against any of the Consortium's subcontractors.

XII. RECORDS AND AS-BUILTS.

1. In addition to those documents set forth elsewhere in this Agreement, the Consortium shall provide to DOT prior to Final Completion a complete set of as-built drawings complying with the specifications of Exhibit G hereto.
2. Fluor Daniel shall maintain the following documents for a period of three years after Final Completion:
   a) All test reports;
   b) Daily Diaries as defined in Article V;
   c) Appraisals, title search information, correspondence, and diaries or worksheets related to acquisition of right-of-way;
3. During the three-year retention period, DOT shall be granted access to those documents upon reasonable notice. At any time during the three-year period, DOT shall have the option of taking custody of the documents.

XIII. CRITICAL PATH METHOD SCHEDULE.

The Consortium shall prepare and maintain a schedule for the Bypass using the critical path method of scheduling. This schedule shall be used to manage the design and construction of the Bypass and shall be in sufficient detail to determine progress, to monitor construction activity and to identify and avoid delays. The Consortium shall provide DOT with a copy of the original project schedule and all significant updates in a paper and/or electronic medium, each accompanied by a management level summary, as may be requested by the DOT representative. Periodic construction meetings shall be held by the Consortium with its consultants and subcontractors to coordinate the work, update the schedule, provide information and resolve potential conflicts.

XIV. COMPENSATION.

1. Definitions.
   a) The "Contract Price" shall initially be Four Hundred Twelve Million Two Hundred Eighty Three Thousand Dollars ($412,283,000.00), but shall be subject to adjustment as set forth below. In the event, and only in the event, that the Permit Modification is rejected, the Contract Price shall be Four Hundred Sixty Five Million Five Hundred Eighty Three Thousand Dollars ($465,583,000.00), and not Four Hundred Twelve Million Two Hundred Eighty Three Thousand Dollars ($412,283,000.00). In the further event that the Permit Modification is granted in part and rejected in part, the parties will agree to a modification to the Contract Price based upon the document titled "Bypass Project Bridge Reduction Cost Decrease" produced by the Consortium and provided to DOT.
   b) The "Owner's Contingency Fund" shall initially be Twenty Four Million Dollars ($24,000,000.00), but shall be subject to adjustment as set forth below, and shall also equal the difference between the Maximum Cost and the Contract Price.
   c) The "Maximum Cost" of the Bypass shall be Four Hundred Thirty Six Million Two Hundred Eighty Three Thousand Dollars ($436,283,000.00). In the event, and only in the event, that the Permit Modification is rejected, the Maximum Cost shall be Four Hundred Eighty Nine Million Five Hundred Eighty Three Thousand Dollars ($489,583,000.00), and not Four Hundred Thirty Six Million Two Hundred Eighty Three Thousand Dollars ($436,283,000.00). In the further event that the Permit Modification is granted in part and rejected in part, the Maximum Cost shall be the Contract Price agreed upon pursuant to Article XIV.1.a above plus Twenty Four Million Dollars ($24,000,000.00).
   d) A "Qualified Cost Reduction" shall be any or all of (i) the amount shown as a saving on the written agreement effecting a "Change," as that term is defined below; or (ii) a saving on the Cost of Acquisition of Rights-of-Way as described in Article VI.
   e) An "Eligible Additional Payment" shall be the actual costs directly associated with any or all of the following: (i) the amount added to the price of

the Bypass as a result of a "Change"; (ii) an overrun on the Cost of Acquisition of Rights-of-Way as described in Article VI; (iii) the Cost of Relocation of utilities as described in Article VII; (iv) occurrences falling within the definition of Force Majeure in Article XV below; (v) "Differing Site Conditions," which term shall refer to unknown physical conditions at the site of an unusual nature differing materially from those ordinarily encountered and generally recognized as inherent in work of the nature provided for herein, including but not limited to, conditions requiring the average pile length for the Bypass as a whole to exceed 60 feet or the existence of antiquities or artifacts at the Bypass site; (vi) intentional interference with or delay (other than for good cause) of work on the Bypass by DOT and/or its Assignee; (vii) injunctions, lawsuits, or other efforts by individuals, groups, or entities to hinder, delay, or halt the progress of the Bypass, provided that such efforts are either premised on alleged wrongs or violations other than repeated or serious environmental or regulatory violations by the Consortium or its subcontractors or are without substantial basis in fact; and/or (viii) the costs in excess of Five Million One Hundred Sixty Nine Thousand Dollars ($5,169,000.00) of meeting storm water drainage requirements; provided that, for purposes of this clause (viii), the Consortium shall require its subscontractors and consultants to provide pricing detail only of that portion of their contract price attributable to such drainage requirements, and DOT's verification of such costs shall be limited to the pricing data so provided.

f) A "Change" shall be any deviation or variation from the scope or design or construction criteria of the Bypass, as defined by Exhibits A, B, C, and E hereto and by the terms of this Agreement generally. No Change shall be implemented without the express written agreement of DOT and the Consortium. DOT's Assignee may not institute or agree to Changes without the written agreement of DOT.

DOT may initiate a change by advising the Consortium in writing of the change believed to be necessary. As soon thereafter as practicable, the Consortium shall prepare and forward to DOT a cost estimate and a schedule impact of the change. The Consortium shall be reimbursed for its actual and reasonable direct costs incurred to prepare such estimate. DOT shall advise the Consortium in writing of its approval or disapproval of the change. If DOT approves the change, the Consortium shall perform the Services as changed. The Consortium may initiate changes by advising DOT in writing that in the Consortium's opinion a change is necessary. If DOT agrees, it shall advise the Consortium and, thereafter, the change shall be handled as if initiated by DOT.

As provided below, a Change as a result of which the parties agree to increase the Contract Price (an "Additive Change") shall increase both the Contract Price and the Maximum Cost by the amount so agreed, but shall not affect the amount of the Owner's Contingency Fund.

g) The term "Permit Modification" refers to changes requested by DOT to Department of the Army permit number 93-2A-105 for the Bypass for the purpose of reducing the cost of the Bypass by reducing the number of miles of bridges required for the Bypass. In the event that the Permit Modification is approved in whole or in part, the Consortium may revise Exhibit A in accordance with the terms of the modification granted.

2. *Adjustment and Payment of the Contract Price.*

a) In full consideration of its services in conjunction with the Bypass pursuant to this Agreement, the Consortium shall be paid by DOT's Assignee the Contract Price, as it is finally adjusted.

b) The Contract Price (and, correspondingly,the Owner's Contingency Fund) shall be adjusted as follows:

(i) The amount of each Qualified Cost Reduction shall be subtracted from the Contract Price (and added to the Owner's Contingency Fund);

(ii) The amount of each Eligible Additional Payment shall be added to the Contract Price (and subtracted from the Owner's Contingency Fund, except for Additive Changes, which shall be added to the Contract Price but not subtracted from the Owner's Contingency Fund). It shall be the responsibility of the Consortium to establish the existence and amount of any Eligible Additional Payment. The Consortium shall provide to DOT adequate documentation relating to the existence and amount of any such Eligible Additional Payment.

c) The Maximum Cost shall not be adjusted, except solely in the event that an Additive Change is agreed upon. In the event of such an Additive Change, and only in that event, the Maximum Cost shall be increased by the amount of such Additive Change.

d) Notwithstanding the foregoing, the Contract Price shall not, as it is finally adjusted, exceed the Maximum Cost, as it is finally adjusted, and the Consortium shall under no circumstances receive more than the Maximum Cost in consideration for its services under this Agreement.

e) Upon execution of this Agreement, the Consortium shall receive from DOT (and not DOT's Assignee), as the Consortium's first payment under this Agreement, Five Hundred Eighty Three Thousand Dollars ($583,000.00), the sum representing the engineering, field work, investigation, and drawing and design necessary for the Permit Modification.

f) For all subsequent payments,the Consortium shall submit applications for payment, substantially in the form attached hereto as Exhibit H, to DOT's Assignee no more frequently than on the first and fifteenth of every month. No application for payment shall be submitted, and no payment (other than the payment provided for in the preceding paragraph) shall be made by DOT or its Assignee unless and until DOT gives its notice to proceed. Each application for payment shall set forth, in accordance with a schedule of values to be developed by the Consortium and approved by

DOT, the percentage of each element of the Bypass completed since the Consortium's immediately prior request for payment.

g) In the event that the Permit Modification is approved, the Consortium's first application for payment shall include Five Million Five Hundred Eighty Three Thousand Dollars ($5,583,000.00), the sum representing DOT's cost of acquiring mitigation property required for Permit Modification, along with the sums committed to engineering, field work, investigation, and drawing and design necessary for the Permit Modification. The Consortium shall, promptly upon receipt of this sum, pay to DOT (and not its Assignee) said Five Million Five Hundred Eighty Three Thousand Dollars ($5,583,000.00) as reimbursement for these costs.

h) DOT shall review each application for payment. Upon approval by DOT of an application for payment, DOT's Assignee shall pay the Consortium for the percentage of the Bypass completed during the period covered by the application for payment, in accordance with the schedule of values developed in accordance with this Article XIV. In the event of a dispute over the percentage of the Bypass completed, and hence over the amount of payment to which the Consortium is entitled, the undisputed amount shall be paid by DOT's Assignee to the Consortium. Payment by DOT or its Assignee shall not preclude or estop either from correcting any measurement, estimate, or certificate regarding the percentage completion of the Bypass, and adjusting future payments accordingly to ensure that the percentage of the Contract Price paid to the Consortium at any given time is to the extent practicable equivalent to the percentage completion of the Bypass.

i) DOT's Assignee shall make each payment (to the extent it is undisputed) within twenty-one days of the receipt of the corresponding application for payment. In the event that a given payment is not made within twenty-one days of receipt by DOT's Assignee of the application for payment, the Time for Completion as defined in Article XVII shall be extended, on a cumulative basis, by 0.20 days for every day by which payment is delayed beyond twenty-one days. It is the intention of the parties that this extension of time shall be in lieu of interest or any other penalty for payments not made within twenty-one days of receipt of the application for payment, and the parties agree that such extension is of equivalent value to interest.

j) In the event that the Contract Price is adjusted pursuant to this Article XIV, subsequent payments to the Consortium shall be adjusted accordingly by prorating the adjustment in the Contract Price over the remaining payments to be made on the work affected by the adjustment, and increasing or decreasing each such payment accordingly. Such adjustments of the payment schedule shall be designed to ensure that the percentage of the Contract Price paid to the Consortium at any given time is to the extent practicable equivalent to the percentage completion of the Bypass.

3. *Final Calculation and Disposition of the Owner's Contingency Fund.*

   a) After Final Completion of the Bypass and after the Consortium's final application for payment has been submitted, the final value of the Owner's Contingency Fund (the "Final Value") shall be calculated by subtracting from the Maximum Cost the final Contract Price (that is, the sum actually paid or due to be paid to the Consortium in consideration of its services);
   b) If the Final Value is greater than Twelve Million Dollars ($12,000,000.00), the Consortium shall receive fifty percent (50%) of the Final Value from the Owner's Contingency Fund, and DOT's Assignee the remainder;
   c) If the Final Value of the Owner's Contingency Fund is greater than Six Million Dollars ($6,000,000.00) but less than or equal to Twelve Million Dollars ($12,000,000.00), the Consortium shall receive thirty-five percent (35%) of the Final Value from the Owner's Contingency Fund, and DOT's Assignee the remainder;
   d) If the Final Value of the Owner's Contingency Fund is greater than zero but less than or equal to Six Million Dollars ($6,000,000.00), the Consortium shall receive twenty percent (20%) of the Final Value from the Owner's Contingency Fund, and DOT's Assignee the remainder;
   e) With respect to each of the immediately preceding subparagraphs (b), (c), and (d), the percentage of the portion of the Final Value to be paid to the Consortium (i) shall be increased by one tenth of a percentage point for each day by which the time from notice to proceed to Substantial Completion of the Bypass is shorter than the Time for Completion, as those terms are defined in Article XVII below.
   f) The distribution of the Owner's Contingency Fund pursuant to this Article XIV shall be made within 45 days after the Final Completion of the Bypass.
   g) Upon distribution of the Owner's Contingency Fund pursuant to this Article XIV, and upon payment of the Consortium's final application for payment, all claims by the Consortium for additional compensation shall be forever extinguished.

4. *Liquidated Damages.*

   The Consortium shall pay to DOT's Assignee Five Thousand $5,000.00 per day in liquidated damages for each day by which the period from DOT's notice to proceed to Substantial Completion of the Bypass exceeds the Time for Completion defined in Article XVII. The total amount of liquidated damages that the Consortium shall be obligated to pay pursuant to this subparagraph shall not exceed the amount that would accrue pursuant to this subparagraph for a delay of two hundred days. The parties agree that this amount constitutes damages and not a penalty. Other than these damages, DOT waives any claims, such as claims for loss of use, that it might have for late completion of the Bypass.

XV. FORCE MAJEURE.

Delays or failures of performance shall not constitute breach hereunder if and to the extent such delays or failures of performance are caused by severe and not reasonably foreseeable occurrences beyond the control of DOT or the Consortium, as the case may be, including, but not limited to: Acts of God or the public enemy; expropriation or confiscation of facilities; compliance with any order or request of any governmental authority other than DOT or a party in privity with it; a change in law directly and substantially affecting performance on the Bypass; act of war; rebellion or sabotage or damage resulting therefrom; fires, floods, explosions, or extraordinary accidents; riots or strikes or other concerted acts of workers, whether direct or indirect; or any similar causes, which are not within the control of DOT or the Consortium respectively, and which by the exercise of reasonable diligence, DOT or the Consortium are unable to prevent. Any expense attributable to such occurrences shall constitute an Eligible Additional Payment as defined in Article XIV above. Any delay attributable to such an occurrence shall be added to the Time for Completion of the Bypass as defined in Article XVII below. Existence of a Force Majeure event does not excuse the payment of money otherwise due and owing hereunder.

XVI. QUALITY CONTROL AND QUALITY ASSURANCE.

The Consortium shall conduct quality control and quality assurance in conjunction with Exhibit I hereto.

XVII. TIME FOR COMPLETION; NOTICE TO PROCEED; SUBSTANTIAL COMPLETION.

1. The "Time for Completion" shall be 1,180 calendar days from the date of DOT's notice to proceed. If, and only if, the Permit Modification as that term is defined in Article XIV, is not approved, the Time for Completion shall be 1,095 calendar days. Additional days may be added to the Time for Completion in amounts equal to any delays to the critical path attributable to:
   a) Force Majeure as that term is defined in Article XV above;
   b) Occurrences beyond the control of the Consortium related to the relocation of utilities, up to a limit of 90 days;
   c) Lengthening of the work schedule due to changes in the scope of the Bypass at the request of DOT, or improper interference in the progress of the work by DOT;
   d) Differing Site Conditions as defined in Article XIV above; or
   e) Injunctions, lawsuits, or other efforts by individuals or groups to hinder, delay, or halt the progress of the Bypass, provided that such efforts are

either premised on alleged wrongs or violations other than repeated or serious environmental or regulatory violations by the Consortium or its subcontractors or are without substantial basis in fact.

2. DOT shall not give its notice to proceed until County X or the County X Transportation Authority has dedicated funds equalling the Maximum Cost of the Bypass less One Hundred Million Dollars for the sole purpose of payments to the Consortium for construction of the Bypass.
3. The Bypass shall be "Substantially Complete" (or shall have reached "Substantial Completion") when DOT reasonably determines that the Bypass may be opened to public traffic.

XVIII. INSPECTION; NO WAIVER; FINAL COMPLETION.

When the Consortium believes that all elements of its work on the Bypass, including all of the requirements of this Agreement (including its Exhibits), have been completed, it shall so notify DOT in writing. Within fifteen (15) days thereafter, DOT shall advise the Consortium in writing of any aspect of the Agreement, the Services or the Bypass that is incomplete or not carried out as provided under this Agreement. Upon completion of necessary corrective action (or as soon as the fifteen (15) day period for such notice has expired if DOT does not advise the Consortium of any such incompleteness or defect within the period), the Bypass shall be deemed to have reached Final Completion.

No inspection, acceptance, payment, partial waiver, or any other action on the part of DOT pursuant to this Article XVIII shall operate as a waiver of any portion of this Agreement or of any power reserved herein or of any right to damages or other relief, including any warranty rights, except insofar as expressly waived by DOT in writing. DOT shall not be precluded or estopped by anything contained herein from recovering from the Consortium any overpayment as may be made to the Consortium.

XIX. OBLIGATIONS UNDER OTHER AGREEMENTS.

Each party represents and warrants that it is under no other obligations that would prevent or hinder it from carrying out its obligations under this Agreement. Each party recognizes that the planning, approval, and construction of the Bypass requires the cooperation of a number of parties other than the Consortium and DOT. Certain aspects of that cooperation are embodied in a series of other agreements dealing with the Bypass. Each of the parties represents that it is familiar with the negotiation and terms of those other agreements as they exist as of the signing of this Agreement, and has considered the effect of those agreements in entering this Agreement.

XX. SUCCESSORS AND ASSIGNS.

DOT shall have the right to assign all or some of its rights and responsibilities under this Agreement to the County X Transportation Authority, or such other group as may be established in the stead of the County Transportation Authority (DOT's "Assignee"). The Consortium hereby acknowledges its understanding that, upon such assignment, the sole responsibility for payment of any or all sums due under this Agreement (other than the $583,000.00 payment provided in Article XIV.2.e) will be that of the Assignee, and that DOT shall have no obligation to make any payments hereunder (except to the extent that DOT shall serve as the agent of Assignee for payment out of Assignee's funds of amounts due hereunder). The Consortium further acknowledges its understanding that any right granted to DOT under this Agreement, including but not limited to rights to inspect the work on the Bypass, control that work, approve Changes, receive title to Rights-of-Way or other acquired land, review documents, or approve plans or designs, shall not be transferred to Assignee by virtue of such assignment, and shall remain the sole right of DOT, unless expressly made assignable by this Agreement.

Otherwise, this Agreement shall not be assignable by either party without the prior written consent of the other party hereto, except that it may be assigned without such consent to the successor of either party, or to a person, firm, or corporation acquiring all or substantially all of the business assets of such party or to a wholly owned subsidiary of either party, but such assignment shall not relieve the assigning party of any of its obligations under this Agreement. No assignment of this Agreement shall be valid until this Agreement shall have been assumed by the assignee. When duly assigned in accordance with the foregoing, this Agreement shall be binding upon and shall inure to the benefit of the assignee.

XXI. INDEMNITY.

1. The Consortium shall indemnify, defend, and hold DOT and its Assignee (as that term is defined in Article XX) harmless from any and all claims, liabilities, and causes of action for any fines or penalties imposed on DOT or its Assignee by any state or federal agency because of violation by the Consortium or any of its subcontractors of any state or federal law or regulation, except to the extent that such violation is the direct result of express instructions given by DOT or a party authorized to give instructions by DOT.
   The Consortium shall indemnify, defend, and hold DOT and its Assignee harmless from any and all claims, liabilities, and causes of action for injury to or death of any person, or for damage to or destruction of property (other than the Bypass itself), to the extent that such injury, damage, or claim results from negligence or recklessness on the part of the Consortium or its subcontractors.

The Consortium shall idenmnify, defend, and hold DOT and its Assignee harmless from any and all inverse condemnation claims related to the Bypass.

2. The Consortium shall be responsible for and obligated to replace, repair, or reconstruct, and to furnish any material, equipment, or supplies furnished by the Consortium which are lost, damaged, or destroyed prior to transfer of care, custody, and control of the Bypass or the affected portion thereof to DOT, however such loss or damage shall occur. DOT assumes all responsibility for such loss, damage, or destruction after such transfer of care, custody, and control to DOT, except to the extent that the matter is covered by the warranty provided under Article XI.
3. The Consortium shall have no obligation to DOT with respect to any damage or loss to property caused by the perils of war, insurrection, revolution, nuclear reaction, or other like perils.
4. Anything herein to the contrary notwithstanding, the parties agree that neither shall have any obligation to indemnify the other or any third-party beneficiary or assignee with respect to any fine, charge, claim, or suit relating to Pre-Existing Waste. Pre-Existing Waste shall mean any hazardous or toxic substance that was present on the site of the Bypass prior to acquisition of the property in question for construction of the Bypass, and which was not placed or deposited there by either the Consortium or DOT. Provided, however, that each party agrees to indemnify the other against such fines, charges, claims, or suits to the extent that the indemnifying party aggravated or exacerbated the Pre-Existing Waste condition through its own negligence, recklessness, willfulness, or malice.

XXII. INSURANCE AND BONDING.

1. Commencing with the performance of its Services hereunder, and continuing until such Services have been completely performed (except with regard to "Builder's Risk" Course of Construction Insurance which shall continue for the period specified in subparagraph (c) below), the Consortium shall maintain standard insurance policies as follows:
   a) Worker's Compensation and/or all other Social Insurance in accordance with the statutory requirements of the state, province, or country having jurisdiction over the Consortium's employees who are engaged in the Services, with Employer's Liability not less than One Hundred Thousand Dollars ($100,000) each accident; and
   b) Commercial General Bodily Injury and Property Damage Liability, including Automobile (owned, nonowned, or hired), Contractual and Contractor's Protective Liability. This insurance shall cover bodily injury to or death of persons and/or loss of or damage to property of parties other than DOT. Such insurance shall be provided in a Combined Single Limit of One Million Dollars ($1,000,000) for any one accident. DOT shall be named as an additional named insured with respect to this insurance; and

c) "Builder's Risk" Insurance protecting the respective interests of DOT, the Consortium, and the Consortium's subcontractors covering physical loss or damage during course of construction and any materials or equipment, including Consortium-owned or leased major construction tools and equipment or supplies furnished by the Consortium for the Bypass while in transit (other than in the course of ocean marine or air transit movement, which is to be provided for pursuant to subparagraph (d) below), while at the jobsite, awaiting and during erection, and until transfer of care, custody, and control of the facilities, or portion thereof, to DOT. This insurance shall be maintained to cover the value of Fifty Million Dollars ($50,000,000.00) for any one occurrence. This insurance shall not cover losses caused by the perils of war or nuclear reaction as defined in the policy of insurance nor shall it cover loss of use, business interruption, or loss of product. DOT and its assigns shall be included as an additional insured. A deductible of Twenty Five Thousand Dollars ($25,000) shall apply to each and every covered loss, except earthquake where a deductible of One Hundred Thousand Dollars ($100,000) shall apply for each loss and other deductibles shall also apply as specified in the contract of insurance; and

d) Ocean Marine Cargo Insurance covering any and all materials and equipment which may be in transit to the jobsite by wet marine bottoms, or by air transportation and/or by connecting conveyances. Such insurance shall be maintained to cover limits at risk.

2. The foregoing insurance shall be maintained with carriers satisfactory to DOT, and the terms of coverage shall be as evidenced by certificates to be furnished DOT. Such certificates shall provide that thirty (30) days' written notice shall be given to DOT prior to cancellation of any policy.
3. The Consortium shall cause to be executed by its ultimate parent, the Consortium's Corporation, a "Parent Company Guaranty" in favor of DOT and its assigns in the form of Exhibit J hereto.
4. The Consortium shall cause each of its contractors or subcontractors to furnish bonds as follows:

   a) One bond to be furnished by each contractor or subcontractor to secure the claims of laborers, mechanics, or material workers employed by such contractor or subcontactor on the Bypass; and

   b) One bond to be furnished by each contractor or subcontractor to guarantee the faithful performance of the work on the Bypass assigned to such contractor or subcontractor.

XXIII. DISADVANTAGED BUSINESS ENTERPRISES.

1. The Consortium shall use all reasonable efforts to attain participation by certified Disadvantaged Business Enterprises ("DBEs") of ten percent (10%) of

the Contract Price for the Bypass. To qualify for credit toward this goal, a DBE subcontractor must be listed in the DOT Directory of Certified DBEs prior to entering a contract for work on the Bypass. The Consortium acknowledges that this directory pertains only to DBE certification, and is not a guarantee or representation of the DBE firm's qualifications. It is the Consortium's responsibility to determine the actual capabilities and/or limitations of these certified DBE firms.

2. The Consortium shall provide DOT with documentation of executed DBE contracts, including the name of the DBE firm, the name of the subcontractor, if any, for whom the DBE will work, the amount of the contract, the type of work to be performed, and an estimated schedule of DBE performance.
3. The Consortium shall comply with the requirements of Exhibit K hereto with respect to use, certification, and reporting of DBE participation.
4. DOT shall have the right to audit all documentation regarding DBE participation in the Bypass.

XXIV. DOT'S AGENT.

DOT shall appoint an individual who shall be authorized to act on behalf of DOT, with whom the Consortium may consult at all reasonable times, and whose instructions and decisions will be binding upon DOT as to all matters pertaining to this Agreement and the performance of the parties hereunder.

XXV. EFFECTIVE DATE.

1. A condition precedent to the enforcement by either party of any term of this Agreement (other than the initial payment to the Consortium of $583,000.00 provided for in Article XIV) shall be the adoption by the People of County X of a referendum substantially in the form set forth in the Ordinance attached as Exhibit L hereto (the "Referendum"). The date of the vote approving said Referendum shall be the "Effective Date" of this Agreement.
2. Notwithstanding the foregoing, the Consortium may at its sole option terminate this Agreement without any further obligation to DOT or its Assignee in the event that County X and/or the County X Transportation Authority fails to dedicate, on or before June 15, 1996, funds equalling the Maximum Cost of the Bypass less One Hundred Million Dollars for the sole purpose of payments to the Consortium for construction of the Bypass.

XXVI. TERMINATION AND CANCELLATION.

1. Should the Consortium become insolvent or bankrupt, or should the Consortium refuse or neglect to supply a sufficient number of properly skilled work-

ers, tools, or material within the Consortium's control, or should the Consortium commit a substantial breach of this Agreement, and should the Consortium thereafter fail to commence proceedings in good faith to remedy such within ten (10) days after written demand by DOT, DOT may terminate this Agreement and enter upon the premises and take possession thereof and at the same time instruct the Consortium to remove from the premises all of its tools, equipment, and supplies, or DOT may take possession of any and all of such tools, equipment, and supplies for the purpose of completing the Services. Upon any such termination, the Consortium shall be compensated by DOT's Assignee for all costs incurred for work on the Bypass then performed, including all direct and indirect costs actually incurred, plus all cancellation charges by vendors and subcontractors, if any. In the event that DOT uses any of the Consortium's equipment or tools, DOT shall return the same to the Consortium in good condition and repair, reasonable wear and tear excepted, and shall pay the Consortium for the use thereof at the Consortium's standard rental rates then in effect.

2. DOT reserves the right to cancel the Services upon ten (10) days' written notice to the Consortium. Should the Services be so cancelled by DOT without legal cause, the Consortium shall be paid for Services properly performed to the date of cancellation and through demobilization, including any cancellation charges by vendors and subcontractors. In no event, however, shall the total payment to the Consortium pursuant to such a cancellation exceed the Maximum Cost as that term is defined in Article XIV.
3. Termination of all or a portion of the Agreement shall not relieve the Consortium of any responsibility it would otherwise have for the work completed, or for any just claims arising from that work.
4. In the event that all work on the Bypass is suspended or becomes impracticable to pursue by virtue of one or more of the events described in Article XVII.1. (a), (d), or (e), for a period that exceeds, in the aggregate, 365 days, the Consortium shall have the option to terminate this Agreement and shall be compensated in accordance with Article XXVI.2.

XXVII. GENERAL PROVISIONS.

1. The Consortium shall be an independent contractor with respect to the Services to be performed hereunder. Neither the Consortium nor its subcontractors, nor the employees of either, shall be deemed to be the servants, employees, or agents of Owner.
2. The Consortium agrees to include, as a term or condition of each purchase order employed by it in the performance of the Services, a patent indemnification provision extending from the vendor under such purchase order to DOT and its Assignee and the Consortium and to render such assistance to DOT and its Assignee as may be reasonably required to enforce the terms of such indemnification by vendors.

3. So long as it has been properly paid for the work in question, the Consortium shall indemnify and hold DOT and its Assignee harmless from all costs, damages, and expenses arising out of any charge or encumbrance in the nature of a laborer's, mechanics' or material worker's lien asserted by parties other than the Consortium in connection with the Services as performed by the Consortium or its subcontractors.
4. It is understood that in performing services of the scope and complexity of the Services to be performed hereunder, it is necessary and inevitable that certain surplus material be purchased and certain temporary construction facilities be provided. The ownership of all such surplus material and temporary facilities (except to the extent that the eminent domain power was used to obtain the property) shall remain, or become, with the Consortium and the Consortium shall be free to dispose of the same at its discretion.
5. The Consortium may subcontract any portion of the Services for the Bypass. In no case shall any such subcontract relieve the Consortium of any of its obligations under this Agreement. The Consortium may have portions of the Services performed by its affiliated entities or their employees, in which event the Consortium shall be responsible for such Services and DOT shall look solely to the Consortium as if the Services were performed by the Consortium.
6. This Agreement shall be governed by and interpreted in accordance with the substantive laws of the State of Y, without regard to choice of law principles.
7. Headings and titles of the various parts of this Agreement are for convenience of reference only and shall not be considered in interpreting the text of this Agreement. Modifications or amendments to this Agreement must be in writing and executed by duly authorized representatives of each party.
8. Unless specifically stated to the contrary therein, indemnities against, releases from, and limitations on liability expressed in this Agreement shall apply even in the event of the fault, negligence, or strict liability of the party indemnified or released or whose liability is limited and shall extend to the officers, directors, employees, agents, licensors, and related entities of such party.
9. In the event that any portion or all of this Agreement is held to be void or unenforceable, the parties agree to negotiate in good faith to reach an equitable agreement which shall effect the intent of the parties as set forth in this Agreement.
10. All notices pertaining to this Agreement shall be in writing and, if to DOT, shall be sufficient when sent registered or certified mail to DOT addressed as follows:

    Deputy Director for Construction, Engineering, and Planning
    State Y Department of Transportation
    Post Office Box 1492
    Anytown, State Y 55555

All notices to the Consortium shall be sufficient when sent registered or certified mail to the Consortium addressed as follows:

The Consortium
100 Corporate Drive
Othertown, State Y 55556-6666
Attention: John Q. Smith

and to:

The Consortium
100 Corporate Drive
Othertown, State Y 55556-6666
Attention: Jane M. Doe

11. This Agreement, with the Exhibits hereto, sets forth the full and complete understanding of the parties as of the date first above stated, and it supersedes any and all agreements and representations made or dated prior thereto.
12. The Consortium shall in no event be responsible or held liable for consequential, incidental, or indirect damages of any nature, however the same may be caused, except as expressly provided in this Agreement.
13. The parties make no representations, covenants, warranties, or guarantees, express or implied, other than those expressly set forth herein. The parties' rights, liabilities, responsibilities, and remedies with respect to the Services shall be exclusively those expressly set forth in this Agreement.
14. In no event shall any failure by either party hereto to fully enforce any provision of this Agreement be construed as a waiver by such party of its right to subsequently enforce, assert, or rely upon such provision.

7

CANADA AND THE CANADIAN ATLANTIC PROVINCES

When the early 1990s brought recession to Canada, creating a C$10 billion budget shortfall, the Ontario provincial government headed by Premier Bob Rae began to look at various methods by which the economy could be stimulated. During a 1993 speech to business-people and MBA graduates of the University of Toronto, Rae publicly announced his government's desire to seek private-sector involvement in future transportation infrastructure projects. His government, together with municipal agencies and the private sector, was planning to invest C$6 billion in infrastructure projects in the province of Ontario. By this infusion of funding, not only would much-needed highway construction begin but water quality would be improved and province telecommunications would be significantly enhanced. And as a result of these expenditures an anticipated 100,000 new jobs would be created over the projected 10-year life span of the capital investment program. Three corporations would be created:

The Ontario Realty Corporation to control the province's inventory of buildings and land holdings

The Ontario Clean Water Agency to operate provincially owned facilities and assist regions and municipalities in planning, financing, upgrading, and operating these facilities

The Ontario Transportation Capital Corporation to finance and build new public transit and provincial highway projects that involve innovation in their financing or delivery system

Traffic congestion was estimated as costing business and industry some C$2 billion a year. Thus, Rae was particularly interested in accelerating the construction of Highway 407, a long-delayed Ontario project for a metropolitan Toronto bypass which would act as a major east–west connector. As a public/private multilane toll road project, Highway 407 would achieve several government goals:

- To construct a highway as rapidly as possible without increasing the tax burden on citizens
- To relieve congestion around, and provide better access to, the greater Toronto area
- To create more employment opportunities and act as a general economic stimulus

At about the same time as the Highway 407 program was being revitalized in Ontario, in Canada's Atlantic provinces two other public/private projects in the planning stages were part of a master plan to link Prince Edward Island to the Canadian heartland.

HIGHWAY 407

The idea of building Ontario's first new highway, a bypass for metropolitan Toronto, was first conceived in 1950. The new highway would alleviate traffic and would ultimately extend to Highway 403. This proposed east–west route around Toronto eventually became part of the draft of the Parkway Belt plan, which was published in January 1967 and subjected to a thorough government review with public hearings. The Parkway Belt West Plan (Figure 7.1), from Highway 403 to Highway 48 including Highway 407, was approved by the province in July 1978.

Funding Highway 407

According to A. Jay Nutall of the Ontario Ministry of Transportation, during the 1985 to 1987 period the average annual expenditure for new highway construction was around C$10 to C$15 million; from 1987 to 1994 approximately C$240 million was authorized in Toronto for construction work completed by 1994. Historically, funding for highway construction in Canada had been funded through the General Fund; any tolls collected from existing toll roads were deposited with the General Fund as well.

Annual allocations from the consolidated revenue fund limited the province's ability to fund substantial projects, so the ministry had to implement a number of relatively small projects over an extended period of time. In fact, from July 1987 to May 1994 the government had awarded a total of 16 separate contracts for work related to Highway 407 such as advance detours and interchange and bridge work. But the construction of the main highway project had yet to begin.

The Capital Investment Act of 1993 established the Ontario Transportation Capital Corporation (OTCC), which would report directly to the ministry of Transportation. The OTCC was responsible for arranging innovative financing and public/

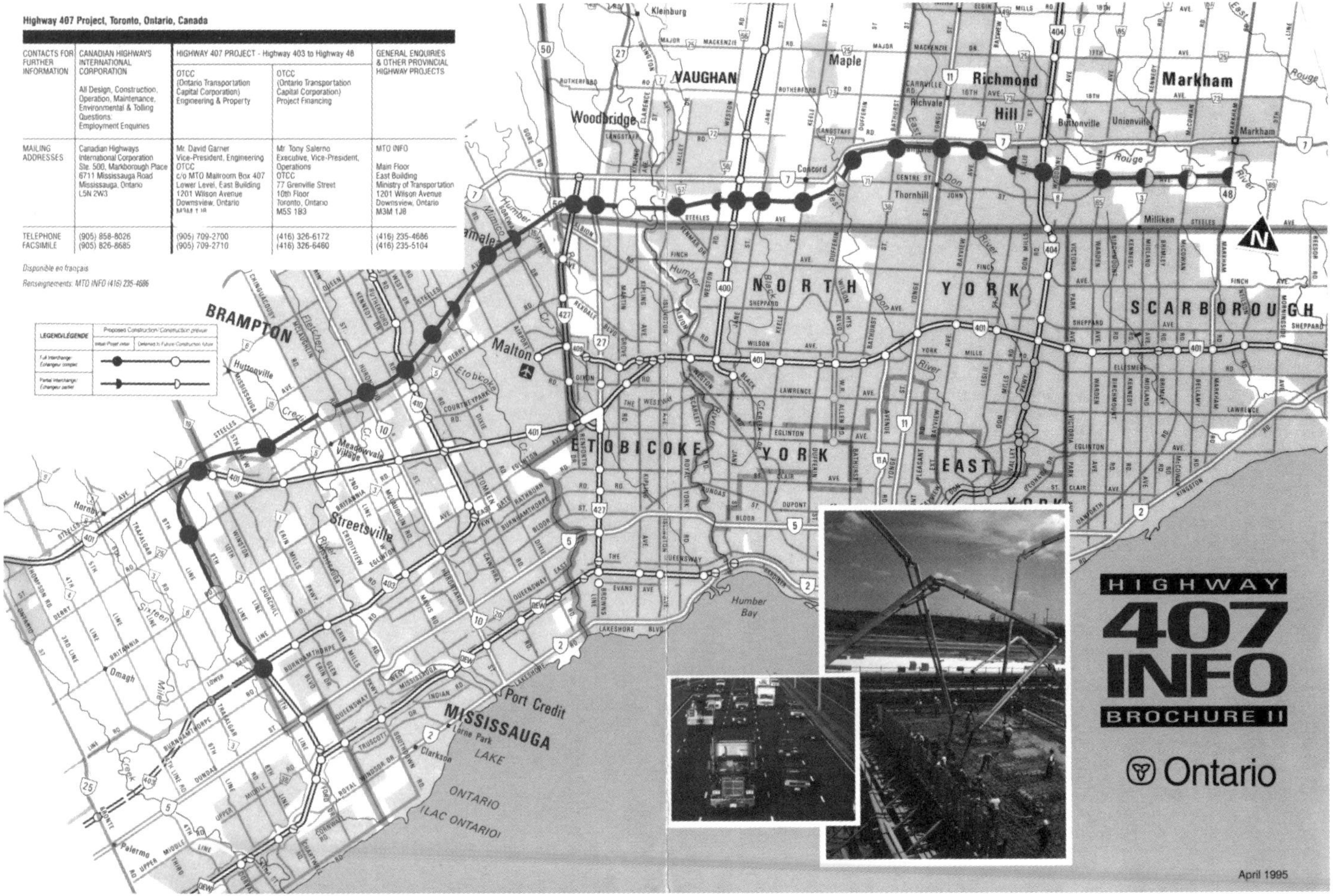

Figure 7.1. The Route of Ontario's Highway 407. *[Source Ministry of Transportation, Canada.]*

private-sector partnerships in the field of transportation infrastructure. The legislation also ensured that the revenue from tolls collected on these public/private road would be kept separate from government general funds so that this revenue could be used to pay for highway construction; it allowed private corporations to collect toll revenues in the event a build, operate, transfer project was proposed. Highway 407 was at the top of the OTCC list.

The Highway 407 Request for Proposal

In response to Premier Rae's expression of interest in private-sector partnerships, the government received early proposals from three firms between February and June 1993. One proposer was rejected for liability in meeting tolling criteria; the other two, Canadian Highways International Corporation (CHIC) and Ontario Road Development Corporation, were given Request for Qualification statements for completion. Both firms were also requested to submit all-encompassing proposals to design, finance, and construct Highway 407 as a toll road.

The Request for Proposal (RFP) was issued September 1, 1993. It contained the current ministry policy as it related to highway construction specifications and design criteria, and suggested project design criteria. Proposers were to include insurance during the design and construction phases only. The selected group would be responsible for all highway operations, maintenance, repairs, and rehabilitation during the concession period. The RFP demanded detailed provisions for

- Project financing
- Design, both architectural and engineering
- Construction
- Operation and maintenance
- Toll technology, with a target of all electronic systems installed by the 1998 highway completion date
- Transfer of title of the project to the Crown at the end of a 35-year concession period

Jobs Ontario The creation of Jobs Ontario, a provincial initiative to stimulate the economy and to create more employment opportunities, had been announced in February 1993. Helping to accelerate the construction of Highway 407 through a public/private-sector partnership was a key goal of Jobs Ontario, so it contained a unique provision: The program would reimburse each consortium that submitted a proposal for the project up to C$1.5 million of the costs incurred in preparing and submitting their individual comprehensive proposals.

Rights-of-Way The Crown had already acquired or would acquire all of the land needed for the Highway 407 project at its own expense. All these lands would be available to the successful bidder. The selected developer would be responsible for relocating all utilities and could obtain assistance from the Crown, which would assume the role of coordinator during any negotiations. Highway 407 also was exempt from all requirements of the Environmental Assessment Act (Ontario Regulation 707/83).

Value Engineering During the proposal stage, two qualified consortiums, CHIC and Ontario Road Development Corporation, offered 200 value engineering suggestions and nearly 60 were accepted by the government, resulting in an estimated C$200 million reduction in the total project costs. Some of the major changes from the initial concept that were offered and accepted were

1. Modification and simplification of the interchange at Highway 401 and 407, eliminating some local road accesses
2. Several interchanges moved and simplified
3. High-speed directional ramps replaced by "loop" ramps
4. Minor modifications made to roadway design standards
5. Three full interchanges and one partial interchange construction deferred until actually needed in future
6. Highway cross section reduced from five lanes in each direction to four with a provision for a HOV buffer
7. Median width constructed with a minimum width of 7.5 meters (24.6 ft)

Tolls and Toll Systems The proposers were to base their proposals on the forecasted traffic volumes prepared by Wilbur Smith Associates that were included in the RFP. The toll technology had to include photo-tracking and identification of the occasional user along with a cash collection system. The criteria were also to include provisions for enforcement, communication of driver information, and the following subsystems: vehicle classification and detection, data collection and analysis, billing and transactions, and administration.

The RFP required that each proposer establish a toll rate that was "fair and reasonable" so as to provide an adequate financial return to the proposer, but it stipulated no specific minimum or maximum level return on investment. The Crown asked that each proposer develop a "creative" approach to toll rate regulation to protect against excessive returns. The proposer would be offered incentives to include HOV lanes and transit services within the corridor, along with incentives for early completion of each phase of construction and overall completion.

Although as of mid-1995 an exact toll schedule had not been established, it appeared that a rate of 9 cents per kilometer (6/10 mi) would be levied on passenger vehicles, equating roughly to C$1 per 11 kilometers (7 mi) trip at peak traffic times. Discounts would probably be offered for off-peak travel and a minimum would be established for very short trips on the toll road.

The Industrial Benefits Provision A unique provision of the RFP was a requirement to include a plan to encourage participation in the project of Ontario and

Canadian businesses. The government wanted local firms to be part of the design and construction process in hopes that they would upgrade their management and technology skills, making them more sophisticated competitors on future projects. Any potential opportunities advanced by the proposer for technology transfer among Ontario and Canadian firms and research centers involved in highway construction would enter into the Crown's selection-making process.

Evaluation of Proposals

The submission deadline of the Highway 407 proposals was December 13, 1993. Ontario premier Bob Rae announced on November 5, 1993, that "work is now underway to select the private-sector group that will design, finance and build Highway 407. It represents a new way of doing business—a government/private sector joint venture that will allow us to deliver the highway faster and at a lower cost."

A financial working group reviewed the proposals, and provided advice to the proposal review board. The final selection was made by the Lieutenant Governor in Council based on the following criteria:

- An assessment of the design, construction, delivery, maintenance, operation, and transfer elements
- The guaranteed maximum price for design and construction
- The overall viability of the business and financial plans
- The anticipated aggregate toll rate and other sources of income
- The number of years required prior to transfer of title
- The proposed regulatory structure, including mechanisms for regulating toll rates, rates of return, and incentive provisions
- The degree to which toll collection and other sources of income support the overall project cost
- An assessment of risk, that is, the proposer's experience and financial viability, where equity participation would originate, the construction and operational risks, the security plans, and the life-cycle costing

Canadian Highways International Corporation (CHIC) was tentatively selected on April 4, 1994. An agreement between the government agency OTCC and CHIC was signed on May 11, 1994. The partnering agreement, a develop, design, build agreement in the amount of C$929.8 million in the form of a guaranteed maximum price contract, was the largest single transportation contract in Ontario's history. The only items not covered by the contract, agreements for operations and maintenance and for toll collection authority, were executed soon afterward. On the signing of the Highway 407 contract, Transportation minister Gilles Pouliot was enthusiastic: "These contracts keep Highway 407 on schedule and create more than 840 jobs. They will ensure that the first section of Highway 407 between Highways 400 and 427 opens in 1996."

The Canadian Highways International Corporation

The Canadian Highways International Corporation is actually a consortium of four leading Ontario-based project management, engineering, and construction firms with an aggregate annual revenue of C$1.5 billion and 10,000 employees. Fellow consortium members have been building infrastructure projects of all sorts in Canada and throughout the world, including the Prague International Airport, the Cikarange cogeneration plant in Indonesia, a nuclear power plant in South Korea, and new port facilities in Bombay and Shanghai, to name a few.

- The lead partner in the group is Monenco AGRA, Inc., founded in 1907, one of North America's largest engineering firms. Monenco also provides the consortium's procurement and construction management services.
- The second member of the group, the Foundation Company, Inc., is a member of Canada's Banister Inc., a diversified construction conglomerate. The Foundation Company is a general contracting firm specializing in civil engineering work.
- Dufferin Construction Company, a division of St. Lawrence Cement, has a great deal of experience in highway construction and is one of Canada's largest bridge builders.
- Armbro Holdings Inc., the fourth member of CHIC, is a diversified road builder and heavy construction contractor.

In September 1994, negotiations began between OTCC and CHIC and the BBMH Group, who were to provide the comprehensive toll collection system for Highway 407. The anticipated cost of the work was estimated to be C$50 million. The BBMH Group was composed of Bell Canada, Bell Sygma, MARK IV Industries, and Hughes Aircraft of Canada Ltd.

The Status of Highway 407

The first phase of the Highway 407 project, the 36 kilometer (22 mi) section from Highway 410 in Brampton to Highway 404, is slated for completion in 1996 and appears to be on target. By the end of 1998, the 33 kilometer (20 mi) section of Highway 407 from Highway 403 in Oakville to Highway 48 will be complete.

In April 1995 the minister of Transportation announced plans to extend the highway in both easterly and westerly directions. The 22.68 (14 mi) western extension will be built as a publicly funded project at a cost of C$250 million and is scheduled to be completed by 1998. Construction on the eastern extension will not commence for several years because an extensive environmental assessment process is required.

As of September 1995, the construction status of Highway 407 was as follows:

C$285 million of construction and engineering work has been completed, representing 31 percent of the overall project.

60 percent of the entire project's engineering has been completed.

5 million cubic meters (6.54 million cubic yards) of rough grading have been finished.

1.9 million cubic meters (2.48 million cubic yards) of fill dirt have been imported.

164,000 cubic meters (214,503 cubic yards) of concrete have been placed.

260 pieces of earth-moving equipment are on site.

The economic accomplishments of Highway 407 have been even more impressive:

1,700 direct full-time jobs have been created.

5,500 direct and indirect jobs have been created to date.

2,800 purchase orders have been issued with a total value of C$120 million.

135 subcontractors are on bidder's lists and more than C$300 million in subcontract agreements have been made to date with a value of C$483 in purchase orders committed.

94 percent of the purchase orders and subcontracts have been to Ontario firms.

The guaranteed maximum price tag for the Highway 407 program is C$929.8 million—C$300 million less than earlier estimates compiled by the Crown. The ministry hopes that the first year of the 43 mile central section of Highway 407 will prove successful and make future extensions of the route attractive to further private-sector participation. The Highway 407 project is another example of what can be accomplished when the private and public sectors combine their skills and create an innovative infrastructure project.

PUBLIC/PRIVATE ALLIANCES IN THE ATLANTIC PROVINCES

Canada's Atlantic Provinces of New Brunswick, Nova Scotia, Prince Edward Island, and Newfoundland are the location of one of North America's most costly and daring BOT projects—the C$850 million, 12.9 kilometer (8 mi) bridge spanning the Northumberland Strait from Jourimain Island, New Brunswick to Borden, Prince Edward Island (Figure 7.2). Another BOT project, Highway 104 in Amherst, Nova Scotia, will eventually tie into the bridge across the Northumberland Strait and link both provinces to the Trans-Canadian highway system. The Highway 104 project is still in the development stage, but the Northumberland Strait bridge project is already underway and is scheduled to open for public use in 1997.

NOVA SCOTIA'S HIGHWAY 104

Nova Scotia's Highway 104, currently running from Amherst to Truro, is the highest volume road in the province. It serves as the main corridor into and out of the province. Road widening to four lanes has taken place on sections of the highway

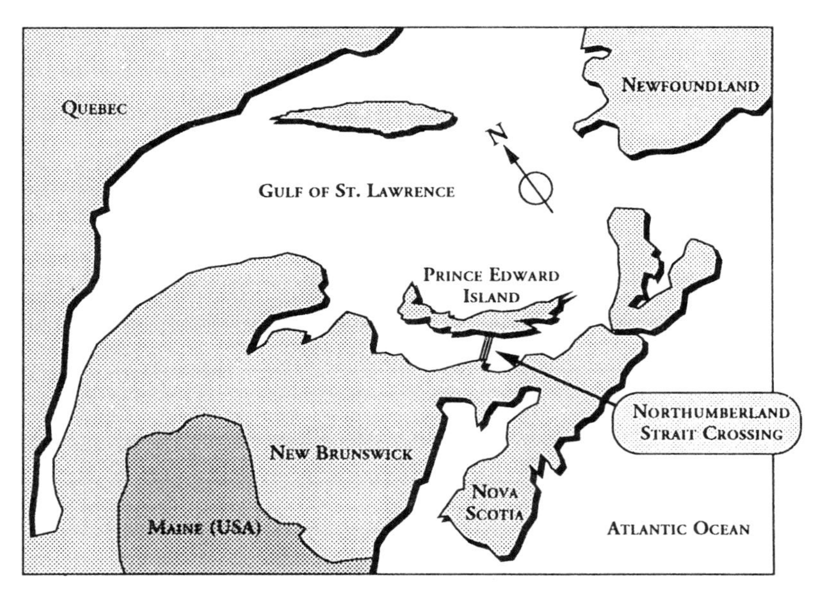

Figure 7.2. The Atlantic Provinces of Canada and the location of the Northumberland Strait Crossing. *[Courtesy Strait Crossing Development, Inc.]*

in the past few years, but the portion between Masstown in Colchester County and Thomson Station in Cumberland County has remained two lanes. The government announced plans to complete the four-lane highway system by creating a Western Alignment joining Masstown with Thomson Station while leaving the old two-lane road in place (see Figure 7.3).

Once the Western Alignment is built, speed limits would be lowered on the old two-lane highway and intercity trucks would be banned from that route. It was anticipated that the accident rate along this portion of the old highway would be substantially reduced.

But the proposed Western Alignment would be attractive to automobile travelers as well because it would save an estimated 30 minutes travel time between Amherst and Truro. A study of projected traffic flow indicated that 57.3 percent of the general traffic flow along the new Western Alignment would be either coming into or leaving the province and only 42.7 percent would remain local truck and auto traffic. For revenue purposes, the total anticipated traffic was projected at 6,000 vehicles per day, 24 percent of which would be trucks; 66 percent of these trucks would be intercity rather than local.

Planning the Western Alignment

Because Canada had limited experience in toll road construction, the government decided to use a specialized market research technique known in the United Kingdom as State Preference or SP to measure the willingness of travelers to use this new toll road. SP, which has been used effectively in Europe, makes use of information collected where the project will be located rather than borrowing information from nonlocal traffic and travel studies. The SP survey gives respondents choices to select for completing a hypothetical journey. When these responses are collected and collated, the researcher will discern the weight respondents attached to their choices. An example of an SP card passed out to local respondents is shown in Figure 7.4. The range of tolls offered to the respondents in the survey were C$2 to C$5 for automobiles traveling one way, and C$3 to C$12 for trucks, also one way. The time savings varied from 8 to 15 minutes using the tolled sections of the highway instead of the existing alignment between Thomson Station and Masstown. Table 7.1 reflects the respondents' preferences for transferring from the old freeway to the new toll road as toll rates increase for both automobiles and trucks.

Although there was a recognizable necessity for the Western Alignment in terms of flow of goods and people, and the need to increase safety and reduce the number

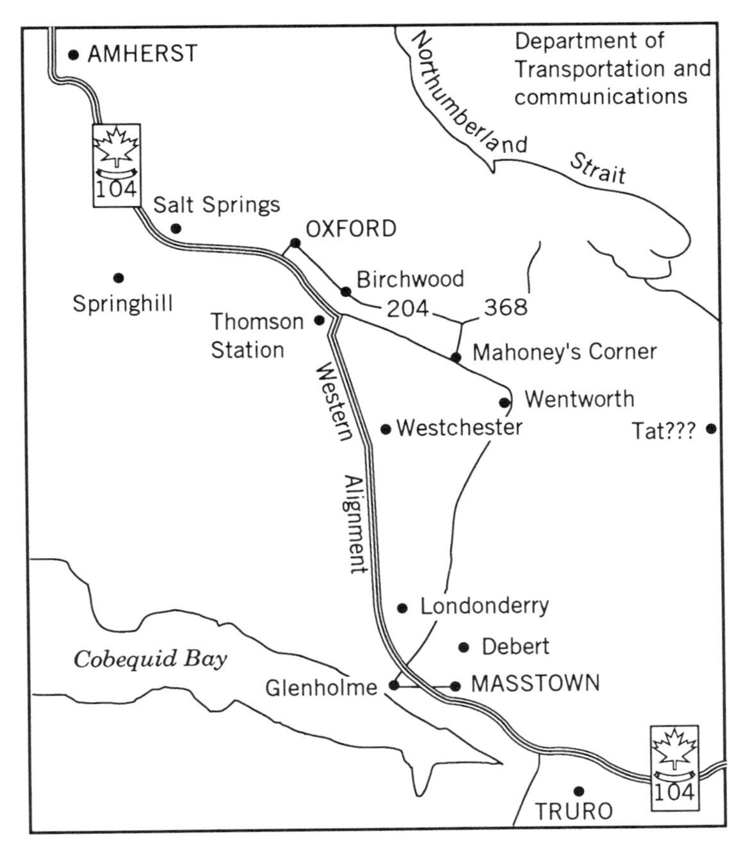

Figure 7.3. The Western Alignment of Highway 104 in Nova Scotia. *[Source Nova Scotia Department of Transportation and Communications.]*

Features of the New Road

Four-Lane Divided Highway	
Toll Charge (Pay cash only)	$3.00 each way
Average Time *Saving*	10 minutes
Distance *Saving*	9 km (5½ miles)

If the New Road Had Been Available Today, What Would You Have Done?

1. Go by new (toll) road *OR*
2. Go by existing highway 104

Figure 7.4. The SP choice card used by Highway 104 planners to test user reactions to a toll road. *[Source Nova Scotia Department of Transportation and Communication.]*

of accidents along the two-lane portion of the highway, no government funding for the project was anticipated until 1999.

Richard Mann, minister of Nova Scotia's Department of Transportation and Communications (DOTC), suggested in 1994 that the province consider a partnership with the private sector just as other jurisdictions in Canada had done to get this new Highway 104 project off the ground. The project would be entirely new construction—not just the widening of an existing highway—and if constructed as a toll road it might prove attractive to a private investment group.

The DOTC, taking into consideration the power of the profit incentive in the private sector, anticipated that the Highway 104 Western Alignment project could be built in 20 months instead of the 52 months it had estimated under conventional bidding processes. It also estimated that the construction costs could be reduced by as much as 21 percent if the project were given over entirely to the private sector.

After placing advertisements about the project in local papers in June 1994, DOTC received sufficient interest in a public/private Highway 104 toll road scheme

TABLE 7.1. Various Toll Charge Options Presented during the User Test Data Program

Traffic Transferring to the New Highway by Toll Charge			
Car Toll (C$)	Truck Toll (C$)	Cars (%)	Trucks (%)
2	3	71.5	58.6
3	6	61.4	50.0
4	9	50.1	41.8
5	12	39.5	33.7
6	15	30.8	26.2

Source: Nova Scotia Department of Transportation and Communication.

that the firm of Arthur Andersen was hired to assist the government in the economic and financial evaluation of such a build, operate, transfer project. The Arthur Andersen study placed a value of C$100 million on the proposed project and determined that it could be completed with a public contribution of approximately C$30 million.

The Highway Western Alignment Corporation

In order for this public/private project to proceed to the next stage, legislation was required to establish a tax exempt corporation that would create a vehicle to finance, construct, and manage the project. On April 20, 1995, such a bill was introduced in the 3rd Session of the 56th General Assembly of Nova Scotia and thus the Highway 104 Western Alignment Corporation was established. The Corporation was not an agent of the Crown, but could borrow money on its own credit and issue bonds and debentures and other securities without recourse against the revenue of the province. The Act authorized the government to negotiate agreements with private developers for the Western Alignment project.

The Corporation was granted the authority to collect tolls for the purpose of paying the expenses and service debt incurred by that entity for construction and maintenance of the proposed highway. To ensure that all revenue collected from tolls would be used to pay for the construction, operation, and maintenance of the project as well as financing costs, the Act exempted the Corporation from paying income tax and other taxes on construction and operations activities. This portion of the Act was looked on not so much as a concession to the private developer but merely as an extension of the tax exempt status of DOTC, the other operator of toll roads in the province.

The Call for Expression of Interest

When the DOTC issued the Call for Expression of Interest for the New Four-Lane Divided Highway 104 on June 1, 1994, seven consortiums responded by June 30. The DOTC inquiry had contained the following stipulations.

1. The Department would purchase the land necessary for the highway right-of-way.
2. The private developer would be responsible for the complete financing and operation of the project until all costs were recovered, at which point title would revert to the province free and clear of all encumbrances. The time frame for this turnover was to be determined by the proposers and included in their proposals.
3. The private developer would be responsible for construction of a 44.9 kilometer (27.8 mi) section of four-lane highway between Masstown and Thomson Station; the developer would be responsible for grade-separated interchanges at Masstown Trunk 4 (present Highway 104) at Glenholm, Station Road, Wentworth, Collingwood Road, and Thomson Station.

4. The operation and maintenance of this section of highway would be the responsibility of the private developer and include interchange ramps in accordance with the standards established for the other highways in Nova Scotia.
5. The private developer would be required to obtain all the permits and approvals necessary for the construction and operation of the highway, including those from regulatory agencies such as Federal Fisheries and Oceans and the Federal Environmental Agency. Approval of the project had already been obtained from the Minister of Environment in accordance with the provisions of the Nova Scotia Environment Assessment Act and the private developer was expected to comply with these requirements.
6. Design standards for the highway were to conform to Nova Scotia DOTC highway design standards and the *RTAC Manual of Geometric Design Standards for Canadian Roads*.
7. The estimate of traffic developers were to consider was

Volume	6,000 vehicles daily
Growth rate	3 percent
Passenger vehicles	80 percent of total
Single unit trucks	5 percent of total
Tractor trailers	15 percent of total

Seven groups submitted documents expressing interest in the project and, through the review and evaluation process, three groups emerged on the short list.

1. SNC Lavalin Inc.

   A group composed of SNC Lavalin, Dexter Corporation, SNC Lavalin Capital Inc., Fenco Shawinigan Engineering Ltd., Wilbur Smith Associates, Scotia McLeod Inc., MacLaren Plansearch Ltd., and Transroute Consultants

2. Atlantic Highways Corporation Inc. (AHC)

   Equity members: Canadian Highways International Corporation, Tidewater Construction Co., Ltd., and Nova Construction Co., Ltd.

   Associate members: Porter Dillon Ltd., Jacques Whitford & Associates

   Specialist members: McDonnell Engineering Ltd., Transroute International, Stewart McKelvey Stirling Scales, Wood Gundy, Inc., Barclays Bank

3. Atlantic Canada Transportation Group

   Developer: Strait Crossing Group, GTM (Canada) Ltd., McNamara Construction

   Consultants: Vaughan Engineering Associates, Ltd., Jacques Whitford & Associates, Jacques Whitford Group, Eastcan Group, KPMG Peat Marwick, Confiroute Corp., Alexander and Alexander, Peat Marwick Thorne, Patterson Kitz

The Request for Proposal

The Highway 104 Western Alignment request for proposal (RFP) was issued on July 17, 1995, a few days ahead of schedule. The milestone dates in the RFP were

RFP Issuance	July 19, 1995
Deadline for written questions at bidder's conference	August 11, 1995
Bidder's conference	August 21, 1995 (week)
Final deadline for written questions	September 8, 1995
Proposal due	September 29, 1995
Notification date	October 30, 1995

The RFP contained the following program objectives:

To arrange for the private sector to carry out the project

To achieve the development of a facility at a reduced cost and in a shorter period than would be possible to achieve by traditional means

To minimize DOTC's financial commitment to the project

To identify innovative financing and construction methods that may be applicable to other DOTC highway projects

To ensure that the facility at the end of the concessionary period is in a good state of repair and preservation and is free and clear of encumbrances

These objectives clearly state the case for a public/private initiative; the private sector can bring a project from inception to completion in substantially less time than a public agency can at less cost and at higher quality levels.

The RFP contained a few unusual provisions as well. Collection of tolls is the sole responsibility of the respondent; however, toll jumping is not an offense under the Motor Vehicle Act. A suggested enforcement method advanced by DOTC was for a developer to install a red and green traffic light at each cash toll booth so that payment of a toll would activate the green light. If the violator did not pay, the light would remain red; if the driver drove through the red light, this would be an offense under the Motor Vehicle Act—a failure to obey a red signal light.

Travelers need to be made aware of any new highway system, and both government and concessionaire would benefit from a well-conceived, properly funded marketing plan for this new toll road. To this end, respondents were required to include a marketing plan with its own budget in their proposals.

In the financial arrangements portion of the RFP, the province ties the length of the concession period to the length of time required to repay the borrowed money. Paragraph 4.1 states that

> the length of the Concession Period will be limited to the length of time needed to repay the money (other than the C$29 million contributed under the SHIP Agreement) borrowed or otherwise made available by parties other than DOTC to pay the cost of

construction of the facility out of toll revenues available after provision for operations and maintenance and all other necessary reserves for repair and rehabilitation work.

The toll rate structure was researched by the DOTC. It took into account the rate per kilometer on selected toll roads in the United States and Canada and determined that a reasonable rate for Highway 104 would be C$3 for cars and C$2 per axle for trucks.

Addenda to the RFP

Subsequent to the issuance of the RFP, four addenda were released, the last of which was dated September 11, 1995.

Addendum 1, released on August 18, 1995, contained the DOTC's responses to questions submitted by respondents prior to close of business on Friday, August 11, 1995:

Q: Will the proposal due date be extended to October 30, 1995?
A: After careful deliberation, no.

Q: If any utilities within the right-of-way are to be relocated, who will pay all associated costs?
A: If the utilities are located within the Western Alignment right-of-way on the date the Omnibus Agreement is signed, the respondent will pay for relocation.

Q: Why does DOTC not legislate the failure to pay a toll as an offense rather than rely on existing provisions of the Motor Vehicle Act?
A: The Legislature has adopted the Act and no amendments to the Act are contemplated.

Q: Will DOTC define what is meant by a "guaranteed maximum price"? Specifically, is DOTC seeking a lump sum guaranteed price or some other form of pricing with the GMP as an upset?
A: DOTC wants a lump sum fixed price.

Q: Request confirmation that the term of the Omnibus Agreement will be variable in order to ensure repayment of all financing and a fair market rate of return to the selected respondent.
A: DOTC reinforces the point that any surplus accumulated in the Corporation belongs solely to the province. The selected respondent must earn its return from (a) the construction contract, (b) the operating contract if the selected respondent elects to operate the project itself, and (c) a return on any debt that the respondent chooses to hold.

Q: Respondent suggests that DOTC weigh individual components of the evaluating criteria and subcriteria.
A: No specific weights have been assigned to the subcriteria; and none of the five criteria components have been assigned a specific weight.

Q: Will DOTC elaborate on the scope of distribution of financial information of the respondents? Will distribution be limited to those with background and experience in evaluation of financial information?

A: Distribution of financial information will be limited to the financial plan evaluation team and members of two committees. No proposal, parts of a proposal, or copies of a proposal will be removed from this secure location until the evaluation is complete.

Addendum 2 was issued on August 25, 1995, and added the requirement of a snowmobile trail network in the right-of-way, crossing the road at two locations. Each crossing was to have an opening 3,000 mm (9.93 ft) minimum width and 3,000 mm minimum height.

The second page of this addendum contained an extremely important change in the monetary contribution by the Canada-Nova Scotia Strategic Highways Improvement Program (SHIP) agreement. The total contribution to be funded by SHIP was raised from C$29 million to C$55 million. The first C$29 million would be made available on April 1, 1996, and the remaining C$26 million would be available on April 1, 1998.

Addendum 3, dated September 1, 1995, dealt with the median options and set a minimum of 22.6 meters (74 ft) between edges of directional travel lanes. DOTC would not accept a median having less than 5.6 meters (18.4 ft) between edges of directional travel lanes and a "tall wall" barrier is required where the median width is less than 12.6 meters (41.3 ft).

Addendum 4, the last addendum, was issued on September 11, 1995, and contained the specific insurance and bonding requirements of both the construction and the operation phases.

Requirements during the Construction Phase

1. All-risk builder's risk insurance
2. Delayed start-up insurance to cover the loss of earnings up to C$10 million per occurrence
3. Contractor's equipment insurance to insure machinery and equipment at replacement cost at C$10 million per occurrence
4. Wrap-up liability to insure all contractors, subcontractors, suppliers, tradespeople, engineers, architects, and consultants working on the project with limit of C$25 million per occurrence
5. Automobile liability at C$2 million per occurrence with a C$25 million overall limit
6. Umbrella and excess liability insurance with a C$50 million overall limit
7. Directors' and officers' liability insurance with a C$1 million limit per claim

8. Wrap-up errors and omissions insurance in the amount of C$10 million each claim
9. Employee dishonesty insurance with a C$1 million limit for each claim
10. Surety bonds during construction in the form of construction guarantee/performance bonds in the principal amount equal to 50 percent of the agreed lump sum price and would include a latent defect clause for two years after substantial completion
11. A labor and material bond in a principal amount of 50 percent of the agreed lump sum price

Requirements during the Operation Phase

1. All-risk insurance, including flood and earthquake coverage up to C$35 million per occurrence
2. Business interruption insurance in the amount of C$10 million, but when combined with all-risk each occurrence would be raised to a C$35 million limit
3. Commercial general liability insurance with a C$25 million limit
4. Umbrella and excess liability at C$50 million per occurrence
5. Directors' and officers' liability insurance at C$10 million limit per occurrence
6. Blanket crime insurance insuring all employees, contractors, and subcontractors with a C$1 million limit on each claim
7. A performance bond in the principal amount of 50 percent of the annualized contract amount for operation and maintenance

Based on the content of these addenda, it does not appear that a final selection will be made before the first quarter of 1996.

THE NORTHUMBERLAND STRAIT BRIDGE

In 1873, Prince Edward Island joined the Canadian Federation and the future economic prosperity of the island depended on a sound and economic "bridge" to the mainland. In the early 1800s the ice boat was the only reliable method of transportation from Prince Edward Island (P.E.I.) to New Brunswick; later in that century the ferry boat became the accepted method of crossing the straits.

In April 1895, George William Howlan, liberal senator from P.E.I., stood up in the Canadian senate and suggested that a "fixed link" be constructed across the stormy Northumberland Strait to connect the island's railroad to the intercolonial rail system at New Brunswick on the mainland. Senator Howlan's proposal was summarily dismissed. In 1965 a causeway was proposed as a replacement for the existing ferry system, but it proved to be too costly for the government to consider;

however, in the mid-1980s a Request for Proposal was issued by Canada's Department of Supply and Services and Public Works for a fixed link across the straits. So about 102 years after this daring scheme was first proposed by Senator Howlan, the fixed link will be in place.

When the Northumberland Strait Bridge is finished, Canada's smallest province will enter the record books as the site of the world's second longest continuous bridge over water. The world's longest bridge over water, in northern Japan, is 14 kilometers (8.68 mi) long; Northumberland Strait Bridge will be 13 kilometers (8.06 mi) in length.

The Fixed Link Advantage

The economy of the P.E.I. region was badly in need of a boost and the proposed bridge would generate more tourism and commercial activity, not to mention the influx of money during the construction phase.

The environmental issues relating to the construction of a bridge across the straits could and would be taken into consideration in both the design and construction of a fixed link. On balance it appeared that the fixed link would be more acceptable than continued ferry operations from an environmental viewpoint. The 75-year-old ferry system, which at this time is sorely in need of modernization, forms the present means of transportation between the New Brunswick mainland and Prince Edward Island. The ferry system's diesel engines are not conducive to air quality, and the ferries are not energy efficient as they burn about 20 million liters (5.29 million gallons) of fuel per year.

However with government funding committed to other uses, neither federal, provincial, nor local agencies could possibly come up with the financing arrangements to fund such a huge project. The Canadian government also recognizes that participation in a public/private venture could give Canadian firms valuable experience that would make them more competitive in the global public/private project arena. All of these reasons provided the impetus for the origination of the P.E.I. bridge venture.

The Call for Expression of Interest

In 1985 and 1986 the government received three unsolicited proposals for a fixed link bridge and began to undertake further feasibility studies. In 1987, 15 consultants were commissioned to conduct 10 studies to determine the economic, structural, and financial viability of the project.

In June of that same year, 12 Canadian companies responded to a government-issued Call for Expression of Interest notice:

Abegweit Crossing Ltd., Cape Breton, Nova Scotia
Banister Continental Ltd., Trimac, Ltd., Toronto, Ontario
Both, Belle, Robb Ltd., Montreal, Quebec

Cartier Construction Corporation, MacNamara Corp., Whitby, Ontario
Consortium Pomerleau-Bouygues, St. George de Beauce, Quebec
Dillingham Construction Ltd., North Vancouver, British Colombia
Monenco Maritimes Ltd., Halifax, Nova Scotia
Northumberland Bridge Builders, Downsview, Ontario
OMNI Systems Group, Halifax, Nova Scotia
P.E.I. Crossing Ventures, Halifax, Nova Scotia
P.E.I. and NB Link Contractors Partners, Montreal, Quebec
W.A. Stephenson Construction (Western) Ltd., Calgary, Alberta

Only seven of the above were deemed qualified at this time and subsequently this list was short listed to three groups: Strait Crossing, Inc., P.E.I. Bridge, Ltd., and Borden Bridge Company.

The Request for Proposal

In 1988 a plebiscite was called on Prince Edward Island, and 60 percent of the population responded favorably to the project. Consequently, the government developed and issued a Request for Proposal (RFP) in which it stated its desire to receive proposals from developers to finance, construct, and operate a toll bridge with a 100-year design life and a 35-year concession period, after which title would be transferred to the government.

A subsidy would be awarded to the winning consortium based on the estimate to operate the ferry service on a net basis. One of the stipulations in the RFP stated that this subsidy, although inflation indexed, would not exceed C$42 million (in 1992 dollars). The developer would obtain revenue from the tolls collected on the bridge during the concession period.

Stage 1 of the three-part proposal process involved seeking interested development teams. Stage 2 required qualified bidders to prepare and submit design criteria for the proposed construction of a fixed link, and stage 3 would be the final selection process that would include review of the proposed consortium's financial plan.

In 1989, the Federal Environmental Assessment Review Office (FEARO) was appointed and the panel's report indicated that the proposed bridge project would be environmentally acceptable. Subsequent public hearings were also held on the subject of environmental impact.

The three short-listed developers—Strait Crossing, Inc., P.E.I. Bridge, Ltd., and Borden Bridge Company—resubmitted their proposals in 1992. On October 7, 1993, a final selection was announced: An agreement was signed with Strait Crossing, Inc.

Strait Crossing, Inc. The Strait Crossing team, which collectively had designed and built thousands of bridges throughout the world, consisted of the following firms:

Strait Crossing, Inc. (SCI) The principals of this firm had a vast amount of experience in the design and construction of bridges, tunnels, recreation facilities, transit systems, and water and waste treatment systems.

Northern Construction Company, Ltd., the Canadian subsidiary of Morrison-Knudsen of Boise, Idaho. M K had extensive experience in a number of large international civil projects.

G.T.M.I. (Canada) Inc., the Canadian subsidiary of GTM Entrepose of Nanterre, France, which is one of Europe's largest construction companies. GMT is one of the joint venture participants in Great Britain's second Severn River Crossing design, build, finance, operate project.

Ballast Nedam Canada Ltd. is the Canadian subsidiary of Ballast Nedam, headquartered in Amstelveen, The Netherlands, and its principal activities are civil engineering, dredging, and general and industrial construction.

Strait Crossing Development Inc. (SCDI) is the developer that will build, operate, and maintain the bridge during the 35-year concession period.

The Strait Crossing project structure is shown in Figure 7.5.

The financial advisors to the consortium are Gordon Capital Corporation of Toronto, KPMG Peat Marwick Thorne of Calgary, and KPMG Peat Marwick, Stevenson & Kellog of Toronto.

Reed Stenhouse Ltd. is the insurance advisor to the developers, and three law firms in Toronto, Calgary, and the Atlantic provinces are the developer's counsel.

The Financial Structure of the Crossing Development

In anticipation of a consortium selection, in June 1993 the government enacted the Northumberland Strait Crossing Act (Canada) setting forth the terms and conditions of a subsidy agreement. This act required the government to designate funds to the Consolidate Revenue to be used for the annual subsidy. The financial structure for the deal was complicated, containing a multitude of cross agreements between private and public sector entities (Figure 7.6).

The federal government's contribution to the project would be in the form of 35 annual payments of C$41.9 million (1992 dollars) to the developer. This represented the federal government's estimate of the cost of the existing ferry service, including operating expenses and the capital expenditures required to keep the system operational. The government estimated that private development of this bridge would save the taxpayers approximately C$250 million (in 1992 dollars), taking into account the maintenance and periodic replacement of government-subsidized ferries and docking facilities.

Strait Crossing Finance Incorporated was formed by the government in 1993. The subsidy agreement would create the keystone around which Strait Crossing Finance would present its C$661 million inflation indexed fully amortized bond issue in October 1993. The bonds were to be used to fund the payment for the construction of the bridge by Strait Crossing Development. Payments on the bridge

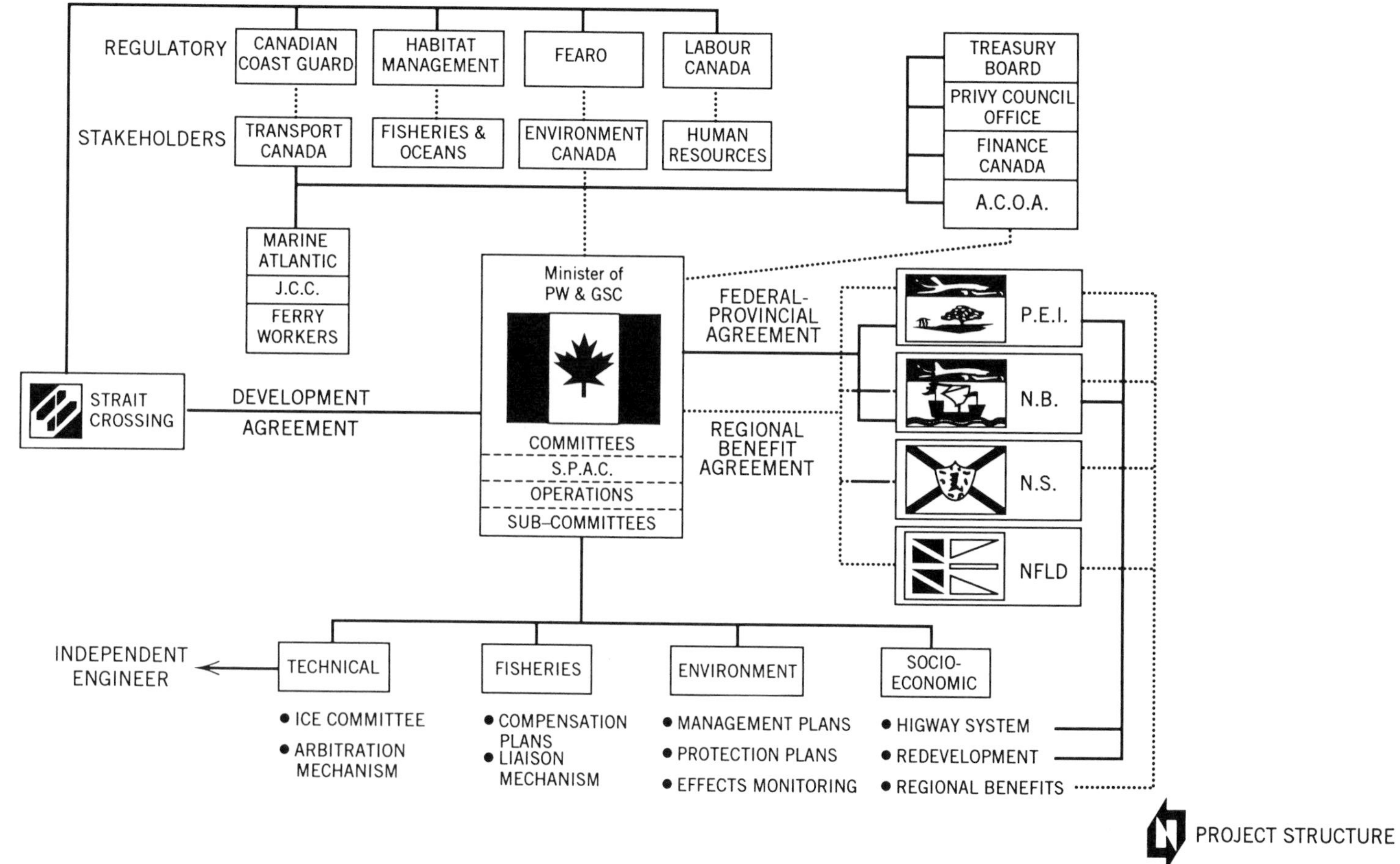

Figure 7.5. The Strait Crossing project structure. *[Courtesy Strait Crossing Development Inc.]*

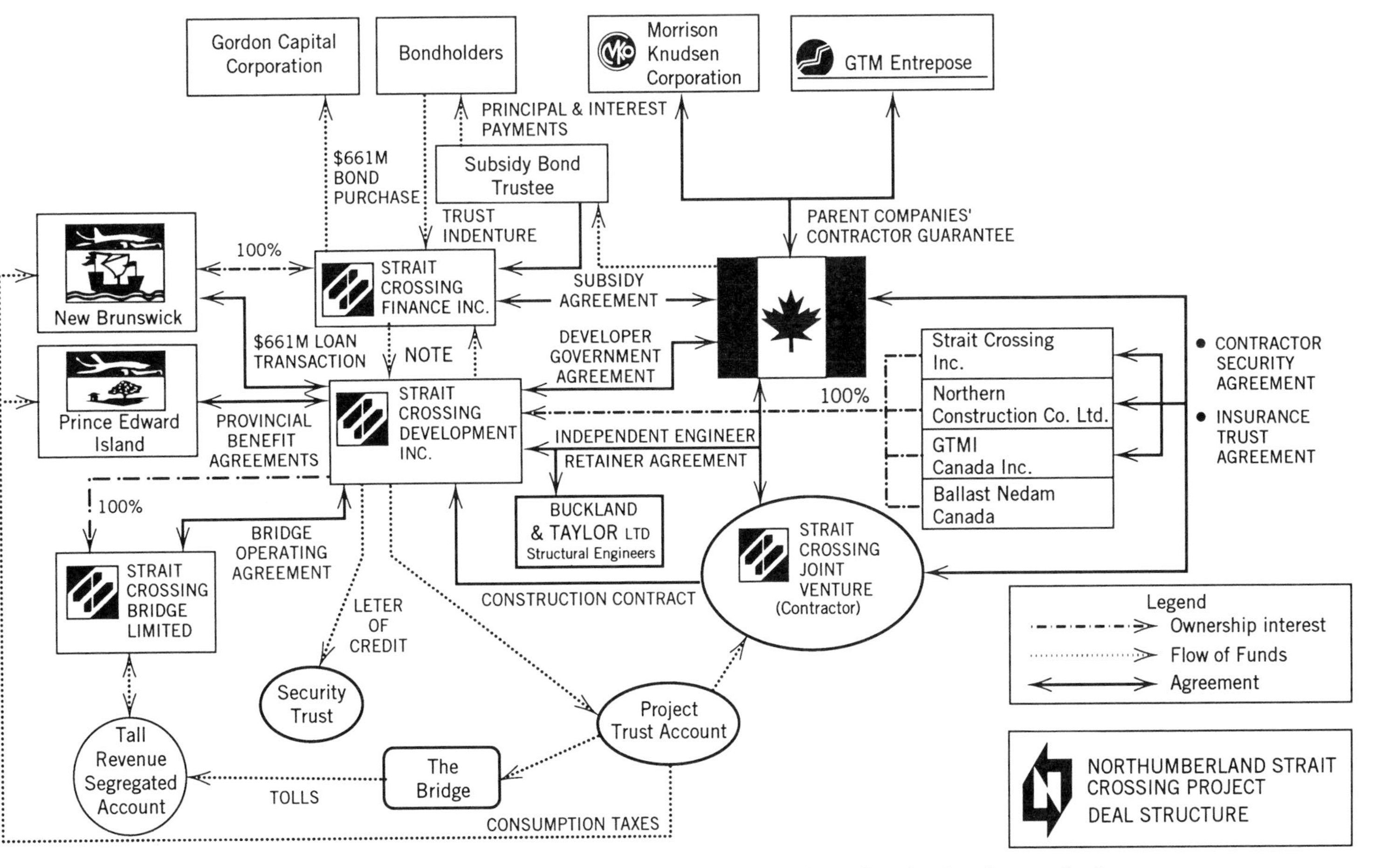

Figure 7.6. The Strait Crossing financing structure. *[Courtesy Strait Crossing Development Inc.]*

bonds will be exactly equal to the annual payment of the government subsidy. Recourse on the bonds would be limited to the security afforded by the assignment of the subsidy agreement.

A 10 percent contingency fund was required by the government and was to be in the form of an irrevocable letter of credit in favor of the government for the duration of the construction period. The developer was also required to furnish other forms of security:

C$200 million performance bond
C$35 million defects assurance bond
C$20 million labor and material bond
C$73.8 million letter of credit
Prepaid insurance policy covering all aspects of construction.

The parent company of each member of the consortium was to provide performance guarantees for the fixed price construction contract.

After the agreements were signed on October 7, 1993, the government's role changed from initiator to overseer. Peter Sorensen, project leader of the Public Works and Government Services Canada, described the government's position this way:

> Its a watchdog role. We're like an insurance team . . . we're here to secure the Government of Canada's interest from a construction, environmental, and social perspective. In 35 years, we'll own the bridge. Now, we've got a good feel for the issues so we can ensure we'll end up with the product we're paying for. But our role is not to be front and center. We want to be like a referee after a good game.

In early 1995 the financial troubles of Morrison-Knudsen, the parent company of Northern Construction, became public. The company posted a C$150 million loss in the last quarter of 1994, and had to raise $125 million to avoid being forced to file bankruptcy. Speculation arose as to the viability of the company and its subsidiary to maintain a continuing role in the Strait Crossing venture. Northern Construction Ltd. was a 45 percent partner in Strait Crossing Development, and the SCI team had confidence Northern Construction. According to Kevin Pytyck, Strait Crossing's manager of contract administration, "Generally, the firm's losses have no bearing on the project. In the event a joint venture partner is not able to meet his security obligations, the other partners assume them. That's bound in the contract with Canada and that's why Canada is not concerned." Pytyck said that they were dealing with the stronger division of Morrison-Knudsen and that if the company were to fail, the consortium would buy out the company's shares and it would "also have to assume their share of risk and guarantees. The project is healthy, it has strong partners, strong participation, it's on track."

Pytyck proved to be correct. Northern Construction Ltd. has carried on with the other members of the team while its parent Morrison-Knudsen continued to sort out its own problems through the balance of 1995.

The Construction

The bridge is designed for a 100-year life span, but its life expectancy should far exceed design life. The approach to the bridge from the New Brunswick side will consists of 14 spans with a total length of 1,260 meters (4,133 ft), and the P.E.I. approach will be made of six spans equalling 540 meters (1,771 ft). The balance of the bridge, spanning the Northumberland Straits, will consist of 44 spans, each 250 meters (820 ft) in length. The total bridge length will be 13 kilometers (7.8 mi). The bridge will rise 55 meters (180 ft) above the water at the center of the channel where the span between piers will be 192 meters (630 ft).

There is a staging area on P.E.I. where production facilities are located to build the 183 main bridge components. The main span box girder is 190 meters (623 ft) long and weighs 7,900 tonnes (8,840 tons). These main bridge components are cast in steel forms with reinforcing steel and then post-tensioned. The connecting drop-in spans measure 60 meters (197 ft) in length. Each segment is "matchcast," a process where a new segment is cast against the preceding one so that it will fit precisely when lowered in place.

Bridge piers are cast in place after the overburden has been dredged from the seabed where it is to be constructed. A cofferdam is built to provide a dry and controlled environment for the pier foundation. Shear keys are drilled into the sea-bed rock; a conical steel ice shield is placed in the cofferdam to be cast into the pier to provide protection from ice damage during winter months when the concrete base is exposed to the elements. These ice shields act as deflectors and allow ice buildups to ebb and flow around them.

The shear-keyed pier base and ice shield are filled with reinforced, cast-in-place concrete, and the cofferdam is removed after an appropriate cure time.

The immense proportions of one of these piers can be seen in Figure 7.7, where the massive pier stands in contrast to the size of the workers who had just completed it.

The precast box pier shaft segments are then erected atop the pier base by a barge-mounted crane; these piers are then post-tensioned to form the complete shaft. A transverse cross section at a pier location is shown in Figure 7.8.

A special ocean-going crane, the MV Svanen, which sailed from Dunkirk, France to P.E.I. in the spring of 1995, was greeted with much fanfare on its arrival. The massive C-shaped catamaran—94 meters (308 ft) long, 65 meters (213 ft) wide, and 100 meters (328 ft) high—will be used to transport the precast concrete bridge components from the casting yards to their final resting place on the piers. The operator of this crane positions the spans by obtaining data from its onboard GPS (global positioning system) which ties into satellites hovering 12,000 miles about the earth. At an installation rate of one segment per day, it is anticipated that Svanen will complete its task in 183 working days.

The pier base crew completed the final pour of the first pier base, P3, on May 4, 1995.

Figure 7.7. The first of several immense concrete bridge piers to be constructed for the Northumberland Bridge project. *[Courtesy Strait Crossing Development Inc.]*

The approach segments are being cast at the New Brunswick facility, and bridge components are being cast on P.E.I. The 170 meter (557 ft) long bridge sections are placed upon the vertical support piers; the crane then places the 80 meter (262 ft) long drop-in sections in between. The segments are tied together by post-tensioning.

Light poles will be installed every 50 meters (165 ft), and closed circuit video cameras will be located on poles every 500 meters (165 ft). Emergency call stations will be placed on the bridge at 500 meter (1,640 ft) intervals.

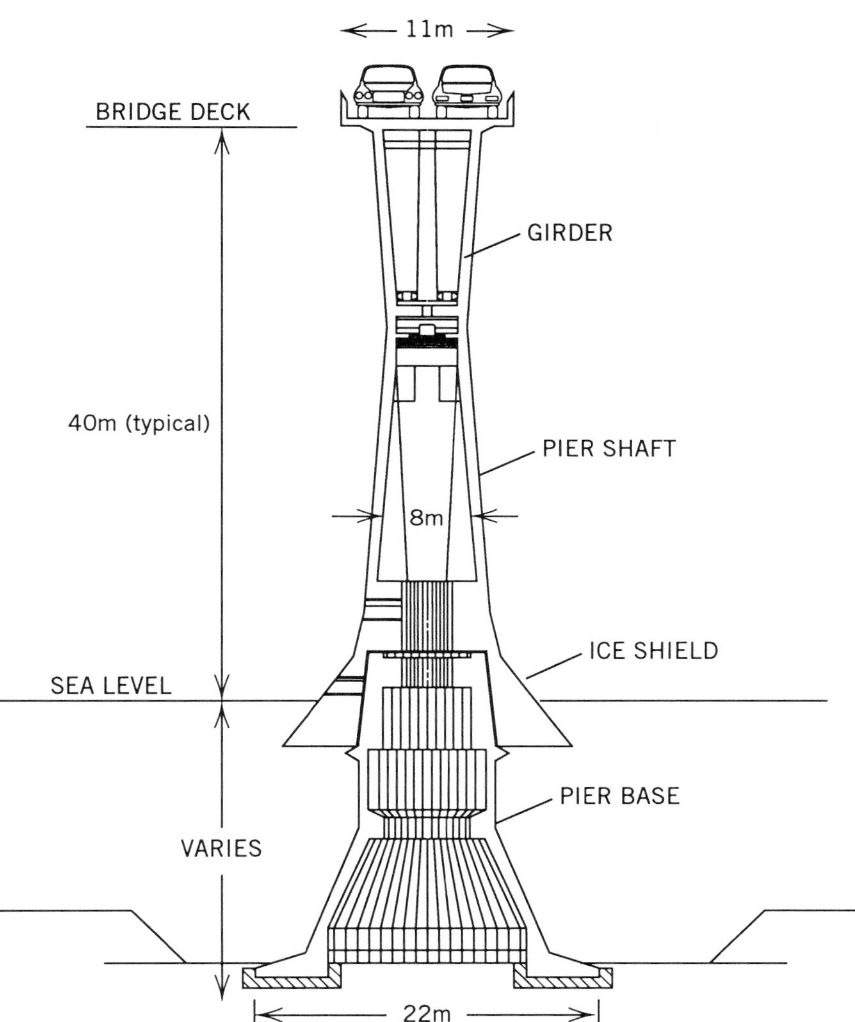

Figure 7.8. A traverse section at a pier, girder, and bridge deck portion of the Northumberland Bridge. *[Courtesy Strait Crossing Development Inc.]*

The bridge's construction has not been without incident. In August 1994 a crane boon failed and landed on a steel cofferdam under construction. Autumn storms in 1994 took their toll on construction when two barges with cranes and hoisting equipment broke from their moorings and were discovered grounded on a sandbar near Amherst Cove. The project's first fatality occurred in December 1994 when a welder fell to his death from the marine jetty at the Borden fabrication yard.

During the severe winter months of 1995 production of the approach bridge segments slowed from one per day to three per week. But as of the fall of 1995, the project was on track and was just about 50 percent complete. The marine jetty had been completed, as was the construction of the shore projection for the P.E.I. approach bridge. Component formwork for the main bridge pier bases, pier shafts, ice shields, and main girders had been erected: All pier bases for piers E1 through E7 had been poured, and pier shafts E1 through E5 were complete.

Project Status

The Northumberland Strait Bridge project has been a shot in the arm to the local economy. In the spring of 1995 approximately 1,000 workers had picked up the pace and were producing seven bridge segments per week. By August, 1,683 Atlantic Canadians were employed by Strait Crossing and another 150 worked for subcontractors on the project. And for many local workers once the bridge is open for business there will be other employment opportunities; these gains may be offset by the 600 ferry workers who will probably be put out of work.

Pytyck, SCI's manager of contract administration, said in early fall 1995 that construction was right on schedule. Paul Giannelia, the project director, shares Pytyck's enthusiasm and optimism—and looks at the Strait Crossing as a bridge to the future.

8

GREAT BRITAIN, HOST TO BOT AND DBFO

Great Britain has become home to the world's most costly BOT project, the Channel Tunnel, and the government has forged ahead on several innovative public/private infrastructure partnerships. For the first time since the 1950s, private-sector financing has been employed in the development of public-sector transportation projects:

- The Birmingham North Relief Road. England's first inland toll motorway, a project being undertaken by Midlands Expressway Ltd., a Trafalgar House–Iretecna joint venture.
- The Queen Elizabeth II Bridge. A 4,260 kilometer (2,663 ft) toll bridge spanning the Thames River east of London, rated as the most heavily trafficked estuary in Europe.
- The Skye Bridge Crossing. A bridge across the straits at Kyleakin to the Isle of Skye, the largest of the Inner Hebrides Islands of Scotland.
- The Second Severn Crossing. A £300 million (U.S.$549 million, 1992) toll bridge across the Severn estuary linking England and Wales.
- The Channel Tunnel Rail Link. A billion dollar venture that will become an integral part of the TransEuropean railway system.
- The Croydon Tramlink. A light rail system servicing suburban London.

In addition, a series of DBFO highway project projects are currently being explored by the Department of Transport, the British equivalent of the U.S. Department of Transportation.

Tenders (bids) for several motorways, referred to in England by the prefix M, and several trunk or A roads have been received by the DOT and are being evaluated as of mid-1995.

The £3 billion (U.S.$5 billion) Channel Tunnel Rail Link—which will complete the final phase of Great Britain's TransEuropean Rail Network by providing high speed train service directly from London's St. Pancras rail station to Brussels and Paris—is a private-sector megaproject in the short-list selection phase as of July 1995. At present the 180 mph Eurostar train operates at top speed only after emerging from the Channel Tunnel at Calais; travel from London's Waterloo International Station is at a more conventional 70 mph (113 kph) and the 20 minute ride through the Channel Tunnel is at a slightly higher speed.

BRITAIN'S NEED FOR PRIVATE TRANSPORTATION PROJECTS

With a forecast indicating the doubling of road traffic in Great Britain by the year 2025, the conservative government in the 1980s took a realistic view of the country's existing motorway and trunk road system.

Motorway travel in England in the past had been free from tolls. Government revenue for highway construction was derived from a number of sources; private vehicle owners paid an annual road tax of approximately £130 (U.S.$221); in addition to this tax there has been a fuel tax. Revenue to the government also accrues from the value added tax (VAT) common to manufactured or items assembled in the United Kingdom.

Car buyers previously had been obligated to pay a one-time excise tax on their new vehicle that averaged £1,000 (U.S.$1,700), but this tax was eliminated several years ago as a measure to stimulate the sale of new cars. Owners of commercial vehicles also pay an annual road tax along with the appropriate fuel and VAT taxes.

Vehicle-related taxes are not specifically earmarked by the government for new transportation projects or the maintenance of existing ones but are treated as another source on which annual budgets depend for their funding. The DOT annually presents its budget to the Exchequer for review and approval in much the same manner as other government agencies do. Faced with the same revenue shortfalls that beset other governments in the early 1980s, Great Britain began to look to the private sector for participation in new transportation projects and the upgrading of older ones.

THE NEW ROADS AND NEW MEANS WHITE PAPER AND THE NEW ROAD AND STREET WORKS ACT

A white paper entitled *New Roads and New Means* (NRNM) was presented to Parliament by the Secretary of State for Transport in May 1989. The NRNM study stated very clearly that the government's objective was to harness the entrepreneurial, financial, and managerial skills of the private sector to further the construction of much-needed highways and bridges in Great Britain.

The New Roads and Street Works Act of 1990 (NRSW) gave ministers the power to initiate these private-sector highway and bridge projects in a manner similar to that used under the existing Highways Act of 1980. It also provided for the charging of tolls on proposed new highways. The application of tolls to existing motorways was also under consideration, but was not yet mandated by this act. Privately financed roadways would be considered by the government when proposed in one of two ways:

1. A private-sector consortium could select, on its own initiative, a particular route for a proposed roadway—one that was technically feasible and would be potentially attractive to users willing to pay a toll for the use of that roadway.

   The government could exercise two options upon receipt of such a proposal: (a) it could enter into an exclusive agreement with that developer without subjecting the proposal to competitive tendering, or (b) if the project was deemed in the public interest, it could invite competitive bids. If initial proposers failed to produce the most competitive bids, they could be compensated, financially, in a prescribed manner outlined in the NRSWA.
2. The Secretary of State could identify a desirable corridor but not a specific route through that corridor and invite competitive bids. Selection would be based on the most attractive route and commensurate financial considerations.

These public/private arrangements were known as DBFO (design, build, finance, operate). They differ from BOT (build, operate, transfer) arrangements in that title to the real estate involved in the project does not pass to the consortium during the concession period, but is leased from the government.

LAND ACQUISITION

If land acquisition for a project is required, it can be accomplished in a number of ways.

The developer can negotiate directly with the property owner, establish a lump sum price for the property involved, and the government will consummate the deal. The developer and the government convert the lump sum acquisition price into a "rental fee" which becomes part of the concession agreement.

In this way, title to the property remains with the government and not in private hands. The government gains by receiving acquired property at a cost that has been subjected to private-sector bargaining without government intervention. The developer can fold the property acquisition costs into the total project costs to be recouped through the anticipated revenue stream. This "purchase-and-lease" arrangement has been deemed the most practical method for parcel assemblage.

In other situations, the government can elect to acquire the necessary parcels directly from the property owners. The total cost of these assemblages is converted into a lease with title remaining with the government and the developer paying rent on an annual basis. Title to lands acquired for public use in England can only be transferred by an act of Parliament.

Compensation for Land Acquired for DBFO Projects

The Land Compensation Acts of 1961 and 1973, the Compulsory Purchase Act of 1965, and the Planning and Compensation Act of 1991 are the four principal pieces of legislation that deal with government acquisition of land deemed to be in the public interest. These acts stipulate that residential property must be purchased from owners at "fair market" value.

Homeowners In recognition that displaced homeowners can suffer personal distress and inconvenience when forced out of their homes, the government provides "disturbance compensation" in the form of moving expenses and other remuneration, distinct and separate from other forms of compensation. To qualify for disturbance compensation homeowners must have maintained the property as their principal residence for at least one year prior to being displaced.

The government also has legislated a provision called "blight notice," which deals with the difficulty of obtaining a fair market value for property when potential buyers are aware of the impending compulsory purchase action. If the owner of residential property attempted to sell that property on the open market but could not find a buyer except at a price below market value, and the seller can present a case that prospective buyers were aware of the government's intention to acquire the property, the homeowner may qualify for blight notice compensation.

Another provision of the act offered assistance to the handicapped. If the homeowner's residence had been modified or adapted for occupancy by a disabled person, the Authority will increase its compensation to include costs to modify a similar dwelling.

Commercial Owners In the case of commercial property owners, not only would fair market value compensation for the property be paid but reimbursement for reasonable expenses or losses incurred as a result of the impending move would be considered by the Authority as well.

Additional causes for compensation include the cost of alterations to equipment and furnishings, and the loss of value of inventory due to a forced sale or depreciation due to relocation.

Tenant Farmers A special provision in the act for tenant farmers includes reimbursement for the value of the unexpired term of their tenancy. There is also compensation for any loss or injury sustained in connection with the forced sale of livestock and the moving of livestock.

Noise Abatement The Department of Transport has regulations dealing with noise abatement for property owners who are not to be relocated and will be subjected to increased noise levels during construction and afterward when the proposed roadway or bridge is operational.

The structures that qualify for noise abatement retrofitting include residences, apartments, senior citizen residences, nursing homes, and residential accommodations in educational establishments. In order to qualify for assistance, these structures must be located within 300 meters (984 ft) of the newly constructed or relocated roadway.

Only certain rooms within these buildings are eligible for government-paid noise abatement work. Living rooms and bedrooms that have "one or more facades" with windows and doors exposed to the project are included. The government definition of a "living room" can include a dining room and a study, but not a kitchen area. Authorized noise abatement work includes secondary glazing over existing windows and the installation of a vestibule and/or double doors at exterior entrances.

THE GREEN PAPER AND USER TOLLS

A May 1993 document entitled *Paying for Better Motorways,* commonly known as the "Green Paper," dealt with toll roads and the manner in which private developers could receive revenue under the DBFO concept. The Green Paper advocated toll roads in order to achieve five basic objectives:

1. Provide a source of financing to improve the motorway network and avoid the traffic congestion that is endemic with economic growth.
2. Provide the impetus to increase private-sector investment in the operation and expansion of the nation's motorway network.
3. Encourage innovative methods to increase the efficiency of the motorway network, such as offering flexibility in toll rates with off-peak usage incentives.
4. Transfer the cost of motorway construction and operation to the users of these roadways.
5. Improve the competitive position of rail and other public transportation facilities, as users of these facilities traditionally pay for those services but motorists have not had to pay to use the roadways.

The Green Paper also addressed the question of the design of highways and related structures; it concluded that bidders on DBFO project must follow the scheme defined by the governments, but would be permitted substantial design and refinement latitude.

The document emphasized that the Department of Transport would seek to transfer all construction risks to the private sector "leaving no scope for claims against the Secretary of State as highway authority." An unusual provision proposed by the Green Paper was that the Department of Transport should consider making a lump sum payment to the developer when it reached substantial completion of the project.

It was thought that this would provide an added incentive for the concessionaire to complete the project expeditiously.

Secretary for Transport John MacGregor addressed the British House of Commons in December 1993 to review the government's plan for promoting new highway construction. According to MacGregor, Britain's motorway network carried 15 percent of all passenger car traffic and 30 percent of the country's truck traffic. Preservation and expansion of the highway system was critical to the nation's economic prosperity. He referred to the three methods of revenue collection proposed by the Green Paper:

1. *Conventional toll collections through the construction of toll plazas and toll booths*. The Green Paper had ruled out this method because conventional tolling leads to congestion at the toll booth and subsequent traffic delays.
2. *The issuance of permits as an interim measure until a fully electronic system can be installed*. This was also discouraged because it was estimated that the cost to provide such an interim program would be more costly than the revenue it produced.
3. *A fully electronic tolling system whereby an electronic tag attached to each vehicle would react to signals as the vehicle passed roadside beacons*. Tolls would be automatically calculated and charged to the user's account.

MacGregor advocated the third option, and suggested to the House of Commons that industry should be invited to demonstrate the practicality of electronic tolling on a trial basis.

The next year, the Department of Transport hosted an April 1994 conference attended by 250 companies from around the world on Britain's proposals for motorway toll technology. MacGregor invited attendees to demonstrate the practicality of electronic tolling on a trial basis using existing technology. With the results of these tests, he proposed requesting that Parliament enact legislation to allow electronic tolls to proceed on the country's motorways:

> Charging on the motorway network will be introduced as soon as Parliament has approved the necessary legislation and as soon as technology is right. Clearly this proposal has unleashed something that is of international interest. There are some remarkable opportunities opening up, both in this country and further afield. I look forward to the benefits that they will bring for the motorist and for society.

However, as of mid-1995 neither a trial basis nor a final decision of the type of electronic tolling had been established. Instead, the government reverted to the use of "shadow tolls" to initiate the DBFO projects.

The Shadow Toll

Shadow tolls are a means to collect revenue based on the actual number of vehicles using the motorway after it is constructed. Vehicle counters, such as induction loops embedded in the roadway, count the vehicles traversing the toll road. Each vehicle that passes over these loops is counted; because these loops can discern the distance between vehicle axles, they can therefore differentiate between automobiles, light trucks, and heavy trucks.

For traditional tolls on new roadways, the proposed concessionaire must develop projected traffic flow indirectly based on demographics of the area and actual traffic flow on the existing nearby roadways servicing the area of the proposed new highway. This data serves as the basis for determining a reasonably priced toll charge and creates the revenue stream for the project's pro forma. When construction is completed and the road opened to the public, tolls are collected by the concessionaire directly from the users of these roadways. Shadow tolls take a slightly different approach. The concessionaire develops the toll rates based on the same data as traditional tolls, but does not collect from the drivers of vehicles as they exit the highway. Per-vehicle tolls based on the actual vehicular count are instead collected from the government.

Critics of the shadow toll say that users feel as though the road is a freeway because they pay no tolls; the government will ultimately pay more for the roads than if they were built with Ministry of Transport funds. Advocates of shadow tolls claim that this system invites developers to propose desirable highways, thereby attracting vehicular traffic and the revenue source required to create a viable project.

Shadow tolls had been proposed to—and rejected by—the Department of Transport for years. But in 1994 Public Transport Minister John Watts suggested that the use of shadow tolls would provide the proper incentive for private investment in new roadway construction. The system is currently being implemented by the government in the preparation of several DBFO projects.

THE SKYE BRIDGE IN SCOTLAND

The Isle of Skye, largest of the Inner Hebrides off Scotland's northwest coast, has the irresistible lure that only a beautiful, unspoiled area can have, and one that recalls the flight of Bonnie Prince Charles to Skye in 1746. The population of Skye in 1990 was 8,920. Kyle of Lochalsh, the mainland approach to the island, has a population of 864. Kyleakin, the small town across the straits can count 510 as its permanent residents.

The ferry service between Kyle of Lochalsh and Kyleakin is operated by Caledonian MacBrayne (Cal-Mac) and operates 363 days a year, and since April 1991 for 24 hours a day. New ferries introduced in April 1991 carry up to 36 cars. During the high season (May to September) the fare structure for automobiles is £4.50 (U.S.$7.65). During low season (October to April) fares drop to £3.70 (U.S.$6.29), and purchasers of a book of 10 tickets receive a discounted fare of £2.12 (U.S.$3.60).

Summer traffic is very heavy and the 4,000 hotel rooms and board and beds on the island are booked solid from May to September. It is then that the ferries can't handle this influx of vehicles and cars at idle speed clog the streets of Kyle of

Lochlash polluting the environment as they inch toward the ferry. A bridge, some government officials reasoned, would reduce pollution, provide for more island accessibility and allow day trippers to come and go at ease. And the bridge would stimulate commercial and light industrial growth in an area where it is sorely needed. It seems small wonder then that a £23.6 million (U.S.$40 million, 1991) bridge would be proposed across the narrow strait where a 24-car ferry now takes 10 minutes to traverse.

The Request for Tenders

In 1975 the Inverness County Council engaged JMP Consultants Ltd. to keep a "watching brief" on the Skye Bridge; the ensuing report indicated that the addition of another ferry would be more cost effective than building a bridge. The second most desirable option was to build a bridge by the year 1993. A subsequent study in 1985 examined the socioeconomic impact of a Skye Bridge and found that 70 percent of the residents were in favor of the structure, with only a small minority in the tourist industry opposed on the ground that the bridge would spoil the island and make it less attractive to tourists.

In the meantime two British construction companies, Trafalgar House and Morrison Construction, suggested to the government that construction of a privately funded bridge was feasible. The Scottish Office Roads Directorate in 1989 advertised for applicants for a DBFO bridge across the strait to the Isle of Skye. These qualification-type applications were to be received by that office not later than noon, December 22, 1989.

Three successful applicants were to be selected and would be issued a tender invitation by early 1990. Submission of tenders was set for the autumn of 1990 with a concession award to be made in December 1990. The applicants would be requested to comply with the following criteria:

1. There were to be two elements to the proposal: the bridge itself and the approach roads connecting the bridge to the roadway network on either side. The Roads Directorate would contribute £6 million (U.S.$10.68 million, 1988) toward the cost of the approach roads which would be adopted on completion and maintained by the government.
2. All costs would be recovered through tolls levied by the concessionaire. If traffic flows were less than projected there would be no adjustment in the 20-year concession agreement, but if revenues exceeded the concessionaire's forecast, the length of the concession agreement could be reduced (sort of a "tails I win, heads you lose" situation).

A proper site investigation was required and the Scottish roads Directorate devised a novel plan to obtain one at no cost to the government. The qualification documents stated that each one of the three short-listed tenderers would collaborate on obtaining a site investigation survey. Costs would be shared among the three, and the successful tenderer would reimburse the other two on receiving provisionary award of the concession.

The government provided the applicants with traffic studies from 1971 through July 1989 and also presented traffic flows for each four-week period for the calendar year 1988 to demonstrate how most traffic was concentrated in the period between May and September. This study revealed that between 1978 and 1988 car and caravan traffic increased 60 percent from 221,400 vehicles to 365,700. Commercial traffic during that same time frame increased 40 percent from 20,300 vehicles per year to 29,500. Fifteen percent of yearly auto traffic occurs in July and the low of 3.4 percent takes place in January.

The Short-List Criteria

The applicants had to meet stringent criteria to make the short list:

- Describe their experience during the past 10 years in the design of the substructures and superstructures of major bridges.
- Describe their experience during the last 10 years in the design of major foundations (not necessarily of bridges).
- Give examples of the major foundations (not necessarily of bridges) they have constructed in the last 10 years.
- Give examples of single carriageway trunk road schemes with an overall cost of at least £3 million (U.S.$4.65 million) that they have constructed in the last 10 years.
- Describe any relevant experience they would expect to bring to bear in the operation of (a) a tolled crossing and (b) bridge maintenance.
- Submit detailed staffing charts and audited financial statements for each member of the group.

The consortium's proposed toll rate could not exceed the current ferry charges by more than the retail price index (RPI) for each year of the 20-year concession period. The fare rates for the ferry vary; 90 percent of 1991 fares collected were for automobiles, for which there are high and low season rates. Commercial vehicles are charged a special rate based on length of the vehicle plus VAT, and pedestrians and motorcycles travel free.

After the Directorate had received and analyzed all the applications, the short-listed tenders selected were Skye Bridge Ltd. (Bank of America, Miller and Dywidag), Skye Crossing Ltd. (Morrison Construction Ltd. and Barclays de Zoete Wedd Ltd.), and the Skye Bridge Consortium (Trafalgar House Construction Ltd. and the British Line Bank). Skye Bridge Ltd. (SBL) was the conditional successful tenderer.

Although the government's contract would be with the developer, SBL was going to enter into a contract with Miller–Dywidag Joint Venture (MDJV) for the construction of the crossing under the terms and conditions listed below:

1. The original tender from MDJV to the developer: £22.25 million (U.S.$37.27 million)
2. During negotiations, several minor amendments were made bringing the sum to: £22.63 million (U.S.$37.9 million)
3. As a result of requirements of the National Trust for Scotland (NTS), modifications were made increasing costs to: £23.65 million (U.S.$39.46 million)
4. The Scottish Office contributed £8.02 million for the approach road and for the NTS modifications, therefore the net capital construction cost to be funded by tolls: £15.54 million (U.S.$26.03 million)

   (Note: All costs are quoted at the tender base exchange rate as of July 1990.)

According to Gary N. MacDonald of the National Roads Directorate's Policy, Finance and Strategy Division, the method of calculating the end of the concession period when tolls will be removed would be based on the total cost of the project to include design, construction, operation, and maintenance costs together with finance charges all expressed at a base price of December 1990, discounted at 6 percent real. The significance of this date is that it marked the end of the last full calendar year prior to publication of the draft toll order in 1991.

According to MacDonald, the final sum referred to as the required net present value (NEV) is £23.64 million (U.S.$39.597 million), and when the aggregate revenue collected (discounted at 6 percent real to December 1990) reaches this total the tolls will be removed.

The maximum length of the concession period would be 27 years, but with projected increases in traffic MacDonald believes that the crossing could be paid for within 14 to 17 years.

This was the deal that was finalized between the government and SBL. Of course, everything depended on the outcome of the public hearings scheduled for January 28 through February 7, 1992.

The Public Hearings

Most residents on both sides of the crossing were in favor of the crossing, and the Highland Regional Council was certainly in favor of it and listed the many advantages that would accrue to the area:

- The bridge would boost the tourist-related economy of Skye because it would make the island more accessible.
- Fish farmers in the area who export their output would gain by reducing the uncertainty of ferry journey times.
- Elimination of queuing benefit residents of Kyle of Localsh and make conditions in their village much more pleasant.

There was also the financial condition of the Cal-Mac Ferry company to consider in this equation. Although Cal-Mac's costs and revenues on individual routes were confidential, their 1990 annual report which was the last one available at the time of the public inquiry revealed that services to Skye and the Small Isles incurred an operating deficit of £168,000 (U.S.$281,400). Cal-Mac receives a subsidy designed to meet the company's deficit incurred while operating a network of services to 23 different islands; the government surmised that this subsidy might have to be increased in 1995.

The objections raised by various individuals ranged from effect on marine life to the inability of large vessels to use the Kyleakin Harbor to ensuring that otters are protected by being diverted away from the approach roads. Sir Iain Noble, speaking for the Royal Fine Arts Commission for Scotland at Skye, was quoted as saying that he had always been "skeptical about the bridge and its cost to the public purse." In his view it was an engineer's exercise and the substantial cost to the taxpayer (of the approach roads, etc.) would prevent realization of more useful projects in Skye and elsewhere. He went on to say that many would regret the loss of mystique of isolation and that tourist traffic might not increase as was being suggested. Sir Iain disputed the claims about the bridge bringing prosperity to Skye and offered that the cheapest solution might be to increase the efficiency of the existing ferry service; he thought that the introduction of larger ferries would turn many townspeople away from the bridge concept.

At the conclusion of these hearings, when all of the evidence had been collected and weighed, the Highland Regional Council recommended that the Skye Crossing should proceed.

The Construction Construction began in mid-July 1992 and the project was completed in September 1995. With the bridge open, the ferries and their 39 employees will be departing. They will be missed by many who treasure the seclusion of prior days. A poem written by Pittendrigh Macgillivray (1856–1938) entitled "Come With Me" seems to convey the feeling of the area—will its mystique remain or be lost to progress?

O come with me where the sea-birds fly
Remote and far by the Isle of Skye
Away with the winds a-sailing!
were dreams are the gifts availing—
Will ye come with me?

THE SECOND SEVERN CROSSING AND APPROACH ROADS

The existing bridge across the Severn River separating England from Wales opened in 1966. In just four years, it carried 275 million vehicles. Between 1980 and 1990 alone, traffic flows had increased by 63 percent and congestion during summer months made travel difficult at best.

The Department of Transport began to study these problems in 1984 and in 1986 announced its intention to build a second bridge some 5 kilometers (8 miles) downstream from the existing bridge. In addition new approach roads were to be built connecting to the existing M-5 motorway (Figure 8.1).

The valley of the Severn estuary is predominantly rural and the environmental impact of a second crossing would be under close scrutiny. The area has been designated a Site of Special Scientific Interest (SSSI) and supports an internationally important bird population. Within the estuary the SSSI recognizes the role of plant and insect life, both of which depend on sufficient quality and quantity of water.

Adjacent to both the Avon and Gwent approach roads, two small grassland areas of ecological interest existed, one at each approach. Because the government recognized that these areas would be lost during construction, the proposing consortiums would be required to provide extensive landscaping and drainage ditches for new wildlife habitats. Studies to ensure the best location for the new bridge took place between 1987 and 1990, and after a series of public hearings the final route was established.

Severn River Crossing PLC

The Department of Transport announced its intention to seek a DBFO project and as such invited four groups to present bids on that basis in May 1989. In 1990 the Secretary of State for Transport announced the successful bidder was Severn River Crossing, a joint venture between Britain's John Laing PLC and France's GTM Entrepose with Cofiroute designated by the joint venture as the toll operator.

The agreement between the government and Severn River Crossing PLC was signed in April 1990 and ratified in October of that year. The project required Parliament's approval and a bill was introduced into the House of Commons. Royal ascent was received in November 1991, and the 30-year concession period began in 1992, signaling the date when the joint venture could begin to collect tolls from the existing bridge.

As part of the bid requirements the successful tenderer must assume operations and maintenance of the existing bridge, and with the ability to collect tolls on that crossing as well as the new one a revenue stream would be established.

The Crossing Design and Construction

The £30 million (U.S.$549 million, 1992) Second Severn Crossing is scheduled for completion in 1996. The total crossing will be 5.1 kilometers (3.16 mi) in length. The design was completed by the engineering firms of Sir William Halcrow & Partners Ltd., headquartered in Swindon, and GTM's subsidiary Société d'Études et d'Équipements d'Entreprises in Paris.

The three viaduct spans, each 98 meters (321 ft) long, required modular deck construction consisting of twin matchcast concrete box girders. Two giant gantries, each 233 meters (764 ft) long, are required to handle the balanced cantilever construction.

The cable-stayed design utilizes four precast concrete cross girders set on cast-in-place concrete pylons. Caissons for the pylons were precast offsite and trans-

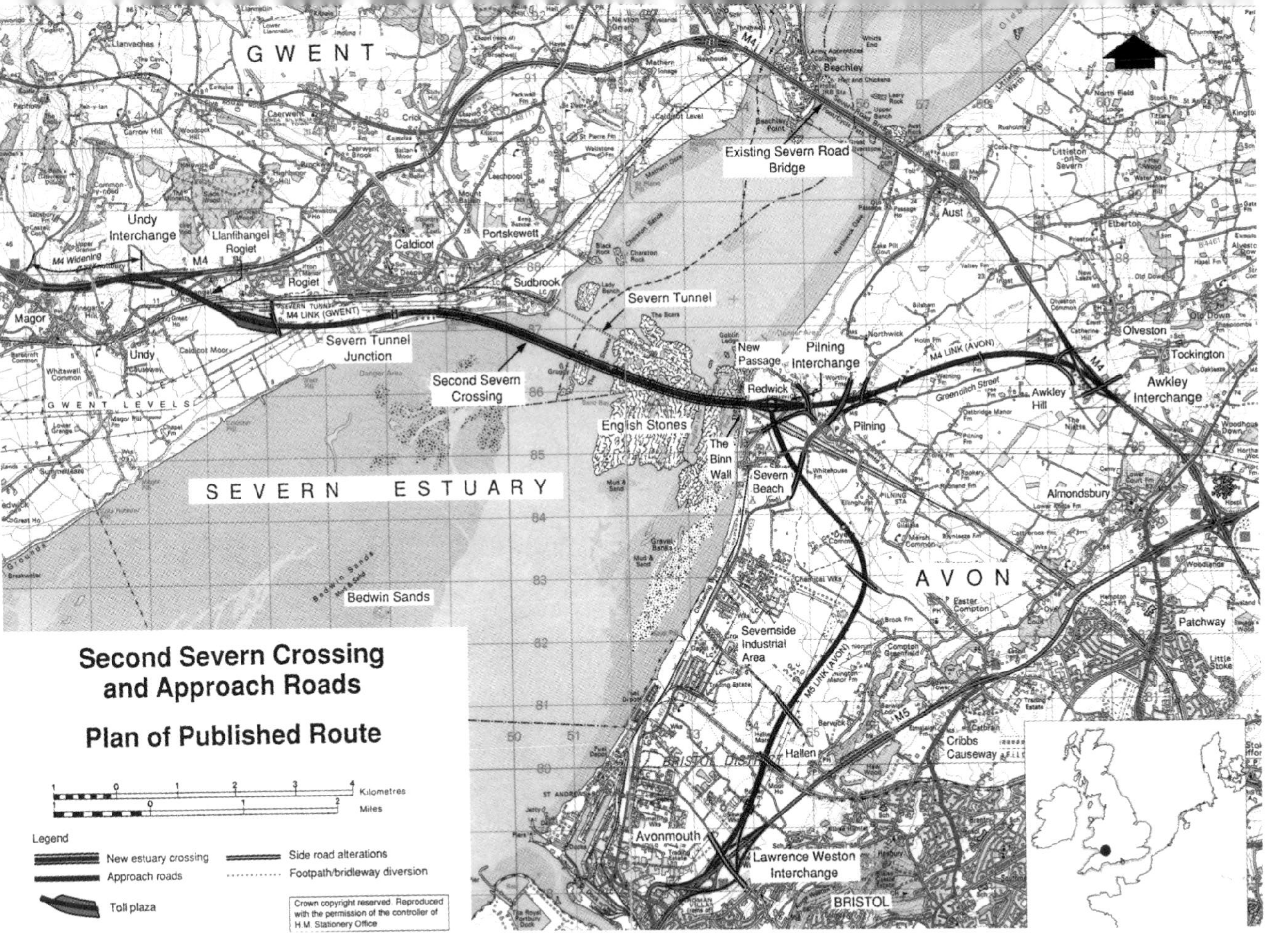

Figure 8.1. Route of the Second Severn Crossing.

ported to the bridge location by truck where they were lifted into place by barge-mounted cranes. The contractor then cast the concrete topping.

Safety Concerns In June 1994 one of the crabs that ran along the top of one of the gantries collapsed under load, causing a 200 tonne (221 ton) concrete unit to crash on the top slab of one girder below. Although minimal damage occurred and the cause of the accident was quickly ascertained and corrective action taken, exposure to construction risks took on added importance.

Weather Delays Weather is a risk faced by many contractors, but Severn Crossing PLC has had to deal with some unusual weather conditions at the estuary site. The contractors were coming out of their third on-site winter as of January 1995; December of 1994 had been especially tough as gale force winds whipped around the site. As the John Laing and GTM project director at the time noted, "The crucial time as far as the weather is concerned is the next three months and the period between October 1995 and April 1996, so we're keeping our fingers crossed."

Wind-Screening High winds plague the Severn River, but motorists will feel safe when the aerodynamically tested windscreen has been set in place. Drivers of slab-sided trucks and boxy vans will especially appreciate this shield which will prevent the bridge from having to close down when winds exceed 45 mph. The new wind-shield, which was thoroughly tested at the government's Road Research Laboratory, consists of 450-millimeter (17.7-in.) thick aluminum panels supported by 3-meter (9.8-ft) high aluminum posts. The panels have a gap every 450 meters to relieve the windloading.

By late March 1995, all foundations had been filled with reinforced concrete, and erection of the piers continued until June 1995. The schedule, weather permitting, anticipates linking the deck from the Gwent pylon and the Avon portion by December 1995. The Second Severn River Crossing joins the already completed Dartford River Crossing and the Skye Bridge in Scotland as the third DBFO venture successfully negotiated by the Department of Transport.

THE NEXT GENERATION OF DBFO PROJECTS

The Developer Survey

The Private Finance Unit of Her Majesty's Treasury conducted a survey on the proposed private finance initiative among the construction and financial community to obtain their views of how bid proposals should be structured in anticipation of preparing future bid documents. On March 1, 1994, the responses to this survey were published.

The Construction Sector

1. Respondents stated the importance of limiting the bidder's short list to a maximum of three to four consortiums.

2. Respondents felt that reimbursement of costs for proposal submissions was warranted given the complex nature of the proposal. These suggestions included reimbursement of 100 percent of the bid (tender) cost if the project was scrapped by the government after all bids had been received.
3. A two-stage bidding process was suggested by a number of respondents. The first stage would be an outline-type proposal whereby respondents would be short listed. The second stage would deal with detailed costs and design criteria.

The Financial Sector

1. A firm commitment to the project on the part of the public sector is essential for the concessionaire to prepare the most competitive proposal.
2. Limited participation in the bidding process is desirable. Many preferred a maximum of three to four tenderers; however, some respondents favored selecting two from a "serious" short list of perhaps six bidders.
3. Lump sum reimbursement might be considered as compensation for the unsuccessful bidders.

Proposals for Four New Projects

On August 17, 1994, John Watts, Minister of Railways and Roads announced the first four DBFO projects for concession periods ranging from 15 to 20 years. The Department also announced its intention to pay shadow tolls to concessionaires based on the number of vehicles actually using the roads.

The Department would be seeking prequalification submissions for the following road schemes:

1. An M-1/A-1 Yorkshire Link Road that bypasses Leeds. The total length of the new roadway would be 18.75 miles (11.63 km) and the budget was listed as £184 million (U.S.$313.5 million).
2. A-1 widening between Alconbury and Peterborough, budgeted at £140 million (U.S.$239 million).
3. An A-419/A-417 trunk road between Swindon and Gloucester, including bypasses for Cirencester, Latton, Stratton, and the dual-laning of the existing road between Stratton and Nettleton in the Cotswolds. The overall length would be 334.4 miles (21 km), and it was budgeted at £34 million (U.S.$57.75 million).
4. A-69 Haltwhistle Bypass construction, plus management and maintenance of the A-69 between Carlisle and Newcastle. Budgeted cost of the project was £9.7 million (U.S.$16.5 million).

In litigious societies, bid quantities varying substantially from actual quantities encountered raises the potential for future claims; however, this does not appear to present a problem in Great Britain. These DBFO offerings did stress that the Department did not have a settled view of the extent of its obligations and allocation of risk; rather, these would be firmed up as the project in question unfolded. The Department advised that it would purchase the land required for the new roadways rather than leave the assemblage and land purchase negotiations between the Department and the DBFO consortium.

The bid packages included project traffic flows both for the year of the proposed opening and 15 projected years, but it was assumed that bid tenderers would perform their own traffic flow assessments as the government had indicated its intention to pay shadow tolls based on actual traffic flow rather than projected.

Some of the packages also included government information of use to specific projects. In the case of the M-1/A-1 Yorkshire Link Road, for example, the Department of Transport included calculations for site cuts and fills indicating that 4.3 million cubic meters (5.7 million cubic yards) of excavate would be created, and 3 million cubic meters (4 million cubic yards) of fill would be required, thereby leaving a surplus of 1.3 cubic meters (1.7 cubic yards). The outline specifications for this project also indicted that a number of potential disposal sites for the surplus excavate have been identified by the Leeds City Planning Department.

The Status of the Projects

Of the 10 proposed DBFO tenders that were scheduled for the 1994, two have not gone forward. The M-1/M-62 Lofthouse Interchange Diversion project has been linked with the M-1/A-1 Yorkshire Link Road and the A-1 Hook Moor–Bramham upgrading work. The A-1 Alconbury–Peterborough upgrading contract 2 has also been grouped in with this contract. Three other tenders have been combined into one, and the A-69 Haltwhistle Bypass job has proceeded to tender as a separate project.

Of the 11 other DBFO tenders planned for release in 1995 and 1996, all are being held in abeyance as of the summer of 1995.

A spokesman for a leading accounting firm in Great Britain accused the government of trying to offload too much risk on the private sector thereby reducing enthusiasm for many of these projects. The government replied that it was not looking to transfer excessive risk, and with no targeted limits set on return on investment or profits the private sector should be very receptive to these private finance initiative projects.

The Environmental Movement Ian Sharman of the Department of Transport said the environmentalist movement in Great Britain has somewhat dampened the private sector's enthusiasm for these projects. While not commenting on the movement itself, Sharman stated that the costs of the delays created by environmental groups such as "Reclaim the Streets" are not lost on prospective developers. One road project, M-11, has been stalled for years as environmentalists occupy house after house scheduled for demolition; the demonstrators occupying these abandoned homesites refuse to leave. The last house to be demolished before construction of M-11 work could commence did not occur until June 21, 1995, once cherry pickers manned by police officers had physically separated protesters from the house's roof.

INFORMATION AND PREQUALIFICATION REQUIREMENTS FOR DBFO ROAD PROJECTS

Ian Sharman of the Department of Transport's DBFO Division indicated in June of 1995 that the government was proceeding cautiously on this private-sector DBFO highway system. The prequalification assessments by the Department of Transport for potential DBFO participants would be based on their compliance with the following criteria:

1. Providing proof of a technical, financial, and economic track record for projects of a similar nature.
2. Providing proof of experience and expertise in the undertaking and procuring of design, planning, construction, maintenance, and operation of projects of a similar nature.
3. Providing a track record of an ability to finance project of a similar nature.
4. Documentation of an understanding of the skills required for major highway construction work as well as the maintenance and operation of same.
5. Demonstration of experience in the method by which design will be approached and how construction, quality control, safety, and effective management will be controlled.
6. Providing evidence of an ability to coordinate and work with the Department's traffic network.

As the proposed bidders would be a consortium or group, the Department required specific information on the composition of the consortium or group:

- A description of the group, the name of each member, and their relationship to other members of the group.
- Detailed descriptions of the major projects undertaken by each member of the group over the past five years, including participation in limited recourse projects, concession-type projects, and road building and maintenance projects.
- Lists of the educational and professional qualifications of each member of the consortium's managerial staff, with clear identification of which individuals in each organization will be responsible for project implementation in each of its phases.
- The percentage each member of the group will likely attain.
- Details of how the internal relationship between group members could change during the design, construction, and operation phases of the project.
- Latest audited statements from each member of the group.
- Statement of annual dollar volume on previous road and/or DBFO type projects during the past three financial years.
- Statement of any contingent liabilities or losses that would require disclosure in accordance with International Accounting Standard 10.
- List of any contracts where there has been a failure to complete; or where there

have been claims for damages; or where damages have been covered within the last five years; and where the value of said contracts or the amount of damages was greater than £4.85 million (U.S.$8.5 million).

- Descriptions of arrangements whereby the financial interest of any contractors who are members of the group will be separated from the DBFO group.
- Information regarding prospective bidders' plans to arrange financing for both equity and debt (though a financing plan is not required).

THE TRAFFIC CONTROL AND COMMUNICATIONS PACKAGES

When the DBFO tender was issued for the M-1/A-1 Yorkshire Link Road (Lofthouse to Bramham) project in early 1995, the invitation to bid on the traffic and communications work was offered shortly thereafter in February 1995.

According to Stephen Raggett of the Department of Transport's Highways Agency, this and other traffic and communications DBFO contracts are under negotiation as of August 1995. Because these requests for proposals contain performance specifications, no decision had been made relating and communications technology and techniques to be employed by the government.

The System Requirements

A model contract accompanied the M-1/A-1 Yorkshire Link Road traffic and communications RFP. Selected sections from this contract provide an indication of the type of systems being considered by the Department of Transport.

The entire traffic and communications system will encompass:

Emergency telephone and other communications facilities.

Automated systems for traffic monitoring, incident detection, and signaling.

Closed-circuit television surveillance.

Motorway matrix signals.

Variable message signs, including enhanced message signs.

Fog detection and reporting equipment.

A maintenance and support system must provide for unforeseeable events such as severe weather, extended periods of power loss, equipment failures under warranty, and major equipment losses due to lightning strikes, fire, or accidents.

Repair personnel must be available on a 24-hour, 365-day basis, and response time must be within two hours of a service call. Any repair problems must be corrected within 24 hours after the DBFO becomes aware of the problem.

The Traffic Control System

The measuring equipment for traffic control must be capable of (1) counting the number of vehicles passing over the roadway each hour in each lane; (2) distinguishing between the large trucks referred to by the Department as "high gross vehicles" (HGVs) and other types of vehicles; and (3) monitoring the speed of each vehicle in one of four rate-of-speed categories: under 40 kph (24 mph), 41 to 70 kph (25 to 43 mph), 71 to 100 kph (44 to 62 mph), and over 100 kph (over 100 mph).

The accuracy of the measuring equipment is not to exceed the following limits:

As to the count of number of vehicles in each lane, plus or minus 3 percent.

As to the total number of vehicles in each lane, plus or minus 1 percent.

As to the total number of vehicles in each direction in each speed band, plus or minus 5 percent.

The traffic census equipment must be able to discern 11 categories of vehicles in each direction with the following degrees of accuracy:

Motorcycles	+10 or −20 percent
Automobiles	+1.5 or −1.5 percent
Bus and Coach	+15 or −15 percent
Light Vans	+10 or −10 percent
Seven types of HGV vehicles	+10 or −10 percent (each)

This equipment must operate in accordance with these parameters for not less than 95 percent of the time during each calendar month.

The Department in their RFP indicated their preference for the inductive loop system of vehicle detection and a general performance specification was included:

1. Vehicles shall be detected by means of subsurface inductive loops and suitable traffic counting equipment for counting vehicles recorded by the loop array.
2. The loop array shall consist of loop pairs set a suitable distance apart to ensure the required level of accuracy against an enumerated count.
3. The loop configuration shall ensure that vehicles straddling lanes are counted only once.
4. Installation of the loop shall comply with all Department specifications.

THE CHANNEL TUNNEL RAIL LINK (CTRL)

Although highway DBFOs may be proceeding cautiously, the government's desire to improve high speed rail lines is shifting into overdrive. In 1986 the nationalized British Rail Company, concerned about the possibility of substantially increased traffic when the Channel Tunnel opened, concluded that it had sufficient capacity to handle international traffic, at least for the foreseeable future. One year later, in August 1987, the Kent Impact Study questioned BR's view that no new construction or equipment was needed until well into the twenty-first century. This study indicated that on the completion of the Channel Tunnel constraints would indeed be placed on existing facilities considerably before the time suggested by British Rail in 1986.

A subsequent study by British Rail in July 1988 concurred with the Kent Impact Study, and BR published a report expressing the need for extra routes and terminal capacity. A selected route was suggested by British Rail in 1989, but in 1990 the Secretary of State for Transport was not satisfied that the best solution was at hand. BR had by this time formed a new subsidiary, Union Railways, to develop a new route in detail within the corridor defined by the government; in March 1993, the Secretary of State announced the route that would be selected for further study. The final route of the CTRL is shown in Figure 8.2.

On November 11, 1993, the Secretary of State for Transport announced the government's intention to transfer a portion of the national railway system to the private sector. In order to accomplish these goals and begin to denationalize the railway system, ownership of British Rail and United Railway Limited Lines would require an act of Parliament, and a Hybrid Bill was introduced to the body in 1994. This type of bill is required when government wishes to obtain Parliament approval for construction of a significant public works project. A hybrid bill is so called because it combines legislation needed by both the public and private sectors. Royal Ascent was received in 1995.

The private sector would be invited to design, construct, finance, and maintain a 108 kilometer (67 mi) high-speed rail service between London and the Channel Tunnel at a cost of £2.7 billion (about U.S.$4.185, 1993), and the ownership of the existing European Passenger Services Ltd. (EPSL) line would be transferred to a private investor. The government request for proposal included new construction, refurbishing of existing rail facilities, and the transfer of title of sophisticated rolling stock and related equipment.

The government announced its three program objectives:

1. To double the existing capacity of four trains per hour (three in the evening) available for international travel between London and the Channel Tunnel.
2. To reduce the journey time of these services between London and the Channel by a half-hour to 40 minutes or less. (The present journey from Waterloo Station, London to the entrance to the Channel Tunnel via Eurostar takes about 1 hour and 10 minutes.)
3. To provide greater capacity and reduced journey times for domestic passengers and to regenerate traffic in the East Thames Corridor.

The Basis for the Competition and Reward of the Concession

The successful tenderer on the CTRL project will be granted ownership of the European Passenger Services Ltd. (EPSL), currently a government-owned organization responsible for the operation of international passenger services utilizing the Channel Tunnel. The services to be transferred to the private sector include continued operations on existing lines between the Tunnel and London and beyond, and an

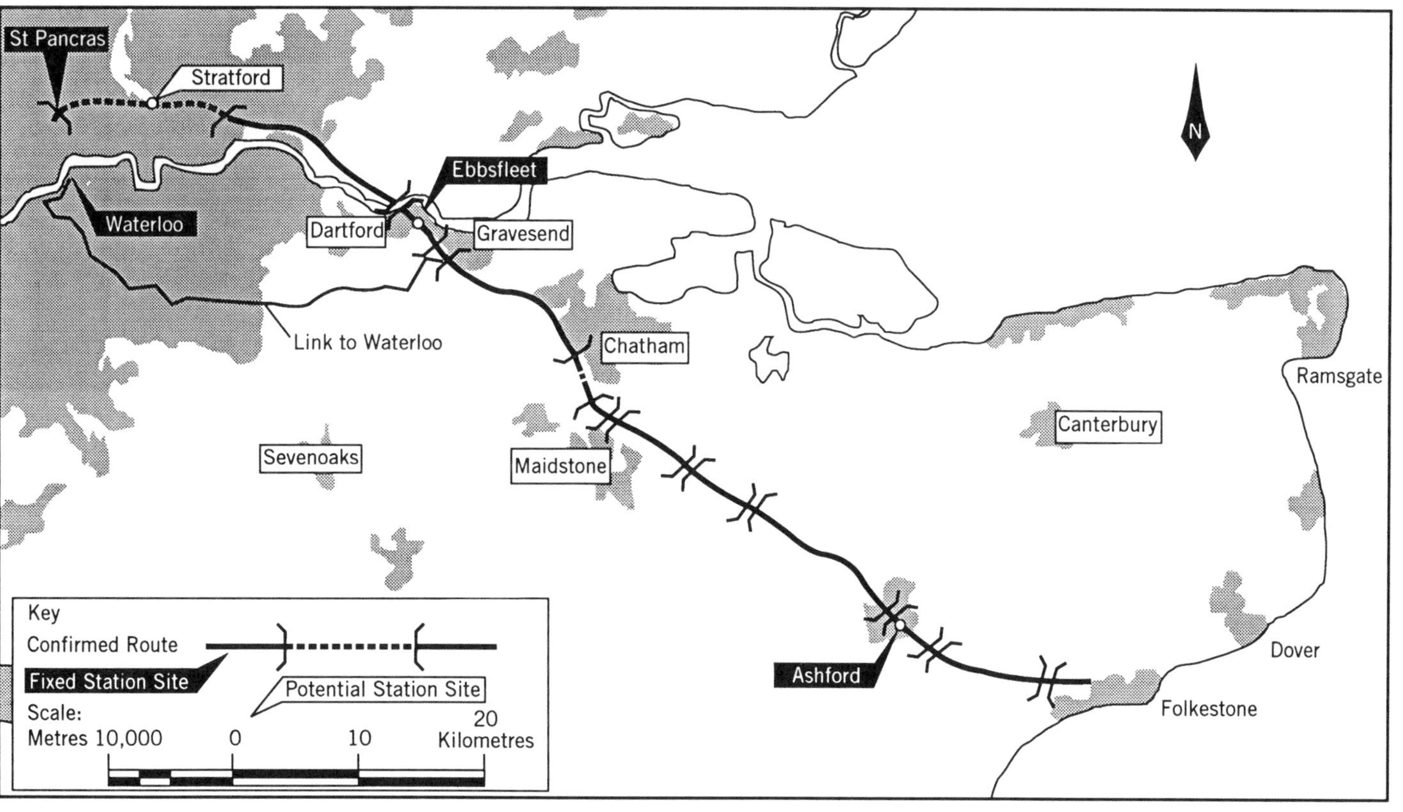

Figure 8.2. Route of the Channel Tunnel Rail Link from London to Folkstone.

obligation to construct new lines which together would form the Channel Tunnel Rail Link (CTRL) network.

The Department, recognizing the costs incurred by developers in preparing a bid as complicated as this one announced in the tender invitation documents that it was prepared to contribute toward the bid costs. Unsuccessful tenderers would receive a contribution from the government if their proposals were of a quality commensurate with the nature of the project. The government would contribute £1.5 million (U.S.$2.55 million) or 33 percent of their costs, whichever is less.

Nine organizations responded to the Department's prequalification invitation. The government would proceed to select the consortium that offered the best deal, requiring the least amount of money from the government and the most risks assumed by the consortium. On June 10, 1994, the following four groups were qualified by the government.

1. Eurorail CTRL: composed of British International Cable Company (BICC), a conglomerate that owns Balfour Beatty, a major British construction firm; GEC; HSBC Holdings; National Westminster Bank; Seeboard; and Trafalgar House.
2. Green Arrow: composed of Hochtief, Costain, and Seimens.
3. London and the Continental Group: composed of Arup, Bechtel, Blue Circle, Halcrow, National Express, Virgin, and Warburg Union.
4. Link: composed of EG, W. S. Atkins, Holzmann, Mowlem, Spie Batignolles, Taylor Woodrow, and Kaiser Engineers.

The four tenderers were requested to submit full proposals to the Department in March 1995, and as of mid-1995 no final selection had been made.

The Project

The Maastricht Treaty envisions a three-stage creation of the European Economic and Monetary Union (EMU) in 1991, 1994, and 1997. Great Britain, having joined the European Union late, was anxious to do its part to further the goals of the treaty. The Maastricht Treaty provides for the European Union to support the financial efforts made by member nations for projects of common interest. The CTRL project, one of the key missing links in the European Commission's proposal for a high-speed TransEuropean rail network, was listed as one of the top 26 priority projects set forth in the European Commission's white paper on growth, competitiveness, and employment.

The European Community had already provided £21 million (U.S.$35.7 million) for transportation studies and funding is expected to increase by 1997.

The European Investment Bank (EIB) has an interest in the CTRL project; its role is to finance projects that contribute to the balanced development of the European Union. Thus, it will not come as a surprise when the successful tenderer on the Channel Tunnel Rail Link project seeks some sort of financial assistance or guarantees from EIB.

The British government listed as its objectives in the creation of this public/private venture:

1. To transfer ownership of the government-owned European Passenger Services Ltd. and Union Railways Limited rail systems to the private sector.
2. To ensure that the Channel Tunnel Rail Link is designed, constructed, operated, and maintained safely and satisfactorily.
3. To minimize the public-sector contribution required and minimize the extent of the risk borne by the public sector in this new railway link. The government was not ready to transfer all risk to the private sector, but was willing to assume certain risks.

EUROPEAN PASSENGER SERVICES LTD. (EPSL)

The future private owners of the EPSL complex will assume operations of the new Channel Tunnel passenger train, the Eurostar. The new Eurostar passenger service is a bright new entrant to the railway system of Great Britain and the continent. The passenger cars are sparkling new, and the futuristic-looking locomotive seems to be straining at the leash when parked at Waterloo Station.

An Average Trip on the Eurostar

Passengers can purchase a ticket in advance or at the ticket booths in the vaulted ceiling addition to Waterloo Station (Figure 8.3). As they wait for the announced train departure, they can pick up a free baggage cart and keep it until their baggage is loaded on the train.

Courteous attendants, fluent in several languages, patiently answer travelers' questions. A small restaurant is near the approach to the trains in case someone wishes to purchase a refreshment before departure.

Once the luggage is stowed in an open baggage compartment on each train or placed in the overhead rack above the seat, passengers settle into comfortable fabric-covered seats while attendants walk the aisle answering questions and offering assistance.

As the train pulls slowly out of the station right on schedule at 8:23 A.M. and rolls out of London toward the Channel, an attendant guides a refreshment cart up the aisle offering cold and hot drinks and various snacks for sale.

The train, moving at 60 to 70 mph, takes 1 hour, 5 minutes to arrive at the Channel Tunnel. Speeding up slightly after entering the Channel Tunnel, Eurostar emerges at Calais, France, about 20 minutes later and begins to pick up more speed as it starts its race toward Paris. An announcement on the car's public address

Figure 8.3. View of Waterloo Station looking toward the ticket booths and the trains beyond.

system notes that passengers may wish to change their watches, advancing them 1 hour for the time zone change.

About 10 minutes after the train has left Calais, the public address system advises passengers in both English and French, "Ladies and Gentlemen, Eurostar has just reached its top speed for the trip: 306 kilometers per hour or 180 miles per hour. We hope you enjoy the trip." Even at this speed there is very little noise—only a moderate high-pitched whine somewhat reminiscent of a turbine.

Not only is Eurostar fast, but it is extremely comfortable: not only is there none of the clickety-clack that U.S. train travelers hear, but there is no swaying from side to side. A half-filled glass of water on a passenger's food tray barely moves at all as Eurostar travels at top speed. And just about right on time at 12:35 P.M. Paris time (11:34 London time), Eurostar pulls into Paris Nord Station.

Cross Channel Traffic

During the past 30 years, cross channel commercial traffic by railship-catamaran services has declined from its postwar peak of 1 million journeys. With the advent of relatively inexpensive flight services and the increase of car/coach ferry services, the use of railways to and from the English Channel has been in steady decline. Taking all methods of transportation into account, however, over the past two decades cross channel travel has seen a threefold increase. As of 1995 more than 60 million people each year made this journey.

With the opening of the Channel Tunnel and the commencement of Eurostar, high speed travel and high quality service to the Continent should increase dramatically over the next five years. The new TransEuropean railway system—which is operated jointly by EPSL, the French national railway (SNCF), and the Belgian national railway (SNCB)—is expected to carry more than 12.5 million passengers a year by 1997 on the London to Paris, London to Brussels route. It is estimated that traffic will grow to 15 million trips by the year 2000.

Services to be Provided

When the full services of Eurostar are operational, there will be one train per hour to each destination for most of the day, and more will be added during times of peak load. Each train will carry 800 passengers.

Some trains will stop at a new international station in Lille, France, affording passengers access to connecting service to the French TGV network.

All three railway companies, EPSL, SNCF, and SNCB, will be integrated under the common Eurostar name. Decisions on frequency of service, staffing of trains, catering and food services, pricing and marketing programs will involve agreement among these three companies. A train leaving Waterloo Station may be composed of French rolling stock, with a British driver and a Belgian crew.

The Asset Base of the New European Passenger Service Ltd. Business

EPSL will own either the freehold or a very long leasehold on the assets, associated physical plants and equipment, exclusive of the freehold land that was valued at £350 million (U.S.$595 million) as of April 1, 1994. These assets include

- Waterloo International Station and the adjacent land to nearby York Road.
- London's North Pole International Depot.
- The relevant lands at Ashford, near the M-20 between London and the Tunnel, where a new international station is planned to be completed in 1995.
- A site in Manchester where a depot is being constructed in 1995 to service some of the EPSL trains traveling north from London.
- Rolling stock consisting of 22 complete sets of trains and 22 diesel and electro-diesel locomotives valued at £370 million (U.S.$629 million), £108 million (U.S.$183 million) of which is being financed through leasing arrangements.

With the announcement of the successful bidder by the Department of Transport at the end of 1995, the CRTL project will be off and running.

The Eurotunnel Relationship

Although the Channel Tunnel Rail Link has no contractual connection with the Channel Tunnel, the two projects are inextricably linked together. Because the Channel Tunnel costs had skyrocketed, Eurotunnel filed an action in mid-1995 to increase its rail user fees by 55 percent. Eurotunnel has requested arbitration in this matter and a decision is expected in late 1995.

The outcome of the arbitrator's decision may have significant effect on the already high proposed cost of the Channel Tunnel Rail Link project, and a detrimental effect on CTRL's anticipated revenue.

Eurotunnel is not only turning its guns on the railroad to seek additional revenue but has filed a court action requesting an end to indirect government subsidies to cross channel ferries and cross channel airlines, subsidies that take the form of exemption from excise duties and sales taxes until the end of the twentieth century. According to Eurotunnel these subsidies amount to about £150 million (U.S.$255 million) per year.

THE CROYDON TRAMLINK

Over the past 40 years, more than 350 light rail systems have been built throughout the world—in cities such as Baltimore, San Diego, Portland in the United States, in Grenoble and Strasbourg, France, and Calgary, Canada. These public systems offer reliable, relatively low-cost transportation alternatives. Great Britain is moving to add another dimension to light rail construction: private sector involvement in which a 99-year concession will be awarded to the successful bidder. The Croydon Tramlink project is a local public transportation scheme linking New Addington, Beckenham, and Wimbledon, all London suburban areas, with the busy shopping and commercial center of Croydon (Figure 8.4).

Outside the central London area, Croydon has the largest shopping center and the highest concentration of office floor space in the southeastern part of Britain. This growing prosperity brought with it more and more traffic jams, and the obvious solution, widening existing roadways, met with large-scale public opposition. The construction of a light rail system seemed to be the most viable people-mover option.

Some 17 kilometers (28 mi) of the proposed line will utilize previous light rail alignment, 3 kilometers (4.8 mi) will be installed on streets running through busy thoroughfares, and 8 kilometers (13 mi) will run over land for which new rights-of-way must be acquired. Thirty-two stops have been planned along its length with an average distance of 900 meters (2,952 ft) in between; however, one stop between Lloyd Park and Addington Palace is at 3,200 meters (1.98 mi).

Each tram will consist of five motorcar sections with a total length of 30 meters (98 ft) and will be known as a Supertram. The cars are electrically powered, utilizing a 750 volt DC overhead supply line. The cars will run on welded track and will be outfitted with resilient wheels that absorb shock and minimize noise and vibration, much like the Metro underground and elevated system in Paris where cars with resilient polymer wheels substantially reduce noise pollution. A study conducted in the French town of Nantes, where a system similar to Croydon Tramlink has been installed, affirmed that these electrically powered cars show relatively low noise levels as measured by decibel readings and as compared to other types of surface transportation (Figure 8.5).

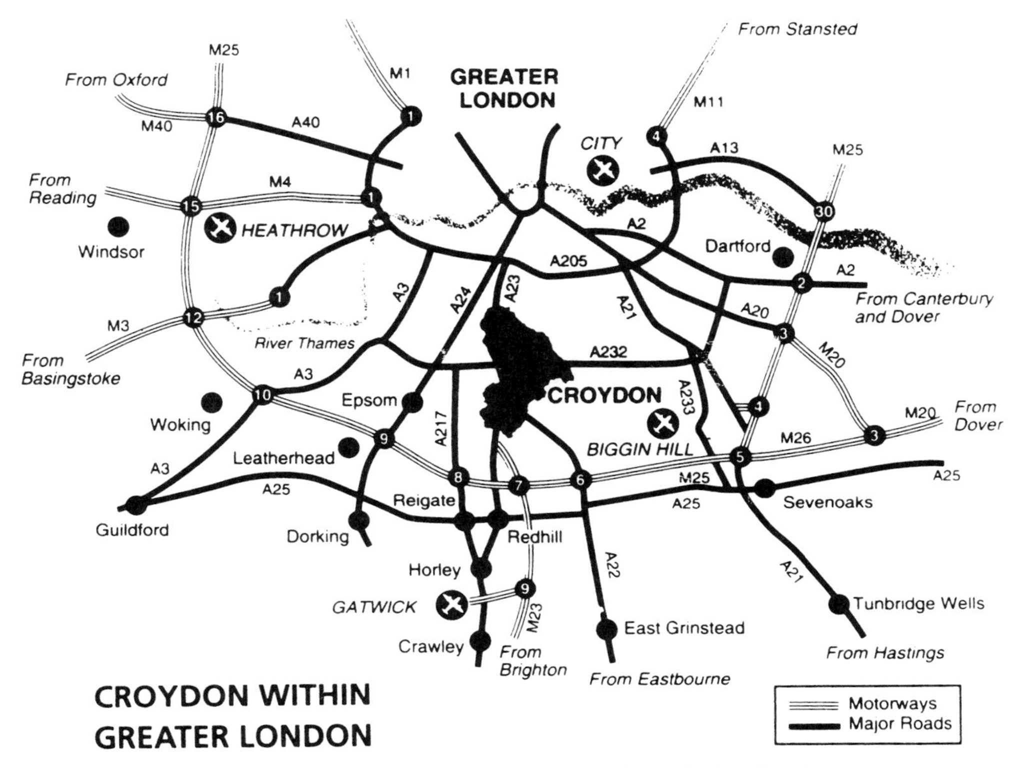

Figure 8.4. The Croydon Tramlink connecting suburban London.

The Notice to Tender

In an unusual move, the government sought consultation with private contractors to prepare the Tramlink's project specifications prior to the issuance of a notice to tender. Contractors Tarmac, AEG, and Transdev were invited to provide technical and commercial light rail expertise to develop the specifications, but were advised by the government that their input would not give them a favored position when bids were ultimately received and evaluated.

In December 1994, the Secretary of State announced that he would be setting aside funds to provide a public-sector grant for this project; however, the size of the grant could not be determined until bids were received and an evaluation of the winning bid indicated the amount of funding required.

On May 30, 1995, the Department of Transport announced its intention to solicit bids for the 28 kilometer (17.5 mi) light rail network. The cost of the project was estimated to be between £150 million (U.S.$255 million) and £160 million (U.S.$272 million).

The project was bid on June 3, 1995, under Britain's Private Finance Initiative as a design, finance, construct, operate, and maintain project over a concession period of 99 years from the date of award. The closing date for bid submissions

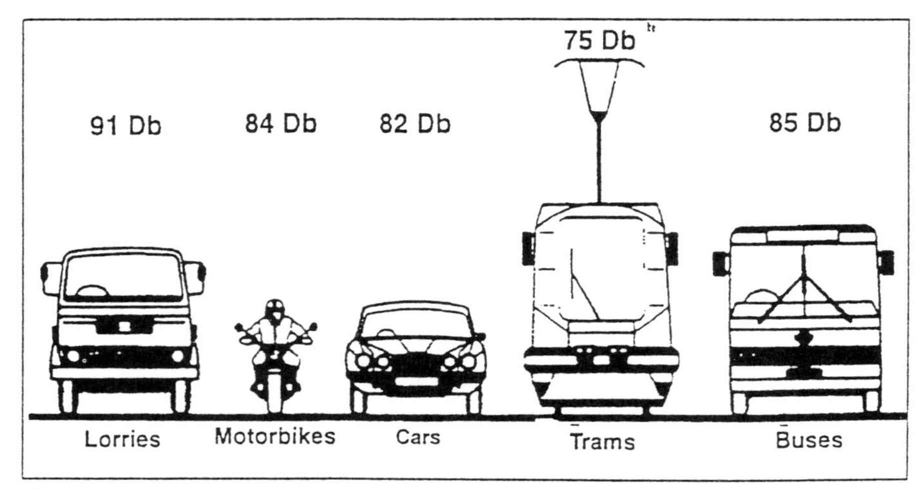

This study made in the French town of Nantes shows relative noise levels in decibels. The readings were taken from a distance of seven and a half metres with traffic moving at 40 kilometres per hour.

Figure 8.5. Comparison of decibel levels of various types of surface transportation.

was September 15, 1995, after which a short list would be created and invitations to tender sent out to the short-listed consortiums. The Department of Transportation planned to evaluate these bids, make its selection, and award a concession agreement in early 1996. The government expected the Tramlink to be operational by 1999.

In order to qualify for the short list, applicants must have included the following information:

- Financial strength and capability of raising the necessary funds.
- Experience in the operation and maintenance of a similar public-type concession.
- Proven ability to develop and manage a successful business.
- Ability to offer equipment, proven in a service environment, that has characteristics similar to the proposed Tramlink.
- Technical capability to manage, design, and construct the works.
- Organizational, managerial, and technical capability to commission a multidisciplinary system and have multiple interfaces with external systems.
- Ability to offer a safe, reliable, and environmentally sensitive system.

A concession award will take into account the following criteria, ranked in descending order of importance:

1. Ability to comply with the mandatory requirements of the project through the 99-year concession period.

2. The financial strength of the bidder and its capacity to raise the required funding.
3. The level of need for U.K. public-sector funds for the project and the degree of risk transfer from the public sector to the private sector.
4. The ability to develop and manage a successful business.
5. The use of proven and established technology to construct, equip, and operate the works.
6. The quality of the service to be provided in excess of the minimum requirements of the performance specifications.

The Light Rail System Advantage

When complete, Great Britain's Tramlink will become a model for other light rail systems throughout that country and the world. Coincidental with the issuance of the request for proposal for the Croydon Tramlink, the state of New Jersey was asking for prequalifying applications by July 31, 1995, from proposers for the first design, build, operate, maintain light rail system in the United States. This $1 billion, two-phase project will provide for a 20.5 mile (33 km) light rail system to service various Hudson and Bergen County communities.

Advocates and supporters of light rail systems justify their choice by pointing out the qualities of other available forms of transportation:

Conventional railway systems. Suitable for carrying large volumes of commuters on long, fast journeys into a major metropolitan area.

Buses. A good means of transportation for use by passengers in low-density areas traveling short distances.

Automobiles. Used most where other means of transportation are not readily available or for what the British refer to as "kiss and ride"—delivering one's partner to the rail station.

Bikes and Walking. Healthy, nonpolluting choices for short distance journeys in good weather.

Light Rail. Provides relatively fast and rapid transportation for large numbers of people where demand is insufficient for construction of subways or heavy railway systems.

Great Britain has embarked on a partnership with the private-sector to regenerate its cities, to link with the European economic community, and to provide its citizens with safe, efficient, cost effective, and environmentally sound transportation systems. Kenneth Clarke, Chancellor of the Exchequer when the Private Finance Initiative was issued in 1993, believes in its merits:

> I believe that the Government's private finance initiative will soon be recognized to be a radical and far-reaching change in capital investment in public services. The private finance initiative offers real benefits: to the private sector it offers more work

> and profit; to the taxpayer an efficient and economical way of delivering more quickly the improved infrastructure and better facilities we all want to see.

The government of the United Kingdom is well on its way to fulfilling this vision.

The Queen Elizabeth II Bridge—also known as the Dartford River Crossing—which opened to the public in September 1991 to resounding success, and the Birmingham North Relief Road, which is nearing construction start, are discussed in more detail in Chapter 9.

9

TRAFALGAR HOUSE, A WORLD LEADER IN PUBLIC/PRIVATE-SECTOR WORK

Great Britain's Trafalgar House Public Limited Company (PLC) is an international, diversified business group based in London. The company was founded in 1956 and since that time has expanded greatly by internal growth and selectively acquiring other companies. In 1986 they acquired John Brown Engineers, ranked by *Engineering News Record* magazine as the top international design firm in 1994. The Cunard steam line with a fleet of eight luxury ocean liners including the Queen Elizabeth II was acquired in 1971 and the Royal Viking Sun cruise ship in 1994. Other acquisitions have included Ideal Homes, one of Britain's largest housebuilders; the Cementation Company, a civil and specialist engineering firm; Cleveland Bridge; Trollope & Colls, a builder and contractor; and several other engineering and construction-related firms.

Turnover for the six months ending in March 1995 amounted to £1.7076 billion (U.S.$2.9 billion), slightly less than the similar period in 1994 when turnover was reported as £1.711 billion. However, the period ending in March 1995 was not so kind as far operating profit was concerned. Trafalgar House reported a six-month operating loss of £14.9 million (U.S.$25.3 million), a substantial change from a profit of $25.6 million posted for the six-month period ending March 31, 1994.

Since March 31, 1994, corporate policy dictated disposal of some noncore businesses, including four hotels and two machinery manufacturers. With top-ranking John Brown and Trafalgar House Engineering and Construction's placement as *Engineering News Record*'s number-one international contractor as of August 1, 1995, the company looks forward to profitable years ahead.

Trafalgar House has a great deal of experience in both build, operate, transfer (BOT) projects worldwide and the design, build, finance, operate (DBFO) projects presently in favor with the government in Great Britain. The public/private-sector projects in which they are currently involved include:

- Tate's Cairn Tunnel, linking the urban areas of Kowloon to the newly developed New Territories of Hong Kong. This project, constructed in 1991 by Gammon–Nishimatsu joint venture, is jointly owned by Gammon Construction Ltd., Trafalgar House, and Jardine, Matheson & Co., Ltd. of Hong Kong.
- The Queen Elizabeth II Bridge (Third Dartford River Crossing) spanning the Thames River linking Dartford in Kent with Thurrock in Essex. Completed in 1991 by Trafalgar House, this bridge is the most heavily traveled estuarial crossing in Europe.
- The Cikampek-Padalarang toll road in Indonesia, a 56 kilometer (34.7 mi) four-lane highway between the cities of Jakarta and Bandung. Trafalgar House is a joint venture participant in this project for which financial arrangements are scheduled to be completed in 1995 for a 1996 construction start.
- The Second Tagus River Crossing in Lisbon, Portugal. This Trafalgar House-led joint venture includes Campenon Bernard SGE and a number of leading Portuguese construction companies. Construction began on March 1, 1995, and is scheduled for completion in 1998 to coincide with the opening of Expo '98.
- The Birmingham Northern Relief road in England. Britain's first privately funded toll motorway is schedule to commence construction in 1997.

Trafalgar House has several other BOT and DBFO proposals in various stages of proposal presentation and is currently a key member of Eurorail CTRL, a consortium that is one of the short-listed bidders on the 80 mile (129 km) Channel Tunnel High-Speed Rail Link. This project, which is estimated to cost as much as £4 billion (U.S.$6.8 billion), will be a major link in the TransEuropean rail system operating out of St. Pancras Station to the Continent via the Channel Tunnel.

TATE'S CAIRN TUNNEL PROJECT IN HONG KONG

It is not surprising that one of England's largest construction firms would venture into a BOT project in the current Crown colony of Hong Kong where the British presence has been felt for several hundred years. The Tate's Cairn Tunnel project is a Gammon Construction Ltd. and Nishimatsu Construction Company Ltd. joint venture; Gammon Construction is owned equally by Trafalgar House and Jardine, Matheson & Company Ltd.

The Gammon–Nishimatsu joint venture was technically and financially very strong. The Gammon Group was established in Hong Kong in 1958; co-partner Jardine, Matheson & Company Ltd., founded in 1832, was the first company to be registered in Hong Kong and is now one of Asia's most respected financial institutions and trading companies. Nishimatsu, a Tokyo-based Japanese construction company, was established in 1974 and operates worldwide with annual sales volume of over U.S.$3 billion. Gammon and Nishimatsu had worked together in Asia for approximately 10 years prior to the establishing of this latest joint venture, but the Tate's Cairn Tunnel project, estimated at H.K.$2.15 billion (U.S.$307 million), was the largest they had undertaken together.

Traffic between Kowloon and Sha Tin in the New Territories had been increasing by 12 percent annually since 1977, and the existing Lion Rock Tunnel and Tai Po two-lane highway traffic were near capacity. In 1986, the Gammon–Nishimatsu joint venture submitted an unsolicited proposal to the government of Hong Kong to build, operate, transfer a tunnel linking the urban area of Kowloon to the New Territories beyond (Figure 9.1). This proposal for a new tunnel proved to be very attractive to the government of Hong Kong; however, the government decided that it could not accept the Gammon–Nishimatsu proposal without competitive bidding and issued a Request for Tender in May 1987.

Six consortiums responded to the project brief. After reviewing all the proposals, the government of Hong Kong awarded a concession agreement in February 1988

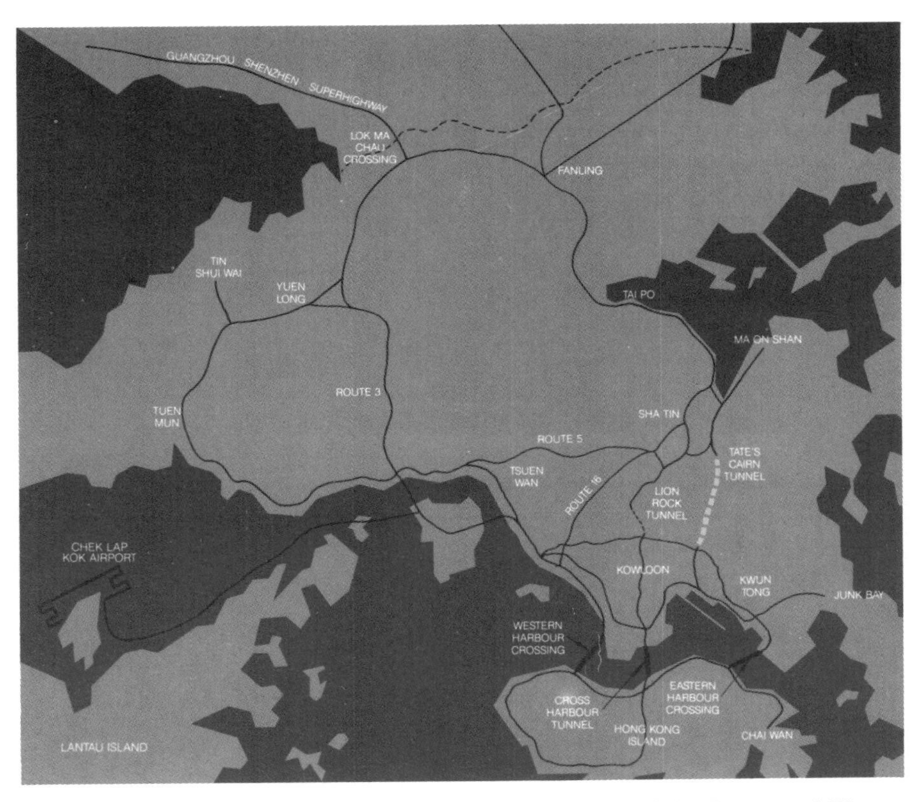

Figure 9.1. The route of Tate's Cairn Tunnel and the routes of the Eastern and Western Harbour Tunnels.

to the Gammon–Nishimatsu joint venture for the design, financing, construction, and operation of the Tate's Cairn Tunnel for a period of 30 years, after which title to the project would revert to the government.

The design team was led by Maunsell Consultants, Asia, and included Parsons Brinckerhoff Asia, Ltd.; Liang Peddle Thorp, a 100-year-old Hong Kong architectural firm; and MVA, a British transportation consultant.

During the final stages of negotiations with the government, the Gammon–Nishimatsu joint venture solicited additional equity partners and the founding shareholder participation was finalized as follows:

Nishimatsu Construction Co., Ltd.	37%
China Resources (Holdings) Co., Ltd.	24%
New World Development Co., Ltd.	24%
C. Itoh & Co., Ltd.	5%
Jardine, Matheson & Co., Ltd.	5%
Trafalgar House PLC	5%

Trafalgar House subsequently relinquished its shares in the Gammon–Nishimatsu joint venture.

China Resources is a general agent in Hong Kong for all Peoples Republic of China import/export corporations. China Resources owns many rail, ocean, and highway transport fleets; wholesale and retail service organizations; supermarkets; and various industrial and consumer goods manufacturing facilities.

New World Development is a publicly owned corporation involved in real estate and hotel investments. New World owns and manages the Hong Kong Convention and Exhibition Centre in collaboration with the Hong Kong Trade Development Council.

C. Itoh & Co., Ltd., established in 1858, is one of Japan's powerhouse trading companies. This company has offices in 85 countries, affiliations with 585 companies, and an annual turnover in the U.S.$100 billion range. The organizational chart of Tate's Cairn Tunnel is shown in Figure 9.2.

The Project

The project involved the construction of twin two-lane tubes through Diamond Hill with approach roads linking the tunnel with existing highways on the Kowloon and Siu Lek Yuen area of Sha Tin. The distance between portals to be 3,945 meters (12,940 ft or 2.45 mi) and for the majority of its length would be driven through sound granite. The width of each tube, as constructed, is 6,750 millimeters (22 ft), and the height to the precast ceiling slab and the overhead light fixtures is 5,100 millimeters (16 ft). On either side of each tube is an emergency walkway. A ventilation duct was created above the concrete ceiling slab and runs the entire length of the tunnel to prevent the accumulation of toxic fumes from forming within the tunnel; in case of fire, emergency fans are activated. Two exhaust towers are installed at quarter points along the tunnel, and electrical power is supplied from two

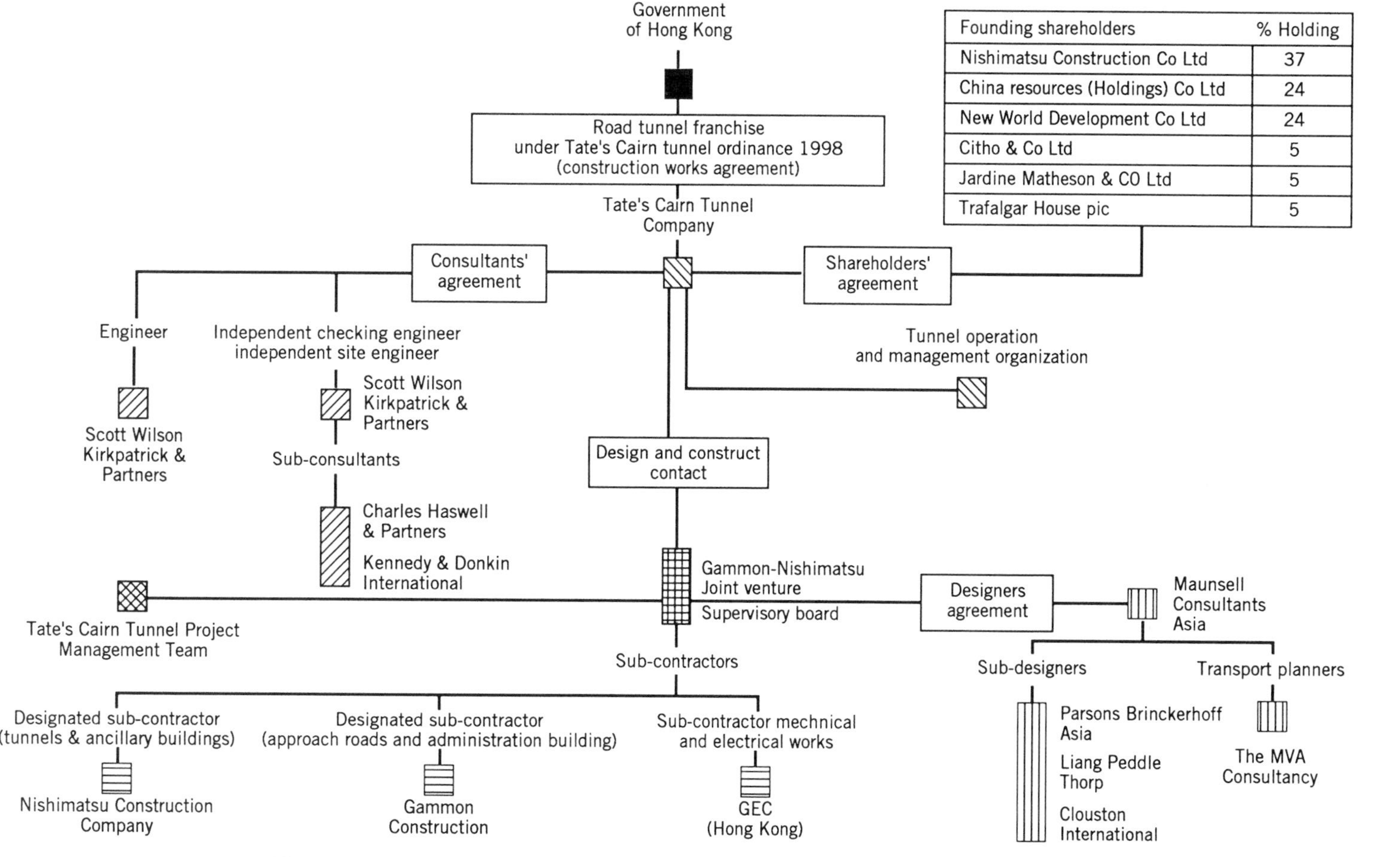

Figure 9.2. Organizational chart of Tate's Cairn Tunnel. *[Courtesy Trafalgar House PLC.]*

independent China Light and Power sources. Should one source fail, the other will be activated. Construction began in July 1988 with the final breakthrough occurring in August 1989. The tunnels opened for business on July 1, 1991, two months ahead of schedule and on budget at H.K.$2.15 billion (U.S.$307 million).

The Revenue Stream and the Financial Arrangements

The toll rate of H.K.$4 was established for private automobiles and H.K.$7 for lightweight commercial vehicles. Medium to heavy duty trucks and buses would pay H.K.$8. Annual increases in the toll rate structure must be submitted to the Governor in Council for approval.

During the first five years of operation a royalty payment of 2.5 percent of the gross operating receipts would be paid to the government; this payment increases to 5 percent for the remainder of the concession period.

Financing of the project included H.K.$600 million (U.S.$85.7 million) in shareholders' equity and a term loan of H.K.$1.55 billion (U.S.$221 million) provided by a consortium of 23 international banks; the lead banks are the Bank of Tokyo Ltd., China Development Finance Corporation (Hong Kong), and the Fuji Bank Ltd. Repayment of the term loan, according to the concession agreement, is to be achieved within 12.5 years from the date the tunnel is opened to the public.

Revenue received from toll collection was 10 percent lower than projected during the first year of operation, ostensibly because travelers were not familiar with this alternate route. But due to lower than anticipated interest rates the consortium was able to meet its projected income levels. During the second year of operation in 1994 the traffic count exceeded projections by 24 percent.

The latest year-end report for the Tate's Cairn Tunnel Company for the year ending June 30, 1994, is not a profitable one. Depreciation of the tunnel is being calculated on a straight line basis whereby the civil expenditure will be written of over the 27 years of the unexpired term of the franchise. The electrical and mechanical components, however, are being written off over a 10-year period.

Turnover for the year ending June 1994 was U.S.$20.8 million versus U.S.$17.98 million for a similar period in 1993. The period of the report indicates a loss of U.S.$18.99; and with an accumulated loss brought forward of U.S.$43.22 million, the accumulated loss carried forward is U.S.$62.21 million. For 1993 the loss of U.S.S$20.59 million resulted in an accumulated loss brought forward at that time to U.S.$22.74 million and a loss carryforward of U.S.$43.33 million.

Financing the Project

The H.K.$600 million (U.S.$85.7 million) shareholder equity was to be used during the first 15 months of the construction period. The term loan of H.K.$1.55 billion (U.S.$221 million) was to finance the remaining period of construction and cover operating expenses during the initial period of operation.

As of June 30, 1994, the Tate's Cairn Tunnel Company had drawn down

H.K.$1,372 million of their total long-term financing agreement which, based on rates of exchange of H.K.$7.72, equates to U.S.$1.77 billion. Interest on bank loans is running about U.S.$8.39 million per year and payments to the shareholders' memorandum account in 1994 were U.S.$9.65 million. This entry is footnoted by Price Waterhouse to call attention to an agreement between shareholders and the government of Hong Kong.

If interest on the bank loans exceeds 8.5 percent per annum before the date of the tunnel opening and 10 percent per annum thereafter, the shareholders will pay the banks for the excess. If the interest on the bank loans is less than 8.5 percent per annum before the date of the tunnel opening and 10 percent per annum thereafter, the company will pay the shareholders the difference.

THE THIRD DARTFORD CROSSING: THE QUEEN ELIZABETH II BRIDGE

The Queen Elizabeth II Bridge is unique in several respects. First of all, it is the busiest estuary crossing in Europe; second, it is anticipated that it will be paid for in 12 years—8 years less than the maximum period of 20 years permitted by the concession agreement.

Pinpoint equity financing was utilized on this project. Peter Goodin, managing director of Dartford River Crossing, was asked to explain this term: "We were able to arrange a series of institutional and bank financing for the £166.6 million (U.S.$283 million) project with only £1,000 (U.S.$1,700) paid in equity capital. That's why we call it pinpoint equity."

The Third Dartford River Crossing of the Thames River is about 16 miles (26 km) east of London and carries M-25 motorway traffic between Dartford in Kent and Thurrock in Essex. The first Dartford River crossing is a two-lane tunnel that opened in 1963 at the cost then of £13 million (U.S.$22 million). Five million vehicles used this tunnel during its first full year of operation and the joint board of Kent and Essex Counties decided that a second tunnel of equal capacity was required. The Second Dartford River crossing was begun in 1972. Due to unforeseen subsurface conditions construction progress was seriously delayed and completion of this tunnel did not occur until 1980 when its cost had risen to £45 million (U.S.$76.6 million).

These two tunnels handled about 30,000 vehicles per day and traffic flowed rather smoothly for a period of time; however, the 1980 start of construction of the M-25 London Orbital Motorway, with two section connecting to the M-11 to the north of the Thames and Swanley/Seven Oaks to the south, would place an immediate strain on these two tunnels when it was completed in 1986. It was estimated that the traffic count would increase to about 80,000 vehicles per day during the summer season, and the need for a Third Dartford River crossing became apparent. Traffic flows in 1991 would actually decrease in part due to people diverting away from the first two crossings because of the hour-long traffic jams that occurred on weekend; however, it was anticipated from the very start, based on traffic surveys conducted by the Kent county council, that a third crossing would be a successful venture.

About this time the government had changed its policy toward new highway construction allowing private-sector involvement. In the fall of 1985, the construction division of Trafalgar House was seeking new contracts and saw an opportunity to create a project. Enlisting the assistance of local landowners and merchant banker Lazard Frères, they presented a scheme to the government to design, build, finance, and operate a bridge across the river. It was envisioned that this project could be financed from toll income derived from its operation over a 20-year period.

The Department of Transport was impressed by the scheme proposed by Trafalgar House but deemed it advisable to seek competitive bids and prepared guidelines for such a tender. Eight consortiums submitted proposals for a Third Dartford Crossing—five were bridge schemes, two were submerged tube tunnels, and one was a twin-bored tunnel. After short-listing three tenderers, Trafalgar House was awarded the contract on September 29, 1986, on condition of Parliament's approval.

Government Approval of the Project

The transfer of ownership of the existing tunnels, assemblage of the necessary parcels of land for the project, and construction of the Queen Elizabeth II Bridge itself required the approval of Parliament. The power to allow a private company to collect tolls would also require government approval; thus, a hybrid bill was introduced in the House of Commons by the Secretary of State for Transport on April 1, 1987. Royal ascent was obtained on June 28, 1988, thus creating the Dartford–Thurrock Crossing Act of 1988. The preamble of the new law stated its purpose:

> An Act to provide for the construction of a bridge over the River Thames between Dartford in Kent and Thurrock in Essex and of associated works; to provide for the Secretary of State to be the highway authority for the highways passing through the tunnels under that river between Dartford and Thurrock and their approaches instead of Kent and Essex County councils; to provide for the levying of tolls, by a person appointed by the Secretary of State or by the Secretary of State [sic], in respect of traffic using the crossing; to provide for transfers of property and liabilities of those Councils to the person appointed and the Secretary of State and for the transfer to the Secretary of State of property and liabilities of the person appointed on termination of his appointment; to provide for the management of the crossing, including the imposition of prohibitions, restrictions and requirements in relation to traffic, and otherwise in relation to the crossing; and for connected purposes.
>
> Be it enacted by the Queen's most Excellent Majesty, by and with the advise and consent of the Lords Spiritual and Temporal, and Commons, in this present Parliament assembled and by the authority of the same.

The Secretary of State was to acquire all land and property for the project using the compulsory Purchase Act of 1965 if it became necessary to obtain the land or property by eminent domain. In turn, the acquired lands and property would be leased to the concessionaire along with any required easements and rights-of-way. The Act set forth the terms and conditions of the agreement between government and concessionaire including operation of the crossing, financial matters, administration of tolls, and regulation of traffic using the bridge and the approach roads. Although certain police powers such as the power to arrest were denied the concessionaire, the right to dun and recover uncollected tolls plus associated administrative collection costs were permitted.

The concession agreement provided for Trafalgar House to purchase the lease of the two existing tunnels for a sum approximating £43 million (U.S.$73 million), build a new cable-stayed 450-meter (1,476-ft) main span bridge across the Thames, and operate all three crossings for a maximum period of 20 years.

The newer of the two existing tunnels which had opened in 1980 had a latent defect—a badly deteriorated roadway surface would require replacement of a major roadway section. This was known when tenders were submitted, and Trafalgar House took it under consideration in their proposal. Peter Goodin said that repair work would commence in early 1996. They planned to saw cut the deteriorated sections out during the evening hours when traffic flow decreases so that they can close some lanes. After those portions of old roadway are removed, they will be replaced by sections of precast concrete. This will act as the base course of a new roadway with expansion joints every 100 meters (328 ft) and will allow reopening of the lanes the following morning. These precast sections would then receive a topping to provide for a smooth-wearing surface. Although the cost of this work would be funded by the government, Trafalgar House's contract with the government requires it to perform the work and supervise the construction.

The Construction of the Bridge

Trafalgar House Construction (Europe) Ltd. was responsible for the design and construction of the foundations and substructure of the Third Dartford Crossing. Their Cleveland Structural Engineering affiliate designed the steel superstructure in association with Dr. Ing H. Homberg and Partner for the cable-stayed portion.

The 2,872 meter (9,420 ft) bridge is composed of three components—an 812 meter (2,663 ft) cable-stayed bridge with a 450 meter (1,476 ft) main span, a 1,052 meter (3,450 ft) approach road on the Essex side, and a 1,008 meter (3,308 ft) approach road on the Kent side. The main cable-stayed structure is an aesthetically pleasing design with twin rectangular section pylons located on each side of the navigational span. The pylons are anchored to reinforced concrete piers that are designed to withstand the impact of a 65,000 tonne (78,325 ton) vessel at 10 knots.

The cable-stayed deck consists of longitudinal steel plate girders at the outer edges of the deck with intermediate longitudinal and cross girders that support an orthotic steel deck with a structural concrete overlay to complete the composite construction. Work on the bridge began in August 1988 and the bridge opened in October 1991, on time and on budget.

Project Funding

An act of Parliament was required to authorize the borrowing of the funds for the Third Dartford River crossing not to exceed £184 million (U.S.$312.8 million); £166.5 million (U.S.$283 million) was borrowed, of which £120 million (U.S.$174.25 million) was provided by the Bank of America. Interest on the syndicated bank loans are floating at a margin of between $\frac{3}{4}$ percent and $1\frac{1}{4}$ percent over prime.

There are four shareholders in the project:

Trafalgar House PLC	50.0%
Kleinwort Benson Ltd.	16.5%
Prudential Assurance Co., Ltd.	16.6%
Bank of America	17.0%

DART-Tag, England's First AVI System

The Dartford Crossing has eight lanes, four northbound and four southbound. There are 14 toll booths at the northern approach to the bridge and 13 booths at the southern approach. Figure 9.3 has views of the bridge and the toll booths from the London side of the crossing. Nine booths in each direction are manned and all lanes are equipped with an automatic vehicle identification (AVI) system known as DART-Tag, but only one lane in each direction is dedicated for AVI usage at this time. DART-Tag is produced by the Swedish firm SAAB Combitech; the toll equipment including the software is made by CSRoute, a French concern.

The Dartford Crossing was the first estuary infrastructure in the United Kingdom to use this AVI system and to date has proven to be very effective. The DART-Tag system allows motorists to pay in advance on a monthly basis for use of the bridge. When motorists' accounts are opened, they are provided with a transponder, approximately 2.5 inches (63.5 mm) square and 0.5 inches (12.7 mm) thick, that mounts inside the windshield. As the vehicle approaches the toll booth, the transponder is activated; the driver's account is debited, and the barrier gate rises to allow it to pass.

When the transponder activates the equipment in the AVI lane, a computer scans the customer's account. There is a set of miniature traffic lights at the AVI lane, and if the driver's account has the proper monetary balance a green light goes on and the barrier rises. If the account is getting low, an amber light appears and the barrier still rises; however, if the account has been exhausted, a read light appears and the barrier remains down. The barrier does not appear to slow down the passage of DART-Tag equipped vehicles as it takes only 7/10 second to open. The dedicated lanes using the DART-Tag system can process 800 vehicles per hour as opposed to the 400 vehicles per hour rate of staffed toll booths. For all lanes, the toll plaza has proven to be capable of processing a total of 11,000 vehicles per hour.

The toll barrier is a small price to pay to ensure payment. Public Affairs Manager Paul Emberley of Trafalgar House Corporate Development said that Dartford River

(a)

(b)

Figure 9.3. (a) London-side view of the Dartford River Crossing. (b) The toll booths for both DART-Tag and regular collection areas. (c) Close-up of DART-Tag toll booth with signal light to indicate account status. (d) Traffic flows smoothly to the entrance.

(c)

(d)

Figure 9.3 *(Continued)*

Crossing Ltd. does have the authority to detain toll jumpers, but does not have the power to arrest them. The local police must be summoned to arrest any vehicle operators avoiding paying the tolls.

A much larger concern involves vehicles with left-hand drive. In Britain, the steering column is on the right; on the Continent, its position is on the left as in the United States. Some of the traffic on the bridge is either coming from or going to the Channel Tunnel, and the traffic with left-hand drive emanating from the Continent cannot use the existing toll booths. These vehicles—mainly trucks—must stop, and the drivers leave their vehicles and walk to the booth to pay their tolls. But at present there is not enough traffic to warrant construction of a special lane for left-hand drive vehicles.

Traffic Flow

There are 146,000 vehicles using the crossing in both direction as of mid-1995. Total volume, from the bridge opening in 1992 to the summer of 1995, was 43 million and annual revenue has been about £48 million. Traffic count at the crossing has been increasing by approximately 7 percent per annum as opposed to a nationwide increase in traffic of 3 percent. Automobiles represent 75 percent of the daily traffic flow with twin-axle trucks contributing 15 percent and multi-axle trucks 9.5 percent.

Currently, 20 percent of all weekly traffic uses the DART-Tag system (up to 30 percent during rush hours); 50 percent of all truckers use the DART-Tag—which is beneficial because commercial toll collecting requires more time than do passenger car tolls. Application for a DART-Tag can be made by mail using one of the convenient account application forms shown in Figure 9.4. A deposit of £30 (U.S.$51) is required of an automobile owner to receive their DART-Tag transponder; owners of commercial vehicles must deposit £125 (U.S.$212). Robert West, secretary of the Dartford River Crossing Company, said there are currently 45,00 tagged vehicles, which translates to £1.475 million (U.S.$2.5 million) in advance deposits. This money is invested promptly to yield maximum return, which is then applied against the term loan.

There are four classes of vehicles as far as toll rate charges:

Motorcycles	30 pence (U.S.$0.50)
Automobiles	90 pence (U.S.$1.53)
Twin-axle trucks	£1.50 (U.S.$2.55)
Multi-axle trucks	£2.44 (U.S.$4.15)

DART-Tag customers receive a 7.5 percent discount rate on these toll charges. Vehicles towing trailers are charged twice the toll, and additional charges are also levied for abnormal loads and for the recovery of broken-down vehicles. According to Peter Goodin, toll rate can be increased annually but only to the extent of the increase in the retail price index (RPI), the British equivalent of a cost of living increase.

ABOUT DIRECT DEBIT

Direct Debit is a simple, reliable and economical way of paying your toll charges. We invite you to sign an Instruction to your Bank/Building Society authorising them to pay us variable amounts from your account at our request. You will appreciate that the Instruction needs to be for variable amounts because you may wish to alter your monthly payment from time to time.

Benefits to You

There are no cheques to write, no paperwork or postage and there is no queuing because Dartford River Crossing arrange payments for you with your Bank or Building Society.

The Direct Debit Guarantee

- This Guarantee is offered by all Banks and Building Societies that take part in the Direct Debit Scheme. The efficiency and security of the Scheme is monitored and protected by your own Bank or Building Society.
- If the amounts to be paid or the payment dates change, you will be told of this in advance by at least fourteen days as agreed.
- If an error is made by us or your Bank/Building Society, you are guaranteed a full and immediate refund from your branch of the amount paid.
- You can cancel a Direct Debit at any time, by writing to your Bank or Building Society. Please also send a copy of your letter to us.

Naturally, you are under no obligation to allow us to direct debit your account and other means of paying tolls continue to be available. However, we urge you to take this opportunity to simplify your payment arrangements by completing the Application Form and returning the enclosed Instruction as soon as possible.

Should you have any queries, please don't hesitate to contact us. You can telephone us on: 01322 280200

'CREDIT LOW' AMBER WARNING LIGHT

The DART-Tag system lane lights are set by default to give customers a 'credit low' amber warning light when the balance remaining on an account falls below £20. If you would like to specify your own credit low warning level please complete the box below and return it with your application form:-

I would like the 'credit low' warning light to operate when my account balance goes below:- £ : 00p (minimum £5)

Signature: ....

DIRECT DEBIT INSTRUCTION

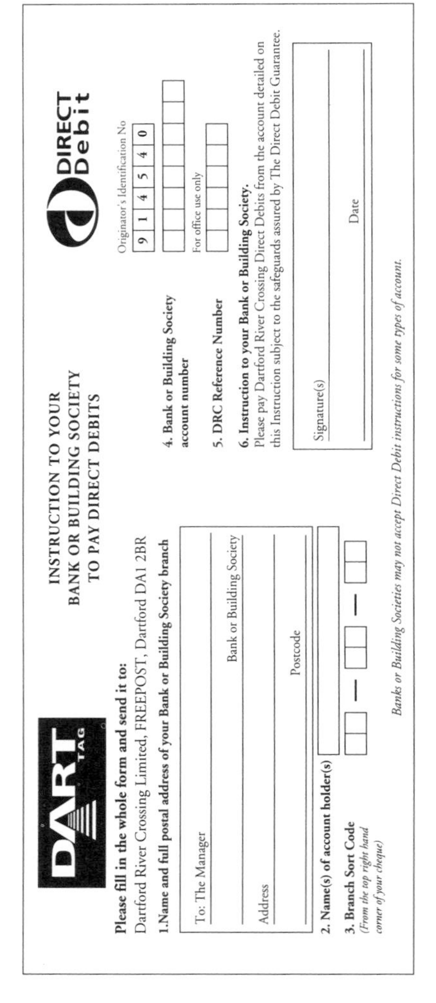

DART TAG

INSTRUCTION TO YOUR BANK OR BUILDING SOCIETY TO PAY DIRECT DEBITS

DIRECT Debit

Please fill in the whole form and send it to:

Dartford River Crossing Limited, FREEPOST, Dartford DA1 2BR

1.Name and full postal address of your Bank or Building Society branch

To: The Manager

Bank or Building Society

Address

Postcode

2. Name(s) of account holder(s)

3. Branch Sort Code *(From the top right hand corner of your cheque)*

Originator's Identification No 9 1 4 5 4 0

4. Bank or Building Society account number

For office use only

5. DRC Reference Number

6. Instruction to your Bank or Building Society.

Please pay Dartford River Crossing Direct Debits from the account detailed on this Instruction subject to the safeguards assured by The Direct Debit Guarantee.

Signature(s)

Date

Banks or Building Societies may not accept Direct Debit instructions for some types of account.

Figure 9.4. DART-Tag application form with toll rates and recommended initial payments. *[Courtesy Trafalgar House PLC.]*

Fleets of Land Rovers are utilized as emergency vehicles in case one of the many television cameras placed along the bridge observes a stalled vehicle or a motorist in trouble. These Land Rovers are also used to escort trucks with hazardous cargoes through the tunnels. When such trucks enter the approach road to the bridge and tunnels, they are shunted off to a special lane where they wait until a convoy of six vehicles is assembled. A Land Rover with flashing lights brings up the rear as the trucks proceed.

In 1990 a 2 million square foot shopping mall opened on the northeast end of the bridge, and people shopping at this huge retail center have added to the traffic flow across the bridge. Initially, 9,000 to 10,000 parking spaces were built at the mall, but this has been increased to 13,000; at times, according to Goodin, all of these spaces are filled. During school vacations or holiday seasons shoppers add to bridge traffic at about 10:00 A.M. and again between 4:00 and 5:00 P.M. when they are returning home from their shopping. Goodin said that another large retail shopping center is planned in the vicinity which will add even more traffic over the bridge.

The Payback Period

Trafalgar House projects that revenue in the year 2000 will be £75 million (U.S.$127.5 million), and with operating and maintenance costs of £15 million (U.S.$25.5 million) profits should be in the £60 million (U.S.$102 million) range. That is why payback is projected to occur long before the 20 years specified in the concession agreement and why Gooding refers to the project as "cash on the hoof."

As of September 30, 1994, the Dartford River Crossing Ltd. company has compiled some healthy financial statements as shown by its profit and loss statement (Table 9.1), cash flow sheet (Table 9.2), and borrowing statement (Table 9.3). In

TABLE 9.1 Dartford River Crossing Limited Profit and Loss Account

	Year Ended September 30, 1993 (£000s)	Year Ended September 30, 1994 (£000s)
Turnover—Continuing operations	40,952	44,849
Cost of sales	(22,382)	(27,641)
Gross profit	18,570	17,208
Administrative expenses	(2,267)	(2,393)
Operating profit—Continuing operations	16.303	14,815
Interest receivable	285	690
Interest payable and similar charges	(17,351)	(15,505)
Profit/(loss) on ordinary activities before taxation	(763)	—
Tax on profit/(loss) on ordinary activities	763	—
Profit on ordinary activities after taxation	—	—
Brought forward	—	—
Carried forward	—	—

The Company has no recognised gains or losses and accordingly a statement of total recognised gains and losses has not been presented.

No reconciliation of movements in shareholders' funds has been presented because there are no such movements.

Source: Trafalgar House PLC. with permission.

TABLE 9.2 Dartford River Crossing Limited Cash Flow Statement

	Year Ended September 30, 1993 (£000s)	Year Ended September 30, 1994 (£000s)
Net cash inflow from operating activities	31,598	33,423
Returns on investments and servicing of finance		
Interest received	266	709
Interest paid	(16,611)	(15,194)
Net cash outflow on returns on investments and servicing of finance	(16,345)	(14,485)
Taxation		
U.K. Corporation tax paid	(273)	—
Investing activities		
Payments to acquire tangible fixed assets	(1,065)	(2,774)
Bridge construction costs	(1,000)	(2,342)
Disposal of tangible fixed assets	11	10
Increase in current asset investments	(6,000)	(4,000)
Net cash outflow from investing activities	(8,054)	(9,106)
Financing		
Repayment of secured bank loan	—	(7,000)
Net cash outflow from financing	—	(7,000)
Increase in cash and cash equivalents	6,926	2,832

Source: Trafalgar House PLC., with permission.

1994, the company had £12 million (U.S.$18.6 million) in cash on hand or in the bank—a hefty £4.8 million (U.S.$7.4 million) increase over 1993. And the banks must also be happy given the interest rates of over 14 percent being paid to secured lenders.

When the bridge has been paid for, operation and toll collection activities will revert to the government; however, the existing act of Parliament also allows the company to continue with the concession for an additional 12 months to build up a fund for maintenance purposes. The managers of the Dartford project predicted that the government would enact some form of legislation to continue the tolls either by awarding another concession agreement or by assuming operations of the crossings itself. But no matter what course the future of the Third Dartford River crossing takes, Trafalgar House will have earned a fair profit on its construction contract and gained valuable experience in the successful operation of a complex BOT project.

TABLE 9.3 Dartford River Crossing Limited Borrowing Statement

Borrowings*	September 30, 1993 (£000s)	September 30, 1994 (£000s)
Debenture loans:		
Secured loan 14.041% 2004	30,000	30,000
Secured loan 14.016% 2008	34,000	34,000
Bank loans:		
Secured repayable in installments by the year 2000 at varying rates of interest	102,500	95,500
Total borowings	166,500	159,500
Less: Amounts falling due within one year	(7,000)	(19,500)
Amounts falling due after more than one year	159,500	140,000

Provisions for Liabilities and Charges		
Deferred taxation:		
Balance as of September 30, 1993	6,231	
Credited to profit and loss account	(1,405)	
Balance as of September 30, 1994	4,826	

Potential Amount of Deferred Taxation†		
Fixed asset timing differences	8,609	5,215
Other timing differences	(384)	(389)
Tax losses carried forward	(1,994)	—
	6,231	4,826

*The debenture loans and bank loans are secured by a fixed and floating charge over the undertaking, property, assets, and rights of the company.

†The full potential liability for deferred taxation has been provided by the Company.

Source: Trafalgar House PLC., with permission.

THE BIRMINGHAM NORTHERN RELIEF ROAD PROJECT (BNRR)

The Midland Expressway Ltd. is a joint venture company formed by Trafalgar House and Iritecna, the Italian trading and industrial company, to build the 43 kilometer (27 mi) Birmingham Northern Relief Road (BNRR) to relieve much of the M-6 traffic congestion around Birmingham (Figure 9.5). This project would be the fourth DBFO project in Great Britain; the first was Trafalgar House's Queen

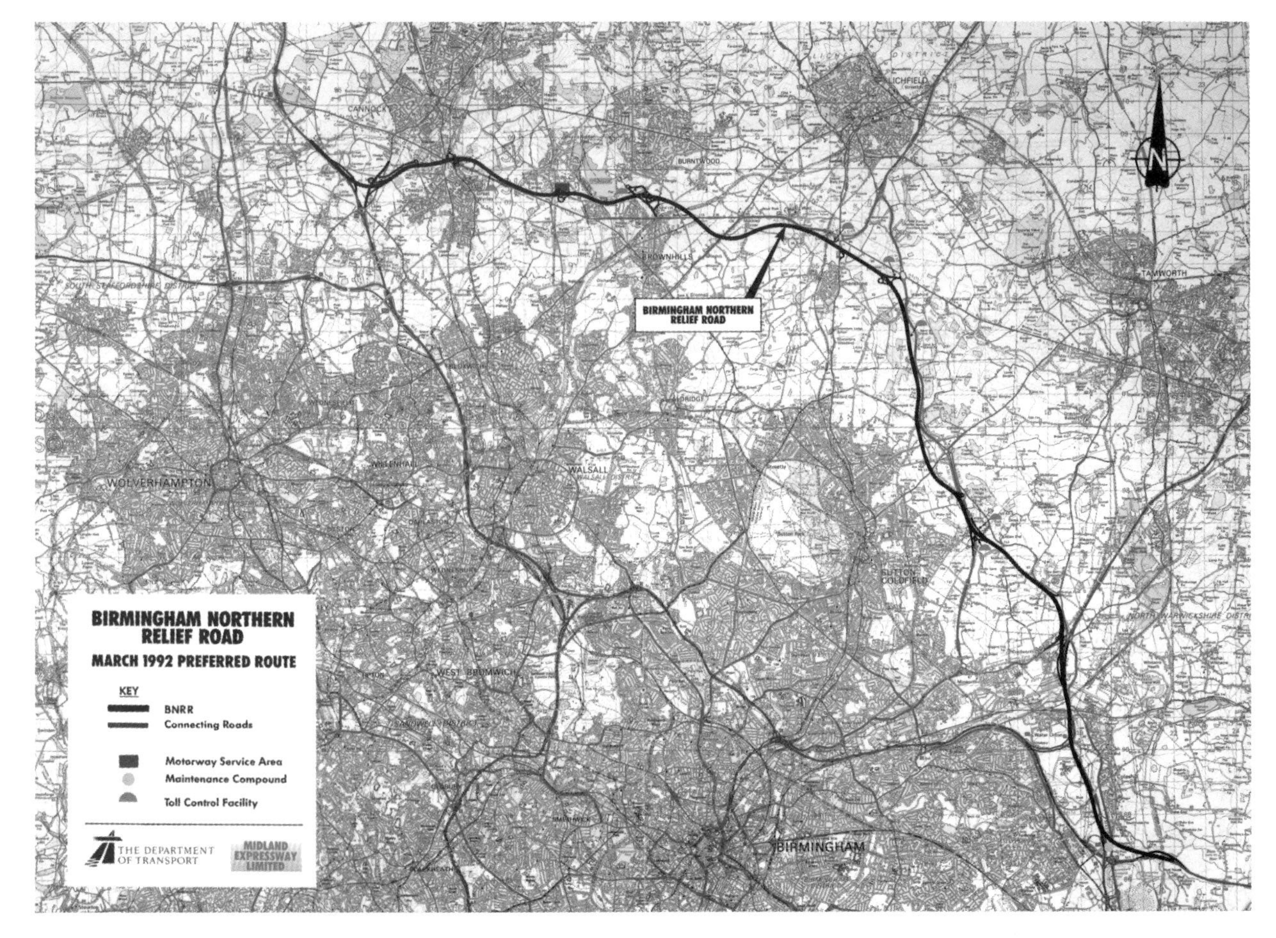

Figure 9.5. Route of the Birmingham Northern Relief Road. *[Source: Department of Transport, United Kingdom.]*

Elizabeth II Bridge, the second was the Second Severn River Crossing, and the third was the £30 million (U.S.$51 million) BOT Skye Bridge in Scotland scheduled for completion in the fall of 1995.

The Need for the Birmingham Northern Relief Road

The portion of the M-6 passing through the communities around Birmingham is one of the busiest sections of motorway in the entire country. Average traffic flows exceed 110,000 vehicles per day, and as many as 130,000 vehicles per day have been accounted for at peak times.

The Birmingham Northern Relief Road project would create a bypass to avoid this congestion that is close enough to the West Midlands industrial and residential areas to give easy access to local trunk roads. Beginning 1984 the government held public meetings to discuss possible routes for a northern bypass around Birmingham, and in March 1986 the Department of Transport announced a preferred routing for a publicly funded highway from the M-54 at Featherstone to the M-6 near Coleshill. But it was not until May 22, 1989, that a private-sector competition to design, build, finance, and operate the Birmingham Northern Relief Road was announced by Paul Channon, Secretary of State for Transport.

Prequalifying bids were submitted in August of that year and a short list of three proposers was released by April 1990. Sixteen months later, on August 12, 1991, the government announced that Midland Expressway Ltd. (MEL), the Trafalgar House–Irtecna joint venture, had won the competition to design, build, finance, and operate the Birmingham Northern Relief Road. The total funding for the dual three-lane highway was estimated at £450 million (U.S.$765 million) when the project was announced in the *Trafalgar House News* corporate newsletter in September 1991. The revised funding in 1995 pounds sterling was estimated at £500 million (U.S.$850 million); the construction phase of the project awarded to Trafalgar House's construction division was estimated at about £300 million (U.S.$510 million).

The concession agreement signed in February 1992 granted MEL the right to design, finance, and operate the BNRR for a period of 53 years, commencing at the start of construction. MEL would be responsible for providing the all required environmental impact studies and establishing the toll rates. These rates could be changed in response to market conditions once the motorway was opened to the public. If rates were too low, it would be difficult to obtain the most advantageous financing because the return on investment would not be attractive to potential lenders. But if toll rates were too high, travelers would avoid the route, preferring to suffer traffic congestion on an alternate route.

An "open" tolling system was proposed in which vehicles are stopped only once when they exit. These toll booths would be equipped with both manual and automatic collection devices. One booth at each location would be devoted to an AVI system, and vehicles would be allowed to pass through the AVI gate at speeds of approximately 30 kph (18 mph). There would be two classes of tolls: £1.70 for automobiles, motorcycles, and light vans; £3 for commercial vehicles. (These toll rates were established in 1991—the 1995 rate of U.S.$1.70 to £1 would result in tolls of U.S.$2.55 and U.S.$5.10, respectively.)

Initially the timetable for a construction start was set for 1995, allowing MEL approximately one year to wade through the environmental studies and obtain the necessary permits. However, the commencement of construction has now been advanced to 1997 primarily because of Great Britain's strong environmental movement. Activist organizations such as Friends of the Earth have voiced objections to the project. Paul Emberley said the fully 60 percent of the original routing of the BNRR has been changed since the initial public consultation. More than 11,000 written objections have been received from citizens in and around the route of the new motorway. When such objections are voiced, the government must appoint an inspector to investigate each matter and render a report to the Department of Transport and the Department of the Environment. Approval must be obtain from both agencies before the objection can be dismissed.

The costs of these delaying tactics have been substantial to the start of construction. Trafalgar House and Irtecna will have invested £25 million (U.S.$42.5 million) in the project, including design and engineering fees; £1 million (U.S.$1.70 million) has gone for environmental impact studies. MEL has 20 full-time employees plus environmental consultants working on this project, and the weekly cost to deal with all of the public inquiries and written objections has been about £20,000 (U.S.$34,000).

The delays associated with obtaining the necessary approvals for the BNRR project have not been lost on the development and investment communities. Emberley said the costs to prepare tenders on BOT and DBFO projects have always been substantial, and after this latest venture the company is taking a cautious stance to submitting proposals on future projects. As an example, Emberly said the preconcession costs, including bridge design, of their successful tender on the Dartford River Crossing was £6 million (U.S.$9.3 million). Their unsuccessful bid on the Severn River Crossing cost Trafalgar House around £1 million (U.S.$1.5 million), and preparation of the Channel Tunnel proposal, which was not accepted, represented a total cost of £5 million (U.S.$7.5 million).

THE CIKAMPEK–PADALARANG TOLL ROAD

Trafalgar House took the initiative once again with its unsolicited BOT proposal to the government of Indonesia. Transroute, the French toll road consultants, conducted traffic studies for a proposed six-lane toll road between Jakarta and Bandung, the largest and third largest cities in Java; the study included a basic route alignment, preliminary design, geotechnical surveys, and estimates of proposed traffic flows. Trafalgar House teamed up with local contractor Citra Lamtoro Gund Persada (CLP) in the summer of 1987 and submitted a joint BOT project for a toll road to the Bina Marga (Department of Highways) and the Ministry of Public Works. The joint venture company was to be known as Citra Gansesha Marga

Nusantara (CGMN); Trafalgar House had a 75 percent share in the joint venture, and the Indonesian firm a 25 percent share. The company created by CGMN to build and operate the toll road was designated the Cikampek–Padalarang Toll Road (CPTR), and in March 1988 the Indonesian government awarded a 25-year concession to CPTR to proceed with the project.

The project would involve construction of approximately 56 kilometers (35 mi) of highway and about 70 structures along the way. Once this highway is constructed and operational, travel time between Jakarta and Bandung will be reduced from four hours to two. The need for the road was amplified by the projections for population and economic growth in the area. The population in this part of western Java is expected to grow rapidly between the end of the 1980s and the turn of the century. Economic growth in that part of the country is anticipated to grow at the rate of 4.5 per annum from 1990 to 2000.

The reason why the proposal was favorably received and the requirement for competitive bids waived was because the scheme had been so well thought out and because Trafalgar House would also provide the necessary financing to design and build this much-needed highway. The British Overseas Development Agency (ODA) was going to provide financial backing for the project as a way of promoting the use of British goods and services. The CPTR project would have the responsibility to use £56 million (U.S.$95 million) of British goods and services, which could include project management cost, design fees, and materials used during construction.

According to Brian Devenish, the Trafalgar House executive in charge of the CPRT project, the Indonesian government has become a small shareholder in the toll road company and is providing CPTR with the land required for the project. The joint venture group CGMN is going to the financial market in 1995 with the anticipation that all equity and debt financing will be completed by the year's end. One-half of the funding will be obtained from the issue of stepped coupon bonds issued in the local capital market, and a supplementary bonus will be paid to bondholders base on achieving a successful revenue stream. Long-term loans are being sought from Bapindo, a state-owned development bank. Trafalgar House and Citra Lamtoro Gung Persada, along with founding shareholders, will invest a total of U.S.$89 million in the joint venture company.

The Revenue Stream

A state-owned company, Jasa Marga, owns the majority of the existing toll roads in Indonesia and will work with CGMN to establish a schedule of tolls. Initial and subsequent toll rates will be contained in a written agreement that will permit increases of at least 2.5 percent to reflect any change in the dollar to rupiah exchange rate. Toll increases can be initiated every two years; however, request for more frequent increases are permitted under certain circumstances per the concession agreement. The plan is to repay all debt by the twentieth year of the 25-year concession agreement.

THE TAGUS RIVER BRIDGE IN PORTUGAL

The existing Tagus River toll crossing, the Ponte de 25 Abril, is heavily trafficked. When it was announced that Lisbon would be host to Expo '98, the Portuguese government stated its intention to receive bids on a BOT project to create a second Tagus River crossing. Not only would this new bridge be required to handle the anticipated traffic generated by the World Expo, but it would complete the Metropolitan Lisbon Northern Highway System and a south-shore interchange that would join the southern highway system to the Setubal Peninsula and on to Spain (Figure 9.6). The government of Portugal anticipated that this new river crossing would also open up land for new development. An ancillary benefit of a new river crossing would be providing an additional emergency route to and from Lisbon in case of a severe earthquake.

The Portuguese Ministry of Public Works, Transportation and Communication issued a request for tender in April 1993, based on the following criteria:

The crossing was to carry six lanes of traffic across 10 kilometers (6.2 mi) of water and 8 kilometers (4.9 mi) of land.

Provisions were to be made to add two additional lanes in the future.

Height of the structure was to be 30 meters (98.4 ft) above the ground at its highest point and 45 meters (147.6 ft) above the active shipping lane. The river has two shipping lanes, and a third, the deepest, would be dredged once the bridge is constructed.

The successful concessionaire would also be required to operate the existing toll bridge from 1996 to the end of the concession period, which was set at 33 years maximum, and the concessionaire would be able to collect tolls at this existing bridge during that period of time.

Lusponte SA, the Trafalgar House–led consortium, was awarded the concession agreement in April 1994. A brochure issued at the time indicated that Trafalgar House and Campenon Bernard SGE would operate as a joint venture (Figure 9.7). The total funding for the project was set at £820 million (U.S.$1.394 billion).

Five Portuguese companies have a 50.4 percent majority shareholding in Lusponte; Trafalgar House and Campenon Bernard SCE, with its Portuguese branch Sociedade de Construcoes H. Hagan, each hold a 24.8 percent share. The detailed shareholder breakdown is

Trafalgar House Corporate Development Ltd.	23.80%
Trafalgar House Construction (Special Projects) Ltd.	1.00%
Campenon Bernard SGE SNC	22.00%
Sociedade de Construcoes H. Hagen SA	2.80%
Bento Pedroiso Construcoes SA	14.84%
Motas & Companhia SA	13.83%
Somague Sociedade de Construcoes SA	13.83%

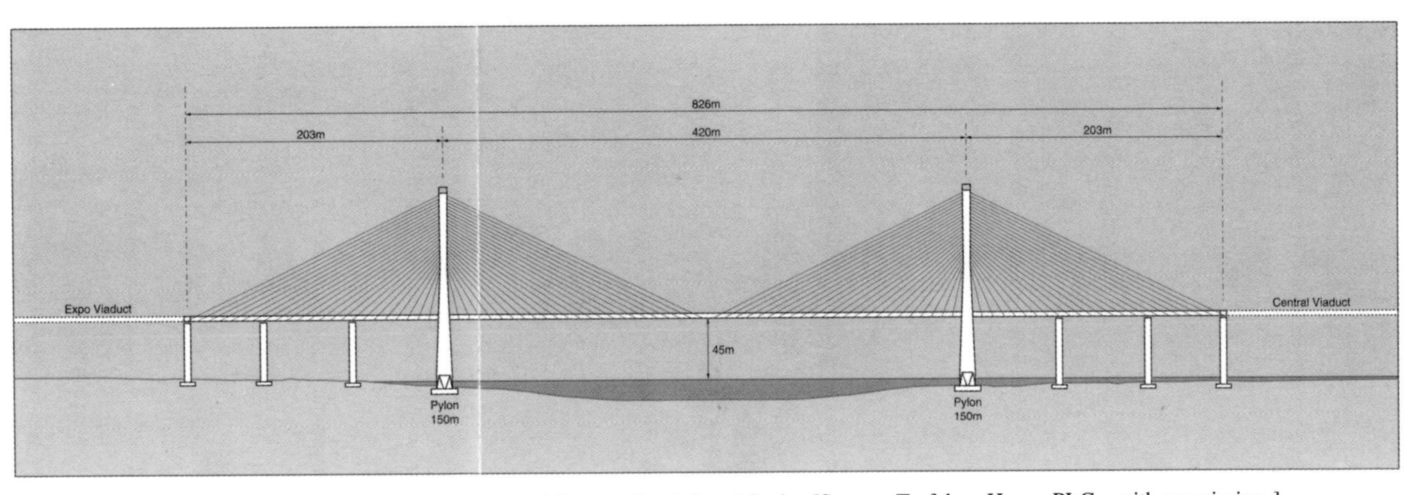

Figure 9.6. The Tagus River Crossing and links to Setubal and Spain. [Source: Trafalgar House PLC., with permission.]

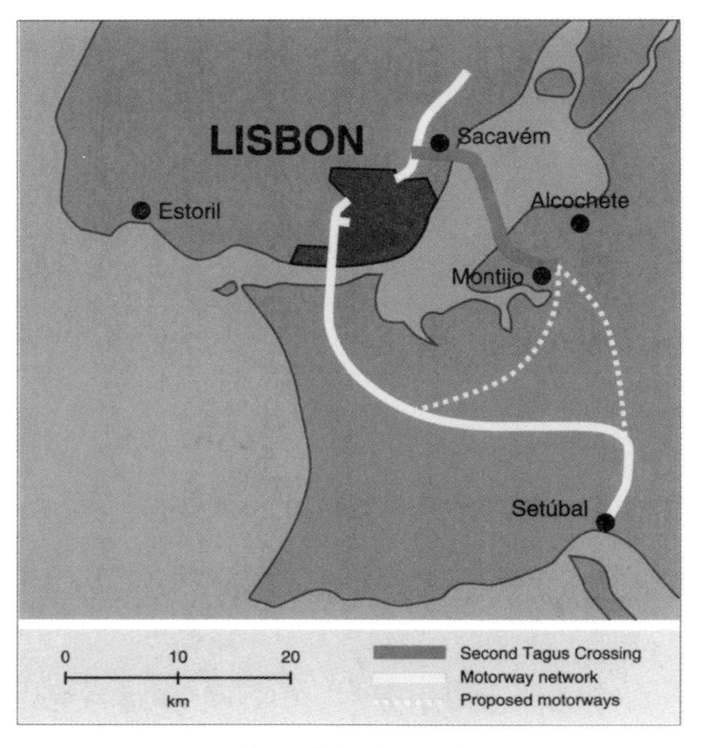

Figure 9.6 *(Continued)*

Teixeira Duarte Engenharia e Construcoes SA	7.50%
Edifer Construcoes Pires Coelho e Ferdandes SA	0.40%

Lusponte has a paid-in capital of 5 billion escudos (U.S.$34,482,000) and two groups of commercial banks, one local and one international, will provide guarantees to the European Investment Bank. The European Union Cohesion Fund will provide a grant of ECU331 million (U.S.$2.274 million) and a loan from the European Investment Bank. The revenue stream from the existing bridge, the Ponte de 25 Abril, will add to the financing sources commencing in 1996. The construction portion of the project is estimated at £550 million (U.S.$935 million) and will be shared by all members of the group.

The maximum concession period is 33 years based on a projected total volume of 2,250 million vehicles traversing the new and existing crossings in both directions commencing January 1, 1996. The existing toll rate for automobiles is 150 escudos (U.S.$1.15) collected in one direction only. The tolls will be increased upon the opening of the new crossing and are expected to be between 2.5 and 3 times the current level.

THE SECOND

TAGUS RIVER CROSSING

For the Ministry of Public Works, Transportation and Communication

Construction joint venture led by

Figure 9.7. Announcement of the Trafalgar House and Campenon Bernard SGE joint venture. *[Source: Trafalgar House PLC., with permission.]*

AN OVERVIEW OF DBFOS AND BOTS

Geoffrey B. Shields is the director of Trafalgar House Finance PLC and has been intimately involved in the Tagus River Crossing for several years, spending considerable time in Portugal to get the project off the ground. When interviewed for this book in June 1995 at the corporation's Berkeley Street headquarters, he discussed his own opinions of the BOT and DBFO concepts.

As far as the DBFO scheme is concerned, it is quite illusory in my view, because all we are doing is deferring capital expenditures by the public sector—at least, that's what we are doing now with this new round of DBFO shadow toll projects, because it is either the government or the taxpayers that pay for the road. Instead of paying through capital expenditures now, the government will pay over the next 30 years.

That, to me, is just playing game with the accounting books. What we should be doing is tolling these roads and making the users pay for them. By going along with the DBFO concept, the government is reacting to anticipated public opinion.

As far as the Tagus River Crossing is concerned, the existing bridge is congested eight hours per day, and on summer weekends motorists spend three to four hours queuing. Another crossing was desperately needed, but it could not be paid for out of government funds. The BOT concept has legitimacy because users will ultimately pay for the project. I do not sympathize with motorists who complain bitterly about paying charges, because if they don't pay, nonusers will pay and effectively cross-subsidize the users.

But these BOT projects are not gold mines. This is something many host governments lose sight of. Just because there is a traffic jam out there today does not mean that there will be an El Dorado tomorrow. In a good case, such as with high traffic volume, etcetera, BOT does present a good opportunity but not a fantastic investment.

The demonstration that I love is to spend a lot of time looking for equity investors. One of the thing about BOT is that people say, "Ah! The contractors are getting construction work and high equity rates of return—you can see the congestion!" But it is difficult getting the equity investment—no one really wants it. There are few takers amongst the financial institutions.

I've been working on this project [the Tagus River Crossing] for four to five years, but it has only been within the last six months that I have received positive responses from the financial institutions as far as taking equity or putting up subordinated debt is concerned.

The banks are fine. The banks have lots of lending capacity, subject to a satisfactorily structured project. It usually means that the promoter or equity investor is trying to find something like 15 to 20 percent of total project costs. This is a huge number. The contractors are struggling to do it. In the contracting industry, profit margins are pitifully thin, so one can't count on generating huge amounts of cash out of the basic business. And the contracting profit can easily be absorbed by the equity requirements.

So when you think about what benefits these projects bring to the construction company, it is fairly much in the form of deferred returns because the contractor is having to plow back into the equity of the concession company and, then, getting an equity return only after the banks are out of the picture.

Trafalgar House brings together an extensive range of skills, and with its team of pragmatic and professional executives such as Geoffrey Shields, it will likely remain a world player in design, construction, development, and the BOT process well into the next century.

10

THE CHANNEL TUNNEL

Much has been written about the Channel Tunnel, a $16 billion BOT project that is one of the outstanding engineering projects of the twentieth century, but this mega project suffered from the same ills that beset contractors and owners on considerably smaller construction jobs.

John Noulton, director of public affairs for the Channel Tunnel Group Ltd., is well acquainted with the project from inception to completion. Noulton was affiliated with RTZ, a study group sponsored by Great Britain's Department of Transport in the early 1970s to review the feasibility of building a channel tunnel. About 10 years later, Noulton would be called on once more to investigate the possibility of such a venture, but this time as part of an Anglo-French group established after a meeting took place between Prime Minister Margaret Thatcher of England and President François Mitterrand of France. Looking back over the years and commenting on the genesis and actual construction of the project, Noulton recalled five basic problems that contributed to the labored birth of the Channel Tunnel:

1. The lending institutions' general distrust of contractors.
2. Lack of a strong owner's representative in the initial stages of the project who would have been able to bridge the gap between the construction group and the financial backers.
3. Lack of sufficient design criteria on which to base a realistic budget and, subsequently, to arrive at a valid lump sum contract price.
4. Exclusion of qualified experts in some disciplines—no transportation experts were on board initially, a field of expertise that would play a major role in the development of the project.

5. Lack of cooperation from both the English and French governments and their reluctance to intercede in such activities as formulation of safety regulations. This would result in overkill requirements that would ultimately add millions of dollars of unexpected costs to the rolling stock.

THE HISTORY OF THE CHANNEL CROSSING

The last time the England was physically connected to the Continent was in 6500 B.C. when a large land mass formed a bridge. Since that time several attempts have been made to build a fixed link between Great Britain and Europe.

Albert Mathieu-Favier, a French mining engineer, is credited with being the first person to seriously propose a tunnel under the English Channel. In 1802 he discussed a plan with Napoleon Bonaparte to bore twin tunnels under the Channel, one emanating from France and the other from England. These two tunnels would rise to the surface at an artificial island to be built mid-channel. Travel in those days was by horse-drawn carriage, and this island would serve as a way station where the horses could be watered, fed, or changed. Mathieu-Favier's plan never proceeded beyond the speculation phase.

Thirty-two years later, in 1834, a plan was advanced by Frenchman Aime Thomé de Gammond who proceeded to develop his scheme by taking soundings at various points in the channel and diving to depths of meters (98 ft) to retrieve samples of soil strata from the seabed. De Gammond's proposal involved the construction of a 34 kilometer (21 mi) twin-tube tunnel constructed of iron ribs encased in masonry. But de Gammond's activities failed to attract any financial backers.

Englishman John Hawkshaw, with his assistant William Low, also became interested in a channel tunnel crossing in 1861 and began to plan a single-bore tunnel to accommodate two rail tracks. However, Low favored two single-rail track tunnels instead of one bore, and this divergence of opinion resulted in his leaving Hawkshaw to join with de Gammond and another Frenchman, James Brullées. Together they developed a rather sophisticated scheme in 1867 that involved building twin tunnels with cross passages to improve ventilation and lessen the piston effect that a train would create as it traversed the tunnel at speed.

Eight years later in 1875, the British and French governments caught tunnel fever and enacted legislation that would permit the formation of a channel tunnel company. As more and more engineering studies were completed thereafter, confidence grew that such a venture was indeed economically feasible. The Association du Chemin de Fer Sous-Marin (the Undersea Railroad Association) concluded that the lower chalk strata under the channel was basically uninterrupted and would be suitable for a bored tunnel. *Engineering News,* the predecessor of today's *Engineering News Record,* published the article "The English Channel" in the April 26, 1879 issue about the proposed tunnel:

> We give some details of the explorations which have been made for the proposed tunnel under the English Channel. M Larousse, the eminent hydrographic engineer

> who name is already well known in connection with the works of the Suez Canal has been charged with the careful survey of the bed of the Channel, with the object of determining the configuration and geological character of the rock of which it is composed. Numerous soundings have already been made, no only in French waters, but also in parts of the Straits near the English Coast: 1,525 observations have been made in this way of the Channel bed district, on the French side of the Channel, and extending to a distance of 17 miles from the coast; from these, 753 samples of the submarine strata have been collected and carefully arranged by means of which the boundary of the calcareous and the argillaceous beds have been determined.

The article went on to relate that, on the basis of these many exploratory observations and soundings, several French engineers had concluded that construction of an undersea tunnel was possible. The *Engineering News* article's report continues with a discussion of this potential tunnel's parameters:

> The tunnel will have a length of 36 kilos and with a traverse a continuous bed of gray chalk. While allowing the necessary incline include to insure the running off of the small quantity of water which will filter into the tunnel, its depth below the seabed will in no part of this course exceed 70 meters. A railway train leaving Paris would enter the tunnel at Sangate, proceed under the Straits and ascend at St. Margaret's Bay, about 6 kilos from Dover.

Tunneling began on the French side in 1881, followed by commencement of work on the British side. Two 2.13 meter (7 ft) boring machines completed a length of tunnel 12 meters (39 ft) in length in 17 hours, using early boring machines as illustrated in Figure 10.1. But the recognition of the military implication of such a crossing resulted in the British government stopping the work in July 1883. Tunneling from France had likewise ceased four months earlier. When all work had come to a halt, 800 meters (2,624 ft) of tunnel work had been completed on the British side.

Toward the end of the nineteenth century several bridge schemes were developed but were rejected as presenting obstructions to navigation and shipping. Engineering know-how continued to advance and tunneling experience greatly expanded; thus, as the twentieth century unfolded, interest revived in a tunnel crossing. However, the world conflict leading to World War I put any cross channel schemes on hold.

In 1921, a 128 meter (420 ft) pilot tunnel was driven from Folkestone, England by a Whitaker boring machine (this machine was rediscovered in its abandoned tunnel in 1990). Supporters of a Channel Tunnel persisted until the outbreak of World War II.

The British Channel Tunnel, a study group that had been established by act of Parliament in 1875, was still on the books after World War II ended. In 1950 its new chairman, Leo d'Erlanger, formed an association with both the American company Technical Studies, Inc., and the Suez Canal Company to form the Channel Tunnel Study Group. The latter association would go through a number of additional organizational changes before the tunnel scheme would become a reality.

Figure 10.1. An example of a nineteenth century tunnel boring machine. *[Courtesy McGraw-Hill's Engineering News Record magazine.]*

British Prime Minister Harold Wilson and French Premier Georges Pompidou brought renewed life to the prospect of a cross channel tunnel when they announced in 1966 that a bored rail tunnel was under serious consideration. The cost of the venture at that time was £365 million (U.S.$620 million 1995 rates). The government of Great Britain did not view this project with much enthusiasm, however; worsening economic conditions in England did not help the situation. And financial backers were of the opinion that a high-speed rail link was required to make the tunnel a profitable venture. With all the problems facing the project, the government lost interest and the Channel Tunnel bill that had been introduced in Parliament in November 1973 was soundly defeated.

In 1979, the Thatcher government drastically reduced expenditures for public projects in Great Britain. A half-hearted public invitation from the Department of Transport to consider a channel crossing resulted in some of the country's largest construction firms professing varying degrees of interest in the project. Tarmac construction took the lead and, with no objections from the Department of Transport, began to develop numerous sketches and work out design drawings for a three-bore tunnel plan. Eric Pountain, Tarmac's chairman, then met with Prime Minister Thatcher to ask if there was a real chance that the Channel fixed-link project would proceed. The answer to his question was an emphatic, No.

But several groups of contractors now were clamoring for the project to proceed because the construction market as a whole was in a depressed state in England during the late 1970s and early 1980s. The builders saw the Channel project as a means of priming the construction pump. So in 1982 the British government initiated a study to investigate methods of financing a channel crossing, and this study concluded that only private funding should be used. The more socialist government of France, on the other hand, was in favor of some involvement, primarily through the participation of the state-owned railroad SNCF.

The British government's thinking remained unchanged until two years later when in October 1984 Colin Stannard of National Westminster Bank wrote to Peter Lazarus of the Department of Transport that a bored tunnel crossing was not only technically possible but, in his opinion, was economically sound and could be privately financed. Stannard's letter had an immediate influence on government thinking. Six months later developers were being solicited to submit tenders on a concession agreement that would include the design, financing, construction, and operation of a fixed-link channel crossing. Several groups responded to the invitation, including the Channel Tunnel Group (CTG) composed of Britain's most respected contractors: Balfour Beatty, Taylor Woodrow, Costain, Tarmac, and Wimpey.

THE FORMATION OF TRANSMANCHE LINK (TML)

The Channel Tunnel Group needed to find a French partner to form the required Anglo-French connection before the closing date for proposal submission, October 31, 1985. Accordingly, they increased their efforts to find a French joint venture team that could help in a concerted effort to obtain a commitment from the respective governments to back the tunnel construction. Government commitment was necessary because legislation would have to be introduced and approved before the concept could progress and become a reality. CTG's search finally succeeded once Sir Nicholas Henderson, a former ambassador to France, joined the group in 1985 as chairman. Sir Nicholas had innumerable contacts both inside and outside the French government; after his meeting with François Bouygue, founder of Bouygue SA, France's largest construction-related conglomerate, the spark was struck.

With Bouygue as prime mover on the French side, four other companies decided to join the project as well—Lyonnaise des Eaux-Dumez SA, Spie Batignolles, Société Auxiliare d'Entreprises SA (SAE), and Société Générale d'Entreprises SA (SGE). These five contractors signed an agreement to join together and pursue the Channel Tunnel project from the French side. Because in France the English Channel is known as *La Manche,* this new organization took the name France Manche.

On learning of the joint venture plans in progress, both the French and British governments made it known that they were not keen on having one private entity representing itself as both owner and contractor. CTG and France Manche therefore agreed to divorce the contracting portion of their plan from the owner/client portion. Translink Contractors became the name of the joint venture company that represented the five British builders; Transmanche Contractors became their opposite number in France. The "client" remained CTG/FM, and it began developing the prospectus around the end of 1986 so that it could commence raising equity. (CTG/FM would continue to search for a more suitable name for itself, and would eventually become Eurotunnel, the current owner.)

If both groups succeeded in convincing their respective governments to proceed with the project and were awarded the concession agreement, the construction work

Balfour Beatty Ltd. Tel: 071-216 6800
One Angel Square, Torrens Street, London EC1V 1SX, UK.
An international construction and engineering company, employing over 18,000 people, with a turnover of £1.8 billion.

Bouygues S.A. Tel: (1) 30 60 47 23
1 Avenue Eugène Freyssinet, 78061 Saint-Quentin-Yvelines Cedex, France.
Operating in some fifteen different business sectors, from construction to the media, Bouygues employ around 90,000 people worldwide.

Costain Engineering & Construction Ltd.
Tel: 071-705 8444
111 Westminster Bridge Road, London SE1 7UE, UK.
A company renowned for tackling some of the most ambitious engineering projects in the world. The latest being, of course, the Channel Tunnel.

Lyonnaise des Eaux-Dumez S.A.
Tel: (1) 47 21 09 12
72 Avenue de la Liberté, 92000 Nanterre Cedex, France.
The Dumez group specialises in major civil engineering projects and urban development, generating a turnover of 87 billion French Francs.

Société Auxiliaire d'Entreprises S.A.
Tel: (1) 46 94 70 00
85 Avenue Pierre Grenier, 92514 Boulogne Cedex, France.
The leading building constructor in France, with over 150 subsidiary companies.

Société Générale d'Entreprises S.A.
Tel: (1) 47 16 35 00
1 Cours Ferdinand de Lesseps, 92851 Rueil-Malmaison Cedex, France.
SGE operate in over 80 countries, and are active in the field of buildings, and road, interior and public works.

Spie Batignolles S.A. Tel: (1) 34 24 30 00
Parc St. Christophe, 10 Avenue de l'Entreprise, 95863 Cergy Pontoise Cedex, France.
Recognised as a world leader in the design and construction of urban and inter-urban railway systems.

Tarmac Construction Ltd. Tel: 0902 22431
Construction House, Birch Street, Wolverhampton WV1 4HY, UK.
Tarmac Construction Limited is one of the world's most experienced international construction companies providing a wide range of construction and design services covering all facets of building and civil engineering.

Taylor Woodrow Construction Holdings Ltd.
Tel: 081-578 2366
Taywood House, 345 Ruislip Road, Southall, Middlesex UB1 2QX, UK.
Taylor Woodrow's construction business covers a wealth of expertise in every aspect of building technology & design, contracting, civil engineering and project management. All on a global scale.

Wimpey Major Projects Ltd.
Tel: 081-748 2000
26-28 Hammersmith Grove, London W6 7EN, UK.
At the forefront of worldwide construction. Wimpey have an outstanding achievement record for multi-skilled building and civil engineering projects across six continents.

Figure 10.2. Profiles of the five British and five French firms that constitute Transmanche Link. *[Source: Eurotunnel.]*

would be carried out through a joint venture agreement, in which each group would share on a 50/50 basis. The name of the joint venture would be Transmanche Link (TML). The joint venture's structure is illustrated in Figure 10.2.

Because CTG/FM wished to enhance the contents of its proposal, it hired a *maître d'oeuvre,* an expert auditor, to review its proposal before it was submitted to the British government. The thought behind this move was to show the government that CTG/FM welcomed an independent observer who would monitor the project throughout its life span. This decision proved to be important, as it weighed in the group's favor and ultimately led to the award of the concession agreement.

THE CROSS CHANNEL PROPOSALS

Although CTG/FM had previously submitted a detailed proposal for the channel crossing, the British government was still considering two other proposals as late as December 1985. Prime Minister Thatcher had a preference for a drive-through tunnel. Such a scheme had been presented in a proposal by a consortium known as EuroRoute. The plan envisioned a cable-stayed bridge extending into the channel from the British and French coasts. Vehicular traffic would travel on the bridge and spiral down to man-made islands, one on each end of the two bridges. An immersed tube tunnel constructed between the two bridges would carry cars and trucks below the shipping lanes. The total length of the tunnel would be 21 kilometers (13 mi). A rail link would also be constructed and would travel through 6 meter (19.68 ft) diameter twin immersed tubes; the operation of trains traveling through these tunnels would be turned over to British Rail (BR) and France's SNCF. The proposed cost of the EuroRoute scheme was considered at that time extraordinarily expensive, estimated at £10.7 billion (U.S.$18 billion, 1995) it was almost double the project cost of the CTG/FM tunnel scheme.

A third proposal also under consideration at that time had been submitted by the Channel Expressway group. This scheme called for two large-diameter tubes to be built through which both rail and vehicles would travel. Estimated at £2.1 billion, this proposal was considerably less costly than that of CTG/FM. However, the Channel Expressway's proposal was not presented with the same degree of professionalism as those of the other tow groups. Also, opponents voiced concerns about motorists suffering from claustrophobia or anxiety attacks as they drove through a 30-mile tunnel.

On January 20, 1986, Thatcher and Mitterrand met in Lille, France and announced that the CTG/FM rail-only proposal was acceptable to both governments, and that a 50-year concession agreement would be issue on ratification of an agreement between the two countries. Subsequently, Parliament introduced a hybrid bill in which the government announced its intention to award a private-sector concession to operate what had previously been a public-sector project. Generally this type of legislation does not require a public hearing, but the Channel Tunnel Bill received about 5,000 petitions. Although the bill was introduced on March 15, 1986, it took about 16 months to obtain royal ascent. By contrast, the French government required less than a week to enact the necessary legislation to allow the project to go forward.

THE CHANNEL FIXED-LINK CONCESSION AGREEMENT

Parliament's Channel Fixed-Link Concession Agreement contained six chapters and four annexes (appendixes):

Chapter 1. Purpose of the Concession
Chapter 2. Construction of the Fixed Link
Chapter 3. Operational Phase
Chapter 4. Common Provisions
Chapter 5. Termination of the Concession Agreement
Chapter 6. Disputes, Laws

Chapter 1 described the project, its purpose, and the events that would trigger commencement of tunnel operations. It also included a section on land acquisition and another on financial requirements. With respect to land, the agreement stipulated that if any land or building required for the project belonged to either the French or British governments, they would be made available to the concessionaire at a cost to be determined later.

Chapter 2 included the creation of an independent project overseer, a maître d'oeuvre. The maître d'oeuvre would be give the responsibility of inspecting the works to ensure compliance with the contract documents. As far as monitoring the design phase of the project, the concessionaires were to submit any proposed modifications to the general characteristics of the fixed link, including any changes to the outline specifications, to the Intergovernmental Commission (IGC), an agency that was newly created for the purpose. If the IGC took exception to any of the *avant projets* (the preliminary project design criteria), it was to lodge its objection within 15 days of receipt of the concessionaire's submission. The IGC would prepare a statement indicating its reason for each objection to these submissions. The concessionaire could request that the maître d'oeuvre review each objection and rejection of the avant projet rendered by the IGC. If the maître d'oeuvre agreed with the IGC, the concessionaire must acquiesce to the decision. If the maître d'oeuvre did not agree with the IGC's evaluation, he or she could mediate the issue between the parties; and if mediation proved unsuccessful and a mutually acceptable solution could not be found within 15 days, the objection would remain in force.

The Operational Phase, Chapter 3, referred to commercial policy, tariffs, operating rules, maintenance and continuity of traffic flow, and security and frontier controls.

The common provisions contained in Chapter 4 relate to the organizational structure of the concessionaire, liability and insurance issues, definitions of force majeure, penalties for breach of contract, duties and taxes, and related topics.

Chapter 5 set forth the reasons for termination and dissolution of property if termination occurred before the end of the concession period.

The settlement of disputes and applicable laws relating to the Treaty were included in Chapter 6.

Annex 1 to the concession agreement was very important in that it contained the definition of the avant projet:

> A 1.71 states that the Avant Projects will comprise a set of documents which will define the Work to be constructed, their objectives and their characteristics, and will provide an explanation of the manner in which they meet such objectives, the feasibility of the Works, and the way in which they conform to safety rules and environmental requirements in accordance with the general obligations set forth in the Concession Agreement.

A 1.72 stipulated that the documents must include:

The design criteria.

The list of principal codes, standards, and regulations.

The justification of the main options selected.

The design criteria and calculation methods used for the main structures and equipment.

The outline calculations for the main structure and equipment.

Plans and drawings indicating the general arrangements for the works, together with principal dimensions.

The outline plans and specifications were being developed for a multibillion dollar project the like of which the world had never seen. Although there was sufficient engineering knowledge and experience to handle the actual tunnel construction, the fixed-equipment design for a tunnel of this magnitude and the railcar design were two key elements that would later come back to haunt both contractors and owner as the design developed and the associated costs seemed to soar out of control.

Provisions in the Agreement for a Second Link

The legislation passed by Parliament contained Clause 34, "Exclusivity and Second Link." Reflecting Margaret Thatcher's design preferences, this clause committed the concessionaires to building a future drive-through link if it was technically and financially viable:

> The Concessionaires recognize that, in due course, the construction of a drive-through link may become technically and financially viable. They undertake as a result to present to the Principals between now and the year 2000 a proposal for a drive-through link which shall be added to the first link when technical and economic conditions for realization of such a link shall permit it and the increase in traffic shall justify it without undermining the expected return on the first link.

Article 34.2 puts a little pressure on the concessionaire to seriously consider the future drive-through link:

> The Principals undertake not to facilitate the construction of another fixed link whose operation would commence before the end of 2020. However, after 2010, and in the absence of agreement with the Concessionaires on the implementation of their proposal for the construction of a drive-through link and as to its timetable, the Principals shall be free to issue a general invitation for the construction and operation of such a link. This new link shall not enter into operation before the end of 2020. Nevertheless, before this date, in the event of a demonstrable lack of quality in the service provided, to be judge according to objective criteria, the Concessionaires shall present to the Principals a proposal to remedy such lack of quality. This proposal may go as far as to involve the construction of a new link and is to be subject to the conditions provided in Clause 34.1

THE DESIGN PARAMETERS

Transmanche Link (TML) had completed a significant amount of design work by the time the Channel Tunnel Act was approved in July 1987. At least this much of the design had been agreed on and established:

- There would be three concrete-lined bored tunnels, each 50 kilometers (31 mi) in length running mostly under the English Channel. Two bores would be 7.6

meters (25 ft) in diameter for train travel. A third 4.8 meter (15.7 ft) bore would be used for services to the travel tunnels.

- These three bores would be linked with cross passages every 375 meters (1,230 ft) to spill air pressure from one to the other and avoid the piston effect of trans traveling through the tunnel.
- Two terminals would be built—one in England at Cheriton near Folkestone in Kent, the other at Frethun in Pas de Calais, France.

Figure 10.3 shows the relationship of the cross tunnels and the service tunnel to the two travel tunnels. Figure 10.4 indicates the path the tunnel takes from Great Britain to France.

Engineering tasks were assigned to teams in both countries. Although differences would inevitable arise, the need for a fully integrated scheme was recognized as paramount. The engineering director would be required to answer to two bosses, Director General John Reeve and Director General François Jolivet. More than one decision would be based on compromise. For example, when the question of the diameter of the service tunnel arose, the British engineers wanted to remain with the size of the original 1970s scheme, the 4.5 meter (14.76 ft) diameter. The French insisted on a 5 meter (16.40 ft) diameter bore. After some protracted back-and-forth arguments, a compromise of 4.75 meters (15.58 ft) was offered by the British in a conciliatory effort. The French refused to accept this. Finally, each side compromised further, giving up 0.06 meters (1.9 in.) and the diameter was set at 4.8 meters (15.74 ft).

THE CONSTRUCTION PHASE

Although some preliminary construction had begun in 1987, the following year was the first full year of tunneling activity, starting in January 1988 at the foot of Shakespeare Cliff, 5 kilometers (3.1 mi) from the White Cliffs of Dover, with the boring of a 5.4 meter (17.7 ft) diameter marine service tunnel.

Cleaning Up the Past

On the British side, the past failed attempts at tunneling continued to cause problems for the modern contractors. One of the important tasks that needed to be addressed was locating the previously drilled bore holes sunk during past tunnel investigations. Unless these holes were located and plugged, subsequent boring operations might cause disastrous inflows of seawater.

Additionally, the contractors had to deal with the Priestley shield boring machine left behind after the aborted 1973 start. The Priestley machine, which was to be dismantled and sold for scrap, was surrounded by pools of oil and grease. U.K. tunnel manager Ross MacKenzie discovered that, contrary to instructions, the demolition subcontractor had been using cutting torches. The result was an oil fire in a pit adjacent to an abandoned 0.5 kilometer (1,638 ft) marine tunnel bored in the 1970s. An train engineer passing the site on a Sunday spotted the billowing smoke,

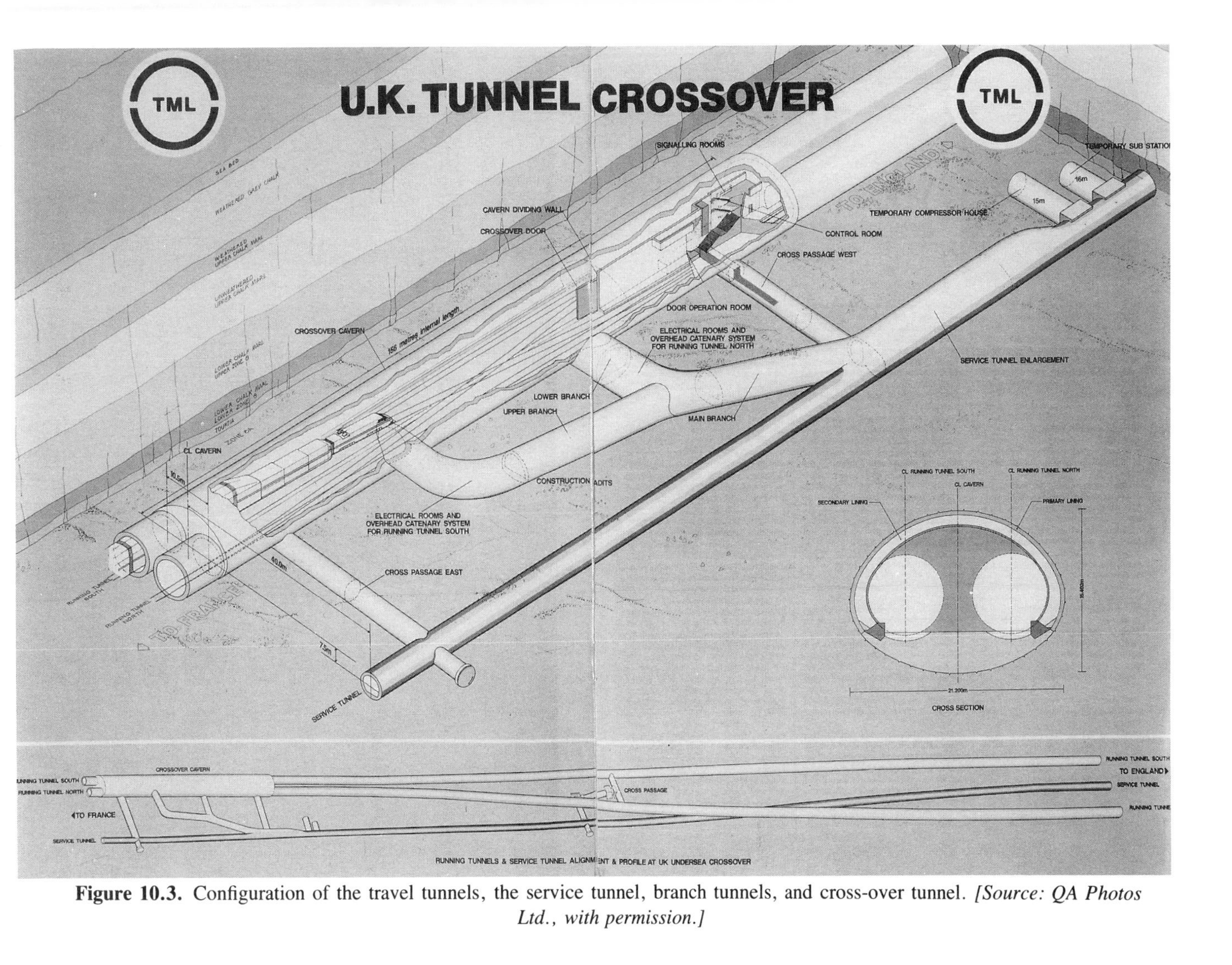

Figure 10.3. Configuration of the travel tunnels, the service tunnel, branch tunnels, and cross-over tunnel. *[Source: QA Photos Ltd., with permission.]*

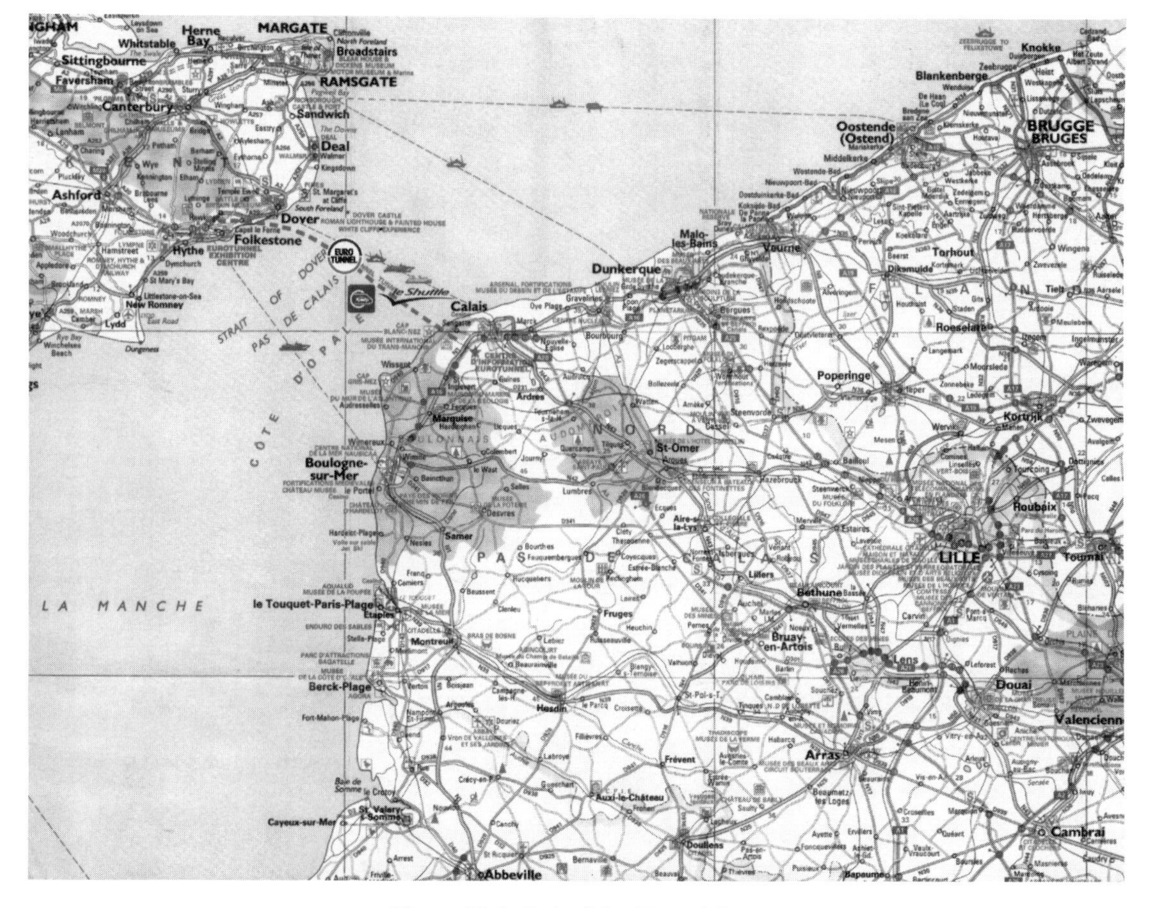

Figure 10.4. Path of the Channel Tunnel.

Figure 10.5. The joyous occasion of breakthrough, demonstrated by one of "Murphy's Marauders." *[Source: QA Photos Ltd., with permission.]*

and reported it to the railway agency. The news that the new tunnel operation was on fire caused quite a commotion, until the true source of the smoke was discovered and the fire extinguished.

Geological Engineering Concerns

Because of the potential problems posed by water, the French commenced work about 1 kilometer (0.62 mi) inland from the coast. They sank a huge 55 meter (180 ft) diameter, 65 meter (213 ft) deep shaft in treated soil so that dry tunneling could proceed.

Some of the approach tunnels, referred to as Adit tunnels, were to be built using the New Austrian Tunnel Method (NATM), which involves excavation, support setting, shotcreting, and rock bolt setting for stabilization purposes. During this process if the surrounding earth is fairly stable, 50 millimeter (2 in.) thick shotcreting and a few rock bolts are all that is required; when the strata is not stable, 200 millimeters (8 in.) of shotcrete must be applied and substantially more rock bolts added. Thus, NATM work requires close supervision and constant attention to changing strata conditions.

Max John, TML's NATM specialist, recalled one occasion when the roof section dropped several inches but was quickly attended to with holes drilled to relieve water pressure and more reinforcement. He commented that, "shear failures of shotcrete are not unknown to me; they do not necessarily mean that the opening will collapse. We are trained to react immediately, to get in there and find out what can be done. There can be no panic—we do not leave the tunnel."

In June 1988, TML encountered an unexpected 3.24 kilometer (2 mi) stretch of wet ground in one of the marine service tunnels. Water began pouring in. The problem was solved by installing the metal rings used to form the tunnel proper, and quickly packing the voids behind the rings.

Because of the potential for noxious fumes, electric-powered locomotives were used instead of diesels to haul materials and equipment in and out of the tunnels. But because of the wet conditions, the electric locomotives had to be sealed against chloride corrosion; the result of this treatment was that the locomotives began to overheat. And after that problem was resolved, it became apparent that these locomotives were losing traction whenever they operated in wet areas, and that their electronics were prey to the salt-laden air. Further modifications were required; in the end, 100 locomotives had to be completely rebuilt.

The Tunnel Boring Machines (TBMs)

Because they anticipated encountering only blue chalk marl, the tunnel boring machines (TBMs) employed by the British were straightforward open face shield tunnel machines. The TBMs used to create the service tunnels were built by the American firm Robbins Company of Kent, Washington. The traveling tunnel TBMs were a joint venture between Robbins and Kawasaki Industries, a Japanese firm with extensive experience in the design and construction of all types of TBMs.

Water had been a major problem for the French when they had briefly started—and abruptly stopped—tunnel work in the 1970s. The French expected to encounter major faults in the chalk and water during their boring operations, so they designed their TBMs accordingly. The French TBMs has cutter chambers sealed to 145 psi (999 kPa) to withstand the pressures exerted by the channel waters above. Because the machine chambers were sealed, the spoil created by the boring operations had to be discharged through a lock to the atmospheric pressure in the tunnel so that it could be carried away for disposal.

Tunneling Progress

Tunneling progress was slow at first. By December 1988, only 856 meters (2,807 ft) had been bored. This increased to 1,816 meters (5,956 ft) by June 1989. The press dwelled on the slow progress, much to the frustration of the engineers and workers who were literally charting new ground every day, encountering obstacles and quickly overcoming them.

By December 1989, 2,210 meters (5,956 ft) had been driven, and by February 1990, 3,116 meters (10,876 ft). TML was by then hitting its stride, reaching a record 5,172 meters (16,964 ft) by April 1990 before backing off to an average of 3,465 meters (11,367 ft) for the balance of the year. Breakthrough occurred in the U.K. land service tunnel in November 1989, and in the MRT South tunnel in September 1990. All were joyous occasions as can be seen by one of Murphy's Marauders in Figure 10.5.

The Tunnel Alignment Problem

The method by which tunnel alignment was calculated as each side raced toward the other was based on a series of control sightings previously established over the years by the Ordnance Survey and the Institut Géographique National, as well as by the use of a global satellite positioning (GPS) system.

For the purpose of tunnel alignment, the British used a series of brackets bolted on the left side of one of the service tunnels. These brackets were spaced at 75 meter (246 ft) intervals. During the first part of 1988, a tunnel boring closure tolerance of 5 millimeters (0.2 in.) was projected, but when quadrilateral traverses were made to new stations on the right-hand side of the service tunnel, error of up to 635 millimeters (25 in.) were being reported. If this problem were not solved, the two tunnels could be off by several diameters by the time the British and French construction crews were to meet.

Horrified surveying crews went over every calculation and every procedure employed in the surveys. They found that refraction was the culprit: The theodolite sightings (the instrument used to measure horizontal and vertical angles) were being taken close to the cold concrete tunnel surface, and the surveyors had not noticed that warm air from ventilation dust was moving back and forth across the tunnel face, greatly affecting their traverse readings. Thus, a second set of control points was establish on the right side of the tunnel, curing the problem and allowing closure to take place within 500 millimeters (19 in.).

THE TERMINAL SITES

The terminal sites in Great Britain and France provide passengers with all of the amenities available to airline passengers—passport control, customs, security procedures, fast food restaurants, automatic teller machines, and duty free shops.

Folkestone Terminal The design of the Cheriton terminal at Folkestone was initially planned as fitting the rail lines to the site, taking into account the topography and environmental issues. However, local petitioners felt that any planned expansion of the terminal would engulf and overwhelm the adjacent communities, so a redesign was initiated and four years later, a simpler design emerged. Eighteen buildings were constructed on the 900 acre (364 hectares) site; a striking architectural feature of the amenity building is a Teflon tented roof that glows in the dark from the interior lighting. Despite its seeming simplicity, Cheriton's Folkestone Terminal has quite a complex nature (Figure 10.6).

Coquelles Terminal On the French side at Frethun, the Coquelles site, Aéroport de Paris constructed the terminal on 1,700 acres (688 hectares). Prior to the start of construction, 75 percent of the existing site was marshland. Aéroport de Paris overcame these adverse site conditions by creating a drainage system that emptied into a man-made lake with a graceful viaduct system (Figure 10.7).

Figure 10.6. Aerial view of the British terminal at Cheriton (Folkestone). Vehicles line up at mid-right section of photo for loading onto railcars at upper left. *[Source: QA Photos Ltd., with permission.]*

THE ROLLING STOCK

The initial concept for the rolling stock was to have commercial vehicles travel in open high gross vehicle (HGV) railcars with only a roof to protect both vehicle and driver from the weather. After this design was submitted to the IGC (Intergovernmental Commission) safety authorities, it was rejected. The IGC said that no further review would be made until a complete and thorough study was conducted of the smoke and fire detection and fire suppression systems that were to be included in the design of the open railcars.

The HGV railcars presented an additional problem in themselves. Because the occupants could remain with the vehicle during the 25- to 30-minute travel time under the Channel, safety precautions for property and person would have to be thoroughly researched.

Twenty-two months later, approval of the fire detection and suppression systems was granted, but in the meantime an order had already been placed with Italy's BREDA/Fiat Corporation to proceed with the manufacture of the modified railcar design. By the time the IGC gave its stamp of approval, more than half of the HGV railcars had already been manufactured.

Thirty-eight locomotives were constructed by a joint venture of Brush Traction and ABB of Switzerland. Each of these 5.6 MV, electric-powered machines had three–two axle bogies, and each axle was independently powered. Passenger cars were built by Euroshuttle Wagon Consortium—a joint venture of Canada's Bombardier, Belgium's BN, and France's ANF. When the HGV shuttles, locomotives, and passenger cars were combined, the total cost of the rolling stock was £800 million (U.S.$1.36).

Figure 10.7. The French terminal at Frethun referred to as the Coquelles facility. The artificial lake with drainage fingers can be seen slightly left of the center of the project. *[Source: QA Photos Ltd., with permission.]*

The Safest Train System in the World

Britain's Health and Safety Act of 1974, which is similar in many respects to the U.S. Department of Labor's Occupational Safety and Health Act (OSHA), contains many of the safe workplace requirements that were implemented by the builders of the tunnel. But during its first days and weeks of operation, the Channel Tunnel's train system had a number of system problems.

- An automobile caught fire during a demonstration for shareholders run in December 1994.
- An October 1994 electrical malfunction caused the entire train to shut down—with 400 journalists on it.
- Attendants in Le Shuttle, the passenger vehicle train, failed to advise a motorist that movement from the train could set off a car's security alarm system—as it did halfway to the northern coast of France. The owner had to be quickly located to deactivate the piercing alarm.
- A car owner took a photograph of his car on board and his flash camera set off the train's fire alarm system, bringing the train to a stop.

Despite these glitches, the Eurotunnel railway system's philosophy has been that of the Roman orator Cicero, *Salus populi suprema est lex:* The safety of the people is the highest law. The train has a plethora of safety features installed in the rolling stock. According to some experts, these features not only added considerable cost to the project, but went well beyond conventional safety standards and into the realm of overkill.

Safety by Design

The basic design of the fixed-link itself incorporates service tunnels, cross-over tunnels, and other systems designed primarily with safety in mind. The Channel was bored through a relatively impervious layer of chalk under the channel; therefore, waterproofing was not a major concern. Only one section on the French side was bored through a fissured and fractured layer of chalk, and that section did require a waterproof lining.

Wall Reinforcement For the greater portion of their lengths, each of the three tunnels was lined with reinforced precast concrete segmental sections. Not only do these precast sections support the surrounding rock, but they were designed for a 120-year life expectancy. TML could not find a suitable subcontractor to perform this work, so they made the decision to cast these sections themselves. In total, 442,755 segments were cast in 35 different sizes. Even in the face of exacting quality control standard, TML's production failure rate was only 0.6 percent over the 42-month casting period.

Ventilation Ventilation of the tunnels had to take into account the air changes for health purposes and design had to anticipate dealing with emergences such as fire and smoke. Every 375 meters (1,230 ft), short passages link each of the tunnels. This spacing corresponds to the length of the trains using the tunnel, so that any train inadvertently coming to a stop in the tunnel will be adjacent to two of these "escape" passages and near to two more.

Ventilation Cross links were also constructed to house electrical equipment and to carry drain water away from one or more tunnels. In addition more than 250 ventilation tunnels link the rail tunnels every 250 meters (820 ft) not only for proper ventilation, but for passenger comfort in that they relieve the "piston effect"—the pressure buildup created by the trains traveling through the tunnels.

Maintenance Additional cross-over chambers were built to allow uninterrupted work on the tracks. At one-third and two-thirds points underground passageway, the two rail tunnels come together to form six separate sections in the 50 kilometer (31 mi) length. If maintenance or reconstruction is required in any one of these sections, trains can be diverted around the section under repair. This does create a situation where trains traveling in opposite directions pass each other at two locations in the tunnel; but unless a cross-over section is in use, large steel rolling doors separate the rail lines. These doors can also assist in smoke control should a fire develop in the tunnel.

Seismic Design Although no seismic events have taken place in the Channel Tunnel region since 1531, the engineering studies considered the possibility and incorporated seismic design.

Rail Design The continuously welded rails, which have a density of 60 kg/m (40 lb/ft), carry coded signal information that will cause any approaching train to stop if the current is interrupted. The rails themselves are secured to precast concrete ties surrounded by a rubber sleeve. The rail, tie, and boot are encased in a continuous cast-in-place concrete bed. Ultrasonics are employed during schedule maintenance checks to detect any cracks in the rail. If a rail should break, the clips securing the rail are designed to hold the broken ends together until a train safety passes over.

Onboard Sensors Detectors on the train will sound an alarm if an axle bearing overheats. If an object is being dragged by the train, another detector will sense this and sound an alarm.

CHANNEL TUNNEL SYSTEMS

Signaling and Control Systems

Power is supplied from a national power source on each side of the channel, and these two independent power sources are interconnected through DC cables installed in the seabed. Two diesel generators of 5 MVA capacity each were installed during the construction phase and remain in place today connected to busbars at each end of the shaft. If both national grid systems are lost, these generators could provide power for essential services such as lighting, pumping, and signaling.

Most train-related accidents in Great Britain have occurred because train operators have ignored the red danger signals; therefore, automatic train protection (ATP) was of paramount importance to the Channel Tunnel. The ATP equipment installed in the trains traveling through the Tunnel receive data transmissions that control the maximum safe traveling speed at all times. This speed is displayed in the locomotive cab, and if the engineer exceeds a predetermined limit, the train automatically slows down to a safer speed or, in some cases, will come to a halt.

Each control center serving the Tunnel has three rooms: the monitor room where traffic flow and local terminal functions are supervised; the control room where power and physical plant functions are regulated; the Major Incident Room where Eurotunnel personnel can coordinate with local emergency forces if needed.

Telecommunications Systems

Both signaling and telecommunications systems use fiber optic networks that are immune to electrical interference. If a cable is severed, there is no danger of a short circuit. Three such telecommunication cables have been installed in the Channel Tunnel, and the systems can still function if only one cable is left intact. In addition to the redundant fiber optic network there is a telephone system throughout the project that interfaces with the national system in both countries.

Carrying redundancy a step farther, four independent radio systems are active within the tunnel:

System 1. The Concession Radio provides a mobile communication system for staff and their vehicles.

System 2. A track-to-track system permits secure communication between the train engineer and the control center.

System 3. The Shuttle Internal Radio allows the crew of one train to talk to the crew of another train. This system also provides drivers attending cars stowed aboard Le Shuttle with prerecorded music and the ability to receive messages from the staff.

System 4. A tactical radio system can be used by fire, police, and ambulance services should an emergency arise.

The Fire Detection and Protection System

Fire and smoke are killers, particularly in tunnel operations, so the 120 mechanical and electrical rooms in each rail tunnel are equipped with optical and gas ionization fire detectors. When detectors activate, a fire suppressant is released in the area and the local ventilation system shuts down; an alarm sounds locally and in the control center.

Although on Eurostar, the passenger service, riders may be the first to discover a fire, this might not be the case with vehicles on Le Shuttle or freight being carted in open rail cars. Fires cannot go undetected, however; smoke detectors have been installed every 1,700 meters (5,576 ft) in the tunnel and, of course, are installed in the rolling stock.

When a smoke detector is activated, the suspect train will be allowed to continue its journey but at a reduced speed. Trains traveling behind the suspect train would be directed to stop. Dampers in the pressure-relief ducts would be closed to stop the spread of smoke to adjacent tunnels. The object of these safely procedures is to get the suspect train out of the tunnel where the fire can be quickly and effectively extinguished.

If a major fire has started in the tunnel itself, a 250-millimeter (9.8-in.) diameter fire main running the full length of the tunnel that feeds hydrants every 125 meters (410 ft) in each of the three tunnels comes into play.

Dealing with Hazards

The motive power has been designed to keep a train running for as long as 30 minutes after a fire has been detected. This is deemed sufficient time to allow the suspect train to traverse the tunnel. If such a situation should arise, on leaving the tunnel the suspect train would be shunted off to a special siding where the fire could be quickly extinguished.

If the nature of the fire is such that the suspect train can be stopped in the tunnel, it would be split to allow the passengers to exit via the mobile portion of the train.

In Le Shuttle where passenger cars are being ferried, barriers are provided at each end of the specially designed railcars to isolate the suspect car from the rest of the train. The floor section of these railcars incorporates 30 minute fire-rated construction; also, sections of the railcars can be uncoupled from the inside if necessary.

Hazardous goods are banned from Le Shuttle. Of particular concern are recreational vehicles and towed house trailers that may contain bottled LPG for cooking and refrigeration. Regulations require that all compressed gas cylinders in vehicles must be valved off prior to boarding. No more than 50 kg (110 lb) of compressed gas in a container is allowed in the vehicle, and this container must be securely fastened to the vehicle. Vehicles powered by LPG or having a dual-power arrangement are barred from Le Shuttle.

FINANCING THE TUNNEL

Huge construction projects require huge amounts of money and financing them is never a simple task. But when all of the unknowns were present in a project as large and as complex as the Channel Tunnel, lenders required a lot of salesmanship.

In September 1986 after the concession agreement had been awarded to what was to become the TML organization, the 10 contractors comprising CTG and FM put up £47 million (U.S.$71.58 million) as equity financing. This financial arrangement was called Equity 1 and represented the first effort to fund the project. That same month, some 40 banks announced their intention to underwrite a certain portion of Eurotunnel's requirements: Equity 2 secured £206 million (U.S.$313.74 million) in private placement.

Coinciding with the announcement of Equity 2, the first of several financial crises surfaced. The Channel Tunnel Bill which was working its way through Parliament was attacked by the Labour Party and for a while it appeared that the bill would be scuttled. This was enough to send shock waves through the financial community that was considering funding portions of the project.

The Management Dispute and Its Ultimate Resolution

Alistair Morton assumed leadership of Eurotunnel in February 1987 and became co-chairman along with Andrew McDowall, who also served as chief executive officer. Morton faced several formidable tasks involving financial matters, and it was anticipated that the money matters plaguing the project would be resolved with his assistance and guidance.

A contract had to be renegotiated with the British and French railroads in order to produce an attractive revenue stream for that portion of the project. Lending institutions would pay close attention to those negotiations. Further Alistair Morton had to finalize a £5 billion (U.S.$8.2 billion) financing arrangement with 50 banks that was pending at the time of his arrival. Equity 3 took place as scheduled during this period of time, and a French bank agreed to provide the project with Fr4 billion (U.S.$663 million). Equity 4 raised £556 million (U.S.$931 million) in 1990, and Equity 5 was a rights issue.

But when Alistair Morton came on board increased costs in construction and impending claims announced by some contractors placed other burdens on his shoulders. To add to his troubles he had to deal with a senior management team that was caught up in internal squabbling.

A shakeup of top management was in the offing in 1988, and Morton took several steps that antagonized the contractors. Morton had a problem with the maître d'oeuvre, and he frequently communicated directly with the contractor, short-circuiting the maître d'oeuvre. Morton felt that the maître d'oeuvre had too many masters, so he proceeded to chop its staff from 300 to 40, transferring the other 260 to the project implementation division whose task it was to check on TML's work not as an independent inspection team but as the owner's representative. The newspapers considered Morton to be "good press" and fed on his statements that the contractors were out to milk the project and that the British contractors were inefficient and needed more men than the French to get the job done. Although the banks relished Morton's "get tough with the contractor" language because it conveyed the impression of a hard-nosed owner look out for his investors, the contractors were furious.

The matter of contractor claims contributed more to the nagging problem of cost overruns. In a multibillion dollar construction project, when incomplete documents are used to create the lump sum contract price, disputes and disagreements are inevitable. The contract between Eurotunnel and TML included a provision for a dispute resolution board to be composed of five members. The appointments of three French and two British representatives—all experts in their fields—to this board were approved by both parties.

As the project progressed, only 20 matters had been referred to the DRB and most of these matters were resolved informally. At the beginning of 1992, however, three rather large claims were presented by the contractor to the board. One claim involved unforeseen subsurface conditions, another dealt with a request for an increase in the lump sum portion of the contract related to the U.K. terminal building and other surface structures, and the third claim involved the cost of the fixed equipment—such as the ventilation and electrical work—where costs were skyrocketing. The first and second claims were settled by mid-summer 1992, but the fixed equipment dispute would not go away; thus, Eurotunnel sought formal arbitration proceedings in accordance with an arbitration clause in the contract.

Although the contractor's fixed equipment claim had no definitive, final dollar value attached, it involved tens of millions of pounds. According to once source in TML, the contractors wanted to increase the fixed equipment budget by £630 million (U.S.$1.071 billion). Neville Simms of TML said Eurotunnel offered them £1.205 billion (U.S.$2.04 billion), but as he put it was in "rubber notes." Simms said that if the offer had been made in Eurotunnel shares the contractors might have accepted it, but they considered the financial instruments backing up the actual offer too complicated to consider.

The Dispute Resolution Board ruled that TML's claim had validity and determined that the owner should begin to negotiate a settlement with the contractors. Until that settlement was obtained, Eurotunnel would have to advance £50 million (U.S.$85 million) per month to TML, on account. Eurotunnel top management was furious.

Eurotunnel requested that the arbitrator in Brussels intervene and rule on what they considered an absurd decision rendered by the board. After reviewing the case, the arbitrator agreed that the DRB had exceeded its authority and could not order "on account" payment, thereby reversing the board's decision. The adversarial relationship between Eurotunnel and TML advanced another two notches or so.

The matter of increasing the budget for fixed equipment remained in limbo for another year until an agreement was reached by both parties in the summer of 1993. Alistair Morton and TML's Neville Simms signed a protocol agreement in July 1993 that ended the stalemate and eased the tensions between owner and contractor. This agreement required Eurotunnel to provide advance payments to TML based on an increase in the lump sum amount that would be thoroughly documented by the contractor on presentation to the owner. In effect, it was like a cost-plus-not-to-exceed arrangement, and TML would get its money subject to a new rights issue which was going to be partially underwritten by the contractors.

As tempers calmed, the word "cooperation" was introduced in to the vocabulary in both camps. With the management crisis over, for the first time since the start of construction there was a light at the end of the tunnel. In return for this major concession by the owner, TML pulled out all stops, rolled up its sleeves to complete the project by December 1993.

Subsequently, in October 1995, Eurotunnel would announce that it was filing a £470 million (U.S.$800 million) claim against TML on the basis of extra costs incurred the commission and modification of the rolling stock. TML's chairman, Colin Parsons, responding to this accusation, said that the rolling stock had been turned over to Eurotunnel in 1993—along with responsibility for operation and maintenance for the system. Another TML member called the claim a desperate attempt by Eurotunnel to seek any means to obtain much-needed funds, and said that if this claim was pursued it would result in a substantial counterclaim by TML. The owner versus contractor battle is heating up yet again.

The Opening of the Channel Tunnel

On May 6, 1994, Queen Elizabeth II and France's François Mitterrand inaugurated the Channel Tunnel. On May 9, 1994, Gerard Haelewijn, a Belgian driver, gained recognition as the first trucker to emerge from the Calais–Folkestone freight shuttle run. The first Eurostar passenger train trip from London's Waterloo International Station to Paris Nord station began on January 23, 1995.

For the first half of 1995 traffic through the tunnel was on the increase, and 100 freight trains per week were traveling through the tunnel carrying 1,400 to 1,500 trucks on each trip. One million passengers booked space aboard Eurostar, and Eurotunnel predicted that 17 million would have passed through the turnstiles by the year's end. But the debt-laden operating company needs much more. Debt service, now that the tunnel is completed, is running at about £2 million (U.S.$3.4 million) per day. John Noulton said that as of June 1995 no payments have been made to the lenders. Three milestones will trigger payments:

1. When revenues cover direct operating costs.
2. When revenues cover operating and office expenses and corporate overhead.
3. When cost flow is positive.

It appears that the second and third milestones will not be attained during this century. Alistair Morton reported a $734 million half-year loss for 1995, and Eurotunnel has postponed paying interest on its U.S.$12.5 billion debt. In Paris in October 1995, Eurotunnel shares fell 3.3 percent to Fr7.25 (U.S.$1.47); just one month earlier these shares had been trading at Fr11.35 (U.S.$2.45). But both these figures are a far cry from the issuance price in 1986 of 300 pence (U.S.$4.56) and its high price of 1,000 pence (U.S.$15.50) in 1989.

The massive loss reported by Eurotunnel may enable the firm to begin discussions in earnest with the banks and possibly to arrange a debt-equity swap. Experts say that in order for the company to survive it must wipe out almost £4.5 billion (U.S.$7.6 billion) of liabilities.

Bankruptcy for Eurotunnel is not in the cards, nor is cessation of operations or sale of the tunnel. Sir Alistair Morton takes the strong position that "you cannot sell a hole in the ground for much if you've closed it." But to paraphrase another famous Englishman, Winston Churchill, "Never have so many given so much to so few."

11

KUMAGAI GUMI, JAPAN'S LEADING BOT PARTICIPANT

A BRIEF HISTORY OF KUMAGAI GUMI

Kumagai Gumi, one of Japan's Big Six contractors operating worldwide with consistent annual sales volumes measured in billion of dollars, is a family business.

Santoro Kumagai, the founder of the company that bears his name, was born in 1862 and began his career as a minor civil servant working in the police department in a small town in Fukui Prefecture, located in the north-central region of Honshu, Japan's largest island. Kumagai had an inquisitive and restless mind, and yearned to leave the small town where he'd grown up. Ignoring relatives' advice to stay home, Kumagai traveled to Yokohama and took up stone carving, working briefly for a company that made religious monuments. While the expansion of Japan's railway system was underway, Kumagai ventured into the construction business as a stone mason and obtained a few small railway roadbed contracts in 1891. But it was while he was building a large stone wall for a temple in his home prefecture that he met Bunkichi Tobishima, the head of a growing construction firm. Kumagai joined Tobishima's firm and worked there for the next 10 years as an employee and subcontractor.

Kumagai started his own company after World War I. In the tradition of many Japanese companies, following generations have carried on the family business, and Taichiro Kumagai, Santoro's grandson, is chairman of the board today.

Many Japanese construction companies ventured into overseas markets to maintain or increase their annual sales volumes from the 1960s to the 1980s rather than depend on the ebb and flow of construction activity in their domestic market. Kumagai Gumi saw an opportunity to create its own market niche in the area of BOT, and has become one of the leading proponents of BOT in Southeast Asia. Successful projects include the following:

The Perisher Valley Skitube. Kumagai Gumi's first BOT joint venture project with Australia's Transfield Pty. Ltd. was also their first BOT project. The rack-rail train system in New South Wales was capable of transporting 3,000 people per hour to Australia's largest ski resort. During the ascent to the ski area the Skitube Railway traverses a 3.7 mile (5.9 km) tunnel that Kumagai Gumi bored out of solid granite. The project, completed in 1988, also included a train station, hotel, and restaurant, along with ancillary facilities.

The Eastern Harbour Crossing. Kumagai Gumi's next BOT project was a U.S.$442 million tunnel project in Hong Kong, which commenced construction in 1986.

The Sydney Harbour Tunnel. Kumagai Gumi returned to Australia to continue its BOT work in 1987 with the U.S.$540 million Sydney Harbour Tunnel project. The company again teamed for the project with Transfield Pty Ltd., Australia's largest contractor.

The Second Stage Expressway. The BOT concept was becoming increasingly attractive to many capital-starved Asian counties and Kumagai Gumi negotiated a BOT contract with the government of Thailand to build the U.S.$1.5 billion Second State Expressway in Bangkok. (This project, which is described in more detail in a subsequent chapter, is an example of one of the most costly risks associated with BOT—a host government that reneges on its agreement and confiscates the project from the concessionaire on completion.)

The Western Harbour Tunnel. In 1994, the company's joint venture with Nishimatsu Construction Co. Ltd. was the successful tenderer for a new U.S.$790 million companion to Kumagai Gumi's first BOT tunnel job in Hong Kong. Construction on the project is slated for mid-1997.

Kumagai Gumi learned early on that it "must act globally and think locally"—a corporate philosophy practiced wherever they operate overseas. Kumagai Gumi recognized Hong Kong as a fertile construction market and opened an office there in the 1970s. They appointed Yu Ching Po, a well-respected Hong Kong businessman, the deputy chairman and managing director. Po has been instrumental in obtaining work for Kumagai Gumi.

Kumagai Gumi opened its New York office in the 1980s and teamed up with William Zeckendorf, Jr., committing about $1.2 billion to real estate development activities in the city of New York. Their most high profile project in that city is the 50-story, 1 million square foot Americas Tower at 1177 Avenue of the Americas.

THE SYDNEY HARBOUR TUNNEL

Crossing the Sydney Harbour

Prior to the construction of the Harbour Tunnel, the Harbour Bridge spanned Sydney's harbor and separated the downtown business area from the suburbs to the north. The bridge, built in 1932, contained six lanes for vehicular traffic, two rail tracks and two tram tracks. The toll rates for the bridge were 6 British pence for cars, 2 p. a head for horses and cattle, and 1 p. for pigs and sheep.

The tram tracks were converted to road lanes in 1959, but the roadway expansion could not keep up with the increased traffic. The annual average daily traffic count increased from 60,000 in 1956 to 180,000 in 1984. By 1986, during peak periods 13,000 vehicles per hour were traveling over the bridge. The government reckoned that, if additional infrastructure could not be built quickly by the end of the 1990s, traffic could conceivably end up bumper to bumper throughout the entire day.

Consulting engineering firm of Wargon Chapman Partners began to study the traffic problem. It appeared that public transportation could do little to solve the congestion. As there were 12 lanes of traffic feeding the bridge from both north and south, one obvious solution would be to add another four lanes to the eight-lane bridge. However, providing additional passage across the harbor had two obstacles, financial and environmental.

- *Financial difficulties.* The high cost of money was one inhibiting factor. If the government increased the fuel tax to produce revenue, the resultant incremental increase in cash flow would take an inordinately long time to accumulate sufficient funds for another bridge or tunnel across the Harbour.
- *Environmental issues.* Any additional crossing would probably be located in the vicinity of the Opera House, a world-famous structure. Nearby residents in Kirribilli chose to live in that area because of its magnificent view of the harbor, and the businesses fronting on Circular Quay depended on tourism for a substantial amount of their revenue. These special interest groups and other concerned citizens could hold up a project for years if they felt that their issues were not being fully addressed.

After considerable study of the traffic situation, Alf Nielsen, managing director of Wargon Chapman, approached two contractors, Australia's Transfield and Japan's Kumagai Gumi, in 1984 with his idea to construct a tunnel under the Sydney Harbour. At that time Transfield and Kumagai Gumi were still building the Perisher Valley Skitube tunnel in the Snowy Mountains, and both were intrigued with the idea of another BOT project. They teamed up with Wargon Chapman and consulted with Westpac Bank on financial concerns to begin addressing the issues they knew the government would stress.

The Proposal for the Sydney Harbour Tunnel

The Australian government has seven objectives in mind when it considers the tender of a BOT proposal:

1. The project must align with government policy.
2. The project must be practical and clearly capable of being brought to fruition.
3. The project has to be financially feasible and not create any unexpected call on public funds.
4. The project must be economically justified.
5. The project must be environmentally acceptable.
6. The project must be of a truly private nature and not call for any government guarantees.
7. The financial arrangements must be such that there be no infringement on the joint borrowing capacity of the federal/state governments or their agencies.

In March 1986, Kumagai Gumi and its joint venture partner Transfield Pty Ltd. presented a proposal to the government of New South Wales to design, finance, construct, operate, and maintain a tunnel under the Sydney Harbour. The Australian Premier requested that a feasibility study be prepared as the Transfield–Kumagai Gumi joint venture began to hold discussions and begin negotiations with the government. The feasibility study was completed in December 1986, at which time the joint venture partners were able to submit a fixed-price, lump-sum agreement that included a provision whereby the joint venturers would assume all construction risks.

The Kumagai Gumi/Transfield team also conducted an extremely detailed environmental study over an 18 month period spanning 1986 to 1987 at a cost of approximately $A2 million. If a tunnel were constructed to funnel off vehicular traffic and thereby increase traffic flow, the study noted that fuel savings of 30 percent could be realized during a typical weekday crossing. And because traffic would flow more readily, air pollution would be reduced in the area—occupants in buildings within 1,000 feet (3,280 m) of the northern pylon of the existing bridge would experience carbon monoxide levels only a third of the acceptable World Health Organization standards.

Although several other proposals for another harbor crossing had been submitted to the government in 1979, they were rejected because they didn't adequately address the issues that were important to the authorities. The proposal submitted by Kumagai Gumi/Transfield was successful because it addressed all seven of the Australian government's objectives:

- It represented an effective solution to the cross harbor traffic problem.
- It was environmentally sensitive, recognizing the need for strict controls during construction and on completion and operation of the project.

- Construction would incur minimal disturbance to the adjacent communities, existing highways, and harbor crossings.
- The economic structure of the proposal made sense to both client and developer.
- All design and construction risks were shifted from the government to the joint venture entity.
- The proposal included a financial package that required funding by the private sector exclusively and did not impact the government's infrastructure budget.

After a protracted period of review and negotiation with the Department of Main Roads, New South Wales, contracts were signed in June 1987 granting the Kumagai/Transfield joint venture the authority to design, construct, own, and operate the Sydney Harbour Second Crossing, valued at approximately U.S.$600 million (1987 rates). An Act of Parliament was presented and approved shortly thereafter, and this new venture was then known as the Sydney Harbour Tunnel Company (SHTC).

The concession period was set at 30 years, and based on completion of the project in 1992 would expire in the year 2022. The entire project was budgeted at $A554 million (U.S.$385, 1986); with interest, principal repayment, and other soft costs, the total amount to be financed would reach $A750, making it the largest privately funded construction project in the country's history to that point. By the time the project was turned over to the NSW government at the expiration of the 30-year concession period, its value would be approximately $A2 billion.

Financing the Venture

Construction costs were to be financed by a long-term bond issue that would be repaid from toll revenues levied on the existing Sydney Harbour Bridge commencing in 1987. When the tunnel was completed in 1992, tolls levied on the new structure would be added to the revenue stream. Because tolls were Australian dollars, Australian financing was sought to avoid any fluctuations in the foreign exchange rate that could impact either the expense or revenue stream.

Sydney Harbour Tunnel Corporation (SHTC) entered into an agreement with the Kumagai Gumi/Transfield joint venture by which the joint venture provided SHTC with a $A40 million performance bond in favor of the Commissioner for Main Roads. The joint venture also provided $A40 million in unsecured loans and an additional $A7 million in equity.

SHTC issued 30-year bonds totaling approximately $A486 million through the Australian firm Westpac Banking Corporation that underwrote the $A486 bond issue, inflation indexed at a rate consistent with market conditions. The bonds contained low cash servicing costs in the early years so that service of the debt would more closely match the income stream from the tunnel's operations.

The Net Bridge Revenue Loan of $A223 million was committed by the Commissioner for Main Roads and was based on projected net revenue received from toll collections during the period 1987 to 1992. The loan would be fully paid off from the accumulated surplus cash produced during the concession period ending in 2022. The $A40 million joint venture loan was to be repaid on August 31, 1992, or project completion, whichever occurred later.

The Revenue Stream The tunnel toll was set at $A1 in 1986 and would be indexed for actual inflation, so it would actually be double by the time the bridge opened six years later. Repayment of the Net Bridge Revenue Loan would come from the tolls collected from the existing Harbour Bridge over the 30-year concession period of the proposed new tunnel. One day after the tunnel was officially opened on August 30, 1992, the New South Wales Road Transport Authority raised the toll on the existing bridge from $A1.50 to $A2 (U.S.$1.14 and $1.52), the same amount charged users of the tunnel.

Construction of the Immersed Tube Tunnel

Because of its expertise in undersea tunnel construction, Kumagai Gumi chose the immersed tube tunnel method for the Sydney Harbour project. The undersea portion would be 3,149 foot (10,329 m) in length and consist of 2 two-lane road tunnels connecting the land approaches at each end. The portion of the work to be subcontracted by Kumagai Gumi was valued a $A280 million, and the projected employment of 1,000 workers would prove a boon to the local economy.

The immersed tube tunnel construction process involves the production of reinforced cast-in-place concrete tunnel sections at a location near the proposed harbor tunnel site. During the precasting process, steel bulkheads are placed at each end of a concrete tunnel section form, and when the casting is complete a water-tight section is created. After all of these precast tunnel sections have been produced, the casting area is flooded to the water level of the adjacent harbor, allowing each tunnel section to float. One by one, the sections are towed to the tunnel location where they will be flooded with water, submerged, and precisely lowered to a predesignated location on a prepared seabed.

Each section is snugged up to the previous one on the harbor floor. The sections are then evacuated of water, after which the steel bulkheads are cut away by divers using cutting torches. This process is repeated until all sections have been submerged, joined together, and pumped dry in place (Figure 11.1).

Immersed Tube Tunnel City The tunnel sections for the Sydney Harbour project were prefabricated in a special dry dock area built at Port Kembla. Ideally, the casting basin for the tunnel tubes should be very close to the area where they will be eventually submerged. Towing a 25,000-ton precast concrete section out to sea is a slow and potentially dangerous operation. Port Kembla was 48 nautical miles from the proposed bridge location, but it was the only available site big enough and close enough to the harbor that could handle the 393 foot (1,289 m) long tube units.

The casting basin was created by removing 78,000 cubic meters (102,016 cubic yards) of spoil from the area. This area then had to be kept dry, so a row of sheet

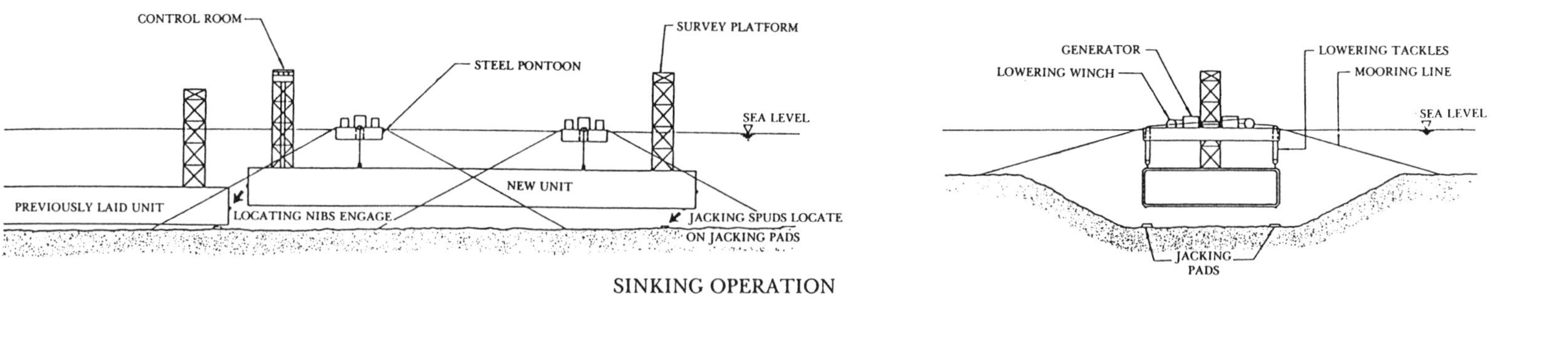

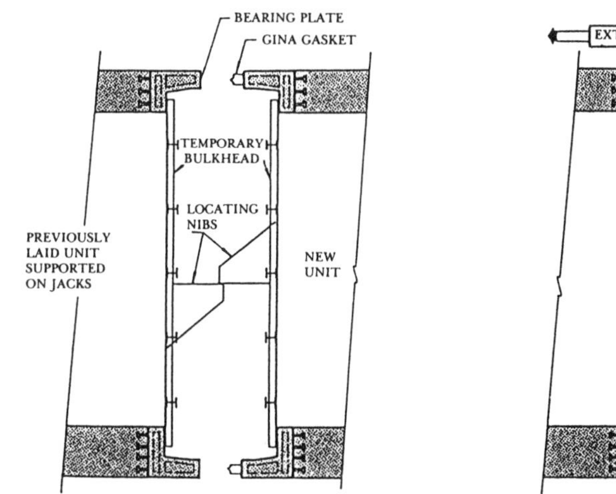

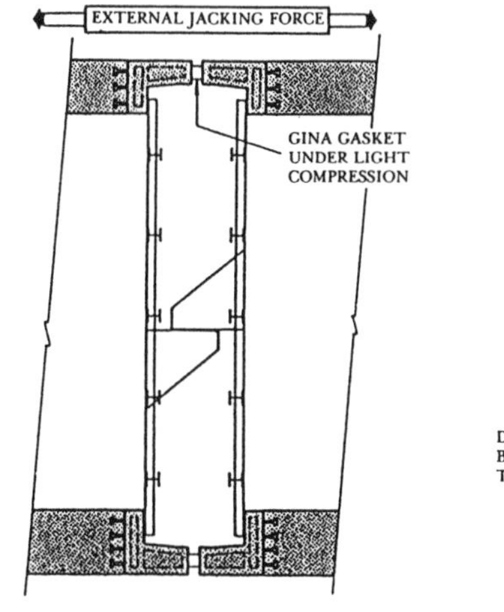

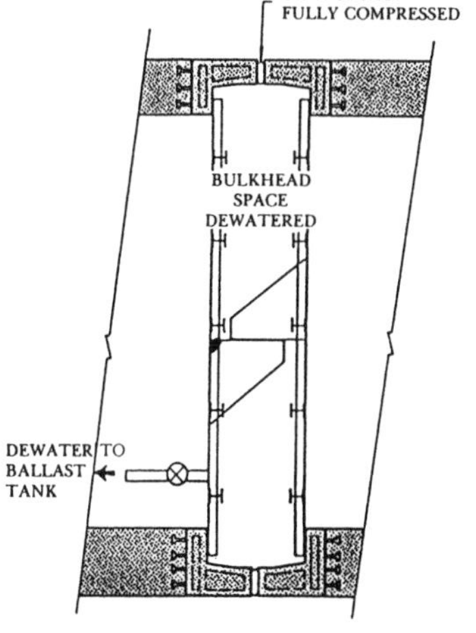

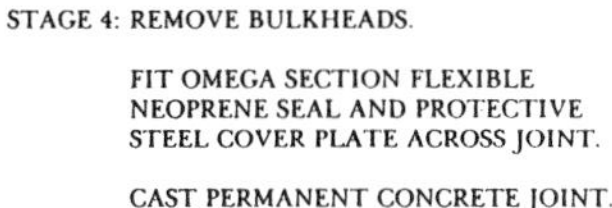
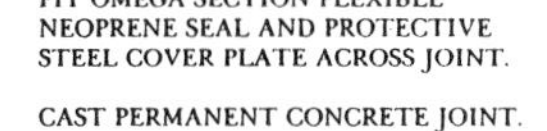

Figure 11.1. The process of joining the immersed tube tunnel sections.

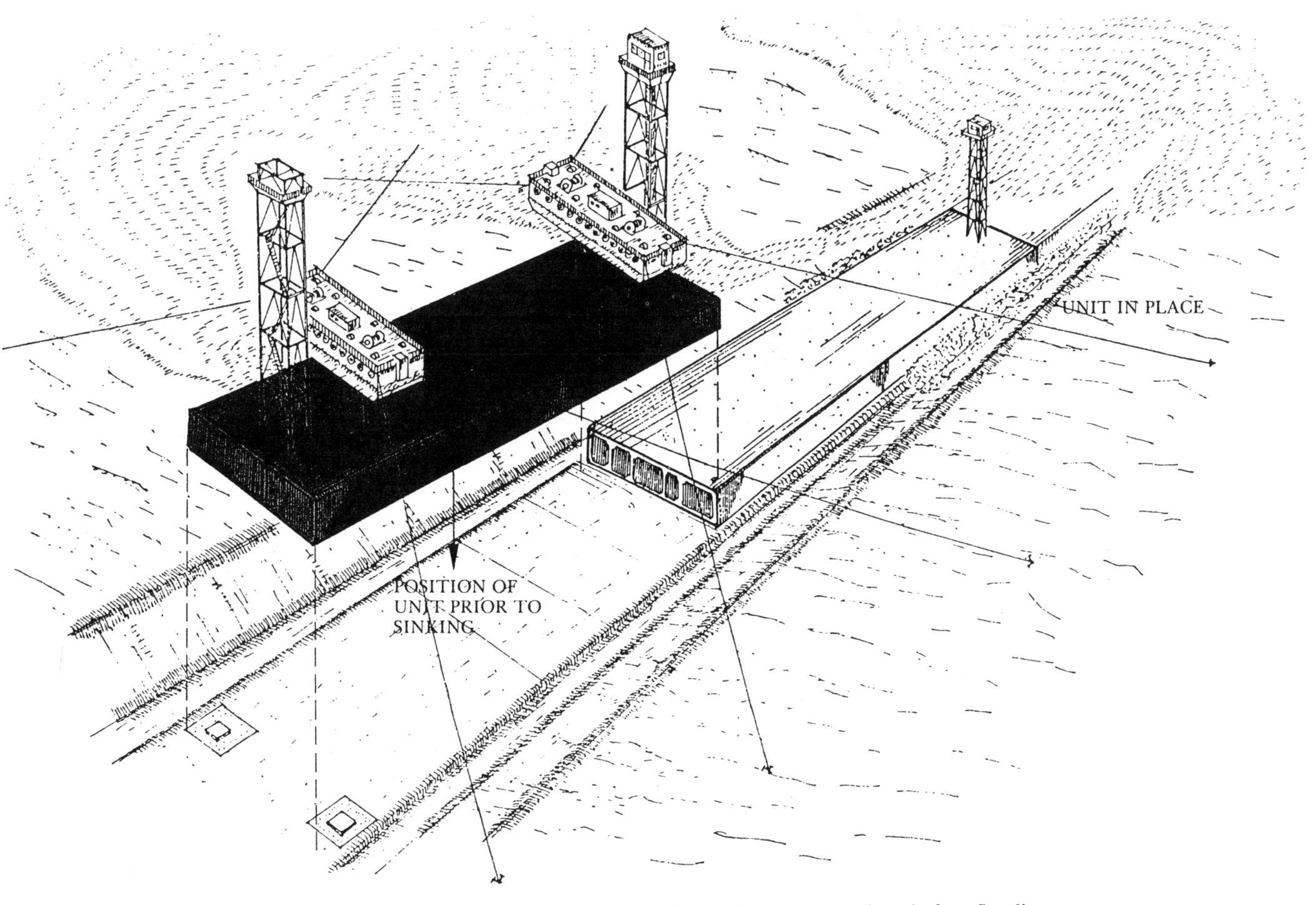

Figure 11.2. Floating alignment towers positioning the tunnel sections before flooding.

piling was driven to bedrock to create a dam between the casting area and the harbor. Also, 300 wells were installed to intercept seeping water.

The bottom of a tube tunnel section must be close to perfect because the tube tunnel will be lowered onto a prepared surface on the harbor floor. Therefore, the floor of the casting basin must be fine-graded to close tolerances. Concrete footings must be poured so that the huge tube sections will bear on them after being cast. A bond-breaker of PVC was placed over this graded area where the tube section would be cast; this PVC sheet would be incorporated into the waterproof seal around the bottom of the finished tube section.

Towing the Sections from Port Kembla Towing the completed tunnel tube sections from the casting basin to the tunnel location proved to be a daunting experience for Kumagai Gumi. If the tunnel sections rode too high in the water, and the seas were rough, these precast concrete structures could develop stress cracks. The joint venture reached the decision that no towing operations would take place if seas were forecast to be higher than 5 meters (16 ft). Video cameras were installed inside the tube sections so that any leaks that developed could be observed. It was decided that during towing operations, the tunnel sections would have no more than 45 centimeters (17.7 in.) freeboard. Two tugboats were required to tow each tunnel section, and their speed was limited to 1.8 knots. It took 24 hours to tow all sections but one (the latter required a 32 hour trip).

Lowering each section into place was also a tricky operation. Two 26 meter (85 ft) high alignment towers were bolted onto each section, one tower at each end of the top surface of the tube (Figure 11.2). Once each section was towed into place and properly aligned on the harbor surface, valves to the ballast tanks were opened and the tube section was slowly lowered onto the seabed, supported and guided as it submerged by cables attached to floating winches. Divers preceded each section to ensure that the precast tunnel sections were being accurately located on the prepared seabed. Shear keys at the joint end of the new section were hooked to the locating nibs of the previously sunken section; the new section was squeezed up against a Gina gasket so that a watertight connection could be made. Each section, once properly placed, secured, and snugged was pumped out so that the steel bulkhead between existing and new sections could be cut away and the watertight passageway extended.

Once all underground sections were in place on the harbor floor, crushed stone was pumped under each section to fill any voids between the seabed and the underside of the concrete tunnel tube. Stone was slowly and evenly placed against the sides and over the top of each immersed tube tunnel section until all underwater sections were enveloped.

The Benefits of the Sydney Harbour Tunnel BOT Project

For the Government and Citizens of New South Wales

A critical piece of infrastructure has been built, one that could not have been provided in a timely fashion using government funding.

If it had been constructed in the conventional manner, the project might have exposed the government to construction cost overruns.

The title to the billion dollar project will revert to the government in 35 years at no cost to the government.

The government has control over construction, operational standards, and toll rates without problems such as cost overruns and the operational concerns associated with these activities.

The government has limited financial exposure.

The toll rates are reasonable and fair.

For Kumagai Gumi and Transfield

The developers obtained a major construction contract—one that was negotiated.

Both companies gained more experience in the formulation and administration of BOT projects.

The project established a close working relationship between two large international construction firms which could result in future joint ventures.

After a preopening-day tour of the project was given to local residents in the interests of good public relations, the Sydney Harbour Tunnel opened to the public in August 1992 amidst great fanfare and publicity (Figure 11.3). Ever since it has continued to meet everyone's expectations.

THE EASTERN HARBOUR CROSSING IN HONG KONG

Hong Kong is composed of three basic areas: Hong Kong Island, the Kowloon Peninsula, and to the north of the peninsula the New Territories (Figure 11.4). In the late 1970s, so many cars and trucks were registered in Hong Kong in the late 1970s that, according to one expert, if all of these vehicles were on the road at the same time they would stretch across the entire highway system end to end.

The traffic in Hong Kong has always been unbearable, and it has been estimated that the value in lost working hours caused by traffic jams amounts to U.S.$2.3 billion—or 2 percent of the country's economic output. But this estimate was based on the cost in lost wages when drivers have to linger in traffic for only 10 minutes. Hong Kong, with the world's highest traffic density, has outpaced the 2.5 percent growth in its road network with a 6.8 annual increase in the number of vehicles on the road.

In the early and mid-1980s, people traveled and commuted across the harbor by car, truck, light rail, ferry, hydrofoil, and speedboat. Weekday vehicular traffic between Hong Kong Island and the Kowloon Peninsula consisted of 1.8 million passengers transported by a light rail system and 120,000 cars crossing the harbor each day. The one existing tunnel, the Cross Harbour Tunnel that had been constructed by Kumagai Gumi in 1979, was filled bumper to bumper with cars and trucks during the morning and evening rush hours.

THE TUNNEL CONTROL CENTRE

A computerised Control Centre with direct links to emergency services operates round the clock to monitor Tunnel use.

- TRAFFIC SURVEILLANCE through a flow detection system built into the road surface and closed circuit television.
- VENTILATION is varied in accordance with traffic conditions.
- LIGHTING is adjusted automatically according to natural light levels at the tunnel entrances.
- FIRE DETECTION includes heat detectors and manual alarms. Motorists can also use the emergency telephones to report fires.
- FIRE PROTECTION is provided by a deluge system on the Tunnel ceiling and is activated from the Control Centre. In addition, fire extinguishers and hose reels are located in cabinets every sixty metres throughout the Tunnel.

SURPRISING FACTS ABOUT THE TUNNEL

- You can listen to local AM or FM radio stations as well as emergency messages via the Tunnel's radio re-broadcast system.
- You can even use your car phone in the Tunnel.

THE TUNNEL AT A GLANCE

- Length: 2.3km
- Capacity: 1,700 vehicles per lane per hour
- Construction Cost: $560 million
- Building Commencement: January 1988
- Workforce: 800 (peak)
- Features: 4 lanes in two separate carriageways.
- Opening to traffic: 31st August, 1992.

HOW THE TUNNEL WILL BY-PASS NORTH SYDNEY AND SYDNEY BUSINESS DISTRICTS

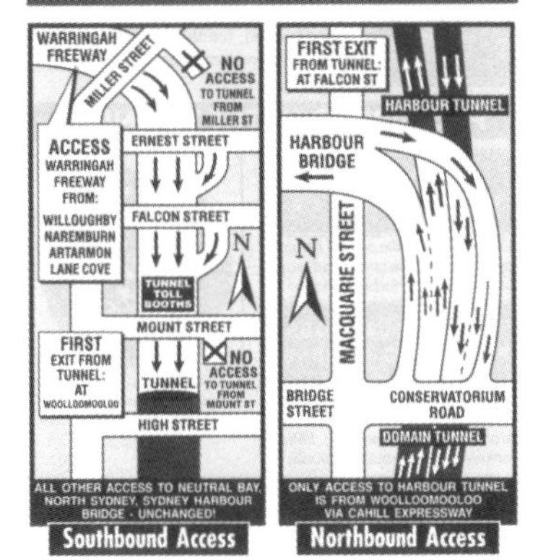

Sydney Harbour Tunnel Company Limited
130 Mount Street North Sydney NSW 2060
PO Box 296 North Sydney NSW 2059
Tel 959 8100 Fax 959 3850

Souvenir Official Opening Programme on sale during Tunnel Walk '92 - Sunday August 30 1992

THE SYDNEY HARBOUR TUNNEL

A users guide.

Figure 11.3. User's guide to the Sydney Harbour Tunnel.

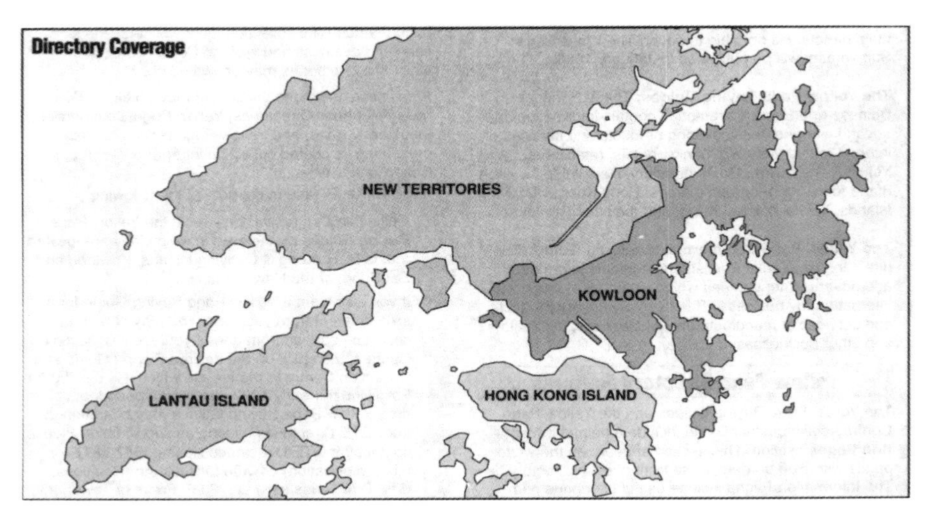

Figure 11.4. The island of Hong Kong, the mainland with Kowloon, and the New Territories. Lantau Island is where the new Chek Lap Kok Airport is being built.

Despite the clear need for infrastructure improvements, in 1984 Hong Kong's economy was stagnant and its ability to finance such projects limited. Kumagai Gumi, perceiving a need that could not be fulfilled by the government of Hong Kong, seized on the opportunity to submit a proposal to the government that could possibly result in a profitable construction contract.

After a rather extensive study of the traffic problem in 1984, Kumagai Gumi prepared a proposal to build a second cross harbor tunnel—this time as a BOT project. The company presented the proposal to the government of Hong Kong, and began plans to team up with Marubeni Corporation, the huge Japanese trading company, if successful in obtaining a contract.

This proposed multibillion dollar BOT tunnel would be one of the first such projects to be built while Hong Kong remained a British Crown colony, but the proposed concession period would extend into period when the Peoples Republic of China (PRC) would assume control of the colonies' government. Thus, in addition to the normal risks associated with a project of this nature, Kumagai Gumi was banking on the assumption that when the British flag was lowered in 1997, the PRC would continue to honor all terms and conditions of the concession agreement.

The proposal for the new tunnel was well received, but Kumagai Gumi was advised that a competitive tender would have to be issued by the government of Hong Kong. With a shrug of the shoulders, the Kumagai Gumi representative accepted this decision, but was of the opinion that the company would ultimately be

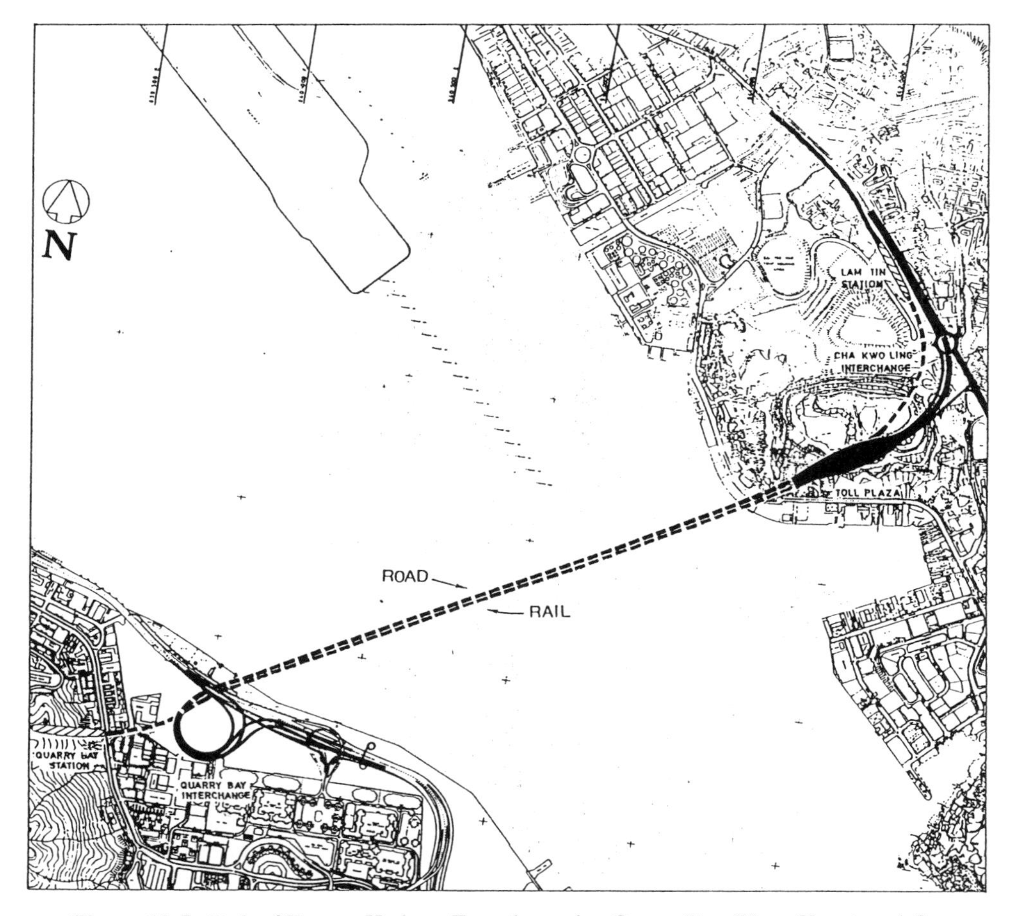

Figure 11.5. Path of Eastern Harbour Tunnel crossing Quarry Bay (Hong Kong) on left.

awarded the contract as it had already completed a concise and thorough study of the proposed project.

Kumagai Gumi was short listed after the proposals generated by the April 1, 1985, open tender were reviewed and evaluated. Even though Kumagai Gumi had been a Hong Kong presence for a number of years, they thought it advantageous to add to the list of their proposed consortium members the China Investment Trust and Investment Corporation (CITIC), a business corporation with the status of a ministry in the PRC. They also invited Paul Y Construction Company Ltd., one of Hong Kong's leading civil engineering firms to join the team, along with Lilley Construction Ltd., a Scottish civil engineering company with extensive mass transit experience.

After considerable negotiations, in December 1985 the government announced its intention to award a contract to the Kumagai Gumi consortium. The Eastern Harbour Crossing Ordinance was enacted on July 17, 1986, providing the necessary legislation to grant the concession. With the formation of the New Hong Kong Tunnel Company Ltd (NHKTC), Kumagai Gumi was ready to begin construction just one month later in August.

NHKTC was responsible for the design and construction of both vehicular and rail tunnels. NHKTC was the corporate entity responsible for operating and maintaining the vehicle portion of the tunnel. A separate corporation, Eastern Harbour Crossing Company Ltd. (EHCC), was responsible for the rail portion; this portion would be leased to the Mass Transit Railway Corporation (MTRC), which would operate and maintain this line along with other existing light rail lines in Hong Kong. The path of the new tunnel and its approach roads is shown in Figure 11.5.

When the tunnel opened in September 1989, 194,000 vehicles passed through in just three months. In 1994, 31 million passenger cars, motorcycles, trucks, and buses traveled through the Eastern Harbour Tunnel.

Financing the New Hong Kong Tunnel

The franchise agreement prepared by the government contained different concession periods for the light rail portion and the vehicular traffic portion: The concession period for the road tunnel would remain in effect until 2016, whereas the rail portion would expire in 2008. In consequence, NHKTC had to arrange separate debt and equity financing for both.

NHKTC required H.K.$2.8 billion (U.S.$360 million); EHCC, the owner of the rail facility, required H.K.$1.6 billion (U.S.$206 million). The total of H.K.$4.4 billion (U.S.$566 million, 1985) for both equity and debt financing was to be arranged by Shearson Lehman Brothers Inc. Of this, H.K.$3.3 billion (U.S.$425 million) would originate from 50 banks, including Barclays Bank, Bank of China, Hang Seng Bank Ltd., Bank of Tokyo, Mitsubishi Bank, Bank Indosuez, China International Trust and Investment Corporation, and Hong Kong and Shanghai Banking Corporation. The remaining H.K.$1.1 billion (U.S.$141.5 million) in equity financing would come from stockholder offerings.

The major stockholders in NHKTC would be Kumagai Gumi and its affiliate companies, with a share breakdown as follows:

Shareholder	NHKTC (%)	EHCC (%)
Kumagai Gumi	69.375	90.0
Paul Y Construction	6.475	
Lilley Construction	4.625	
Marubeni	2.775	
CITIC	9.25	10.0
Government of Hong Kong	7.50	0

Kumagai Gumi anticipated that NHKTC would yield a 16 percent return on investment. The expected rate of return on the rail tunnel would be only 4 percent, but inasmuch as the government had granted them the right to develop property above the rail stations, the rate of return on equity for the entire project would be between 13 and 14 percent.

Toll Structure The proposed tolls (in Hong Kong currency) for the Eastern Harbour Tunnel Crossing were as follows:

Motorcycles, motor tricycles	$ 5
Private cars, taxis, and electric-powered passenger vehicles	$10
Private light buses	$15
Light-goods trucks, special purpose vans	$15
Medium-goods vehicles with gross not to exceed 28 tons	$20
Medium-goods vehicles with gross not to exceed 45 tons	$30
Public and private single-decker busses	$20
Public and private double-decker buses	$30
Each axle in excess of two	$10

Building the Harbor Crossing

To construct the Eastern Harbour Crossing, Kumagai Gumi used the same immersed-tube tunnel method of under-harbor construction that it employed for the Sydney Harbour Tunnel in Australia.

To create the casting area, Kumagai Gumi purchased 13.6 acres (5.5 hectares) of land on the Kowloon side, at the base of a mountain adjacent to the harbor. The 2 million cubic meters (2.6 million cubic yards) of aggregate required to prepare the seabed to receive the tunnel sections would be obtained from rock blasted out of the casting area. (After the crossing project was complete, this dry dock area became a public parking lot.)

The undersea portion of the tunnel consisted of 15 precast concrete sections; five units were 420 feet (1,378 m) long, and 10 units were 400 feet (1,312 m) long. Each of the 15 sections was approximately 32 feet (105 meters) high and 116 feet (380 meters) wide. Each section contained five cells: two for the railway, two for the road, and one for the mechanical and electrical systems.

The total immersed tube section length would be 1,860 meters (6,100 ft); however, the aggregate length of underground tunnel for the railway system would be 5 kilometers (3.1 mi)—4.3 kilometers (2.67 mi) would be underground, and about 701 meters (2,300 ft) above ground. The total length of surface and underground vehicle roadways would equal 8.6 kilometers (5.33 mi).

Among the extensive array of construction equipment that Kumagai Gumi owns are numerous ocean-going floating docks, concrete mixing vessels, tug boats, and other assorted vessels—all of which were available for use during the Eastern Harbour Tunnel project.

A Six-Year Review of the Eastern Harbour Tunnel

In ways other than affording investors a high rate of return, the Eastern Harbour Tunnel project has been very successful. The daily traffic rate has risen from 22,809 in 1989 to 86,023 in 1994. There are presently 31.4 million vehicle trips per year, an increase of 11.3 percent from 1993.

Even in the face of the greater volume of 1994 traffic, accidents and incidents decreased by 8.4 percent—a tribute to the dedicated safety and traffic staff employed at the project. In 1994, 7,833 spot checks were made on vehicles to ensure that those carrying prohibited dangerous cargoes and improperly secured loads were flagged down and prevented from entering the tunnel. Additionally, 509 cars and trucks with excessive emissions were stopped.

A recapitulation of the project from its grand opening through January 1994 reveals a number of issues critical to a successful BOT project. The tunnel, constructed at a cost of H.K.$2.214 billion (U.S.$283.85 million, 1986), was financed by shareholders in the amount of H.K.$750 million ($96.15 million) and bank loans of H.K.$1.464 billion (U.S.$187 million). Toll revenue accounts for 98 percent of the project's revenue and provides cash flow for operating expense, interest expense and repayment of bank loans. The 1994 cash flow was as follows:

Revenue	H.K.$409 million (U.S.$52 million)
Operating expenses	H.K.$95 million (U.S.$1.227 million)
Interest	H.K.$73 million (U.S.$943,000)
Bank loan repayments	H.K.$71 million (U.S.$917,000)
Net cash	H.K.$170 million (U.S.$2.20 million)

Distribution of H.K.$163 million (U.S.$2.11 million) was paid out in the form of dividends in early 1995. This payment represented a simple return of 2.4 percent per annum to shareholder, the same as the last nine years. Although shareholders received little cash in the initial years of the tunnel operation, cash is expected to grow as the project matures. Turnover in total revenue grew by 10.7 percent in 1994 and total income, primarily from tolls, likewise grew by 10.7 percent. (The Western Harbour Tunnel's potential effect on the Eastern Harbour Tunnel's revenues is discussed later in this chapter.)

THE WESTERN HARBOUR TUNNEL PROJECT IN HONG KONG

Hong Kong's New Airport

The existing Kai Tak Airport, where giant jets duck over the skyscrapers to land, is woefully inadequate by today's standards, especially as current estimates of future passenger traffic are expected to be 350 percent higher by the year 2010. Construction began in 1992 on 3,084 acres (1,248 hectares) on an island off Lantau for the new Chek Lap Kok Airport project, which is scheduled for 1997 completion. The dual runways will be approximately $\frac{9}{11}$ mile apart when the airport is operational. In March 1995 the Provisional Airport Authority executed a U.S.$1.3 billion contract for the airport's terminal building. The Kai Tak Airport will provide air travelers in and out of Hong Kong with 24-hour service. An anticipated 28.6 million passengers and 2.4 million tons of freight will pass through its facilities during its first year of operation, but the design capacity of the airport will allow upward of 87 million passengers and 19 million tons of freight per year.

A 35 kilometer (22 mi) Airport Railway system will extend from the airport via West Kowloon to the Central Reclamation area. The railway will whisk visitors from the airport to the Central area in less than a half hour. The present government of Hong Kong is also proceeding on a master plan to provide rapid transportation from Hong Kong Island to West Kowloon by adding a new network of highways to the Chek Lap Kok airport, and from there to the northwest New Territories. A new harbor crossing was an integral part of this master plan, and in February 1992 the government of Hong Kong issued an invitation to bid for the Western Harbour Crossing.

The Invitation to Tender

In 1991 the government of Hong Kong indicated its desire to award a franchise for a BOT tunnel connecting West Kowloon with the Western District of Hong Kong Island by mid-1997. The Western Harbour tunnel would be a key element in the Airport Core Programme, which includes the billion dollar Chek Lap Kok airport and the strategic transport links to it which will aid in the development of North Lantau and the West Kowloon region. The Western Harbour Tunnel would also form an integral part of the Route 3 project connecting Hong Kong Island, West Kowloon, the Northwest New Territories, and the border crossing into the People's Republic of China at Lok Mau Chau. According to the government traffic estimates included in the invitation to bid, the average daily traffic through the tunnel in 1997 would be 85,000 vehicles, increasing to 120,000 in the year 2001 (Figure 11.6).

General Conditions for Submission of Tenders All companies bidding on the tunnel project had to submit six copies of the proposal along with a deposit in the amount of H.K.$1 million (approximately U.S.$129,000). Alternative proposals would be accepted, but the BOT concept, could not be changed. The closing date for proposal submission was set at July 3, 1992, and the government indicated that the short-listed tenderers would be requested to submit a draft construction contract for review. The criteria to be used for final selection would take into account:

- Level and stability of the proposed toll changes. The government indicated that there should be 9 classes of tolls—at the low end motorcycles and motor tricycles, followed by private cars and continuing upward to heavy trucks and public and private buses.
- Methodology for toll adjustment.
- Ability to meet the proposed completion date of mid-1997.
- The financial strength of the bidders and the structure of the debt and equity financing. The government would not take an equity participation in the project

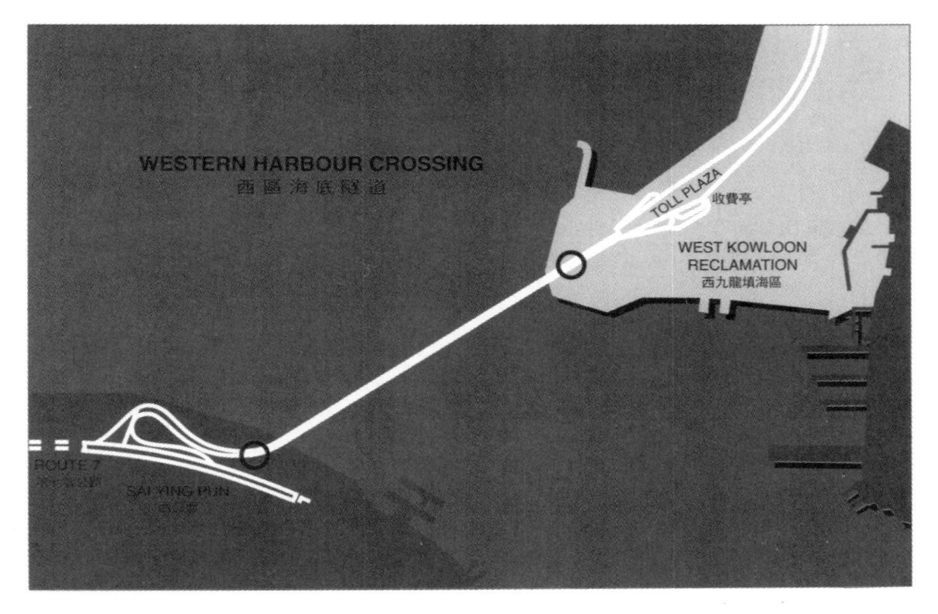

Figure 11.6. Route of Western Harbour Crossing from Sai-Ying Run in Hong Kong to the West Kowloon Reclamination area.

and that all financing must be without recourse. No tax exemptions would be available to the franchisee.

- The quality of the engineering design, landscaping and environmental issues, traffic control, mechanical, electrical, surveillance systems. The requirement for dual three-lane configuration could not be changed, but other conforming schemes using different structural forms or designs would be viewed as acceptable alternatives.
- Operation, maintenance, and inspection procedures.
- Ability to manage and operate the project efficiently.
- Stated benefits of the proposal to the government.

Each proposal had to include a detailed breakdown of capital costs, preoperating costs, operating costs, and projected daily traffic volume for each class of vehicles. The concessionaire would be able to propose toll rates in their tender; however, these tolls could not provide a revenue flow that would result in "excessive rates of return" as determined by the government. The government would retain its right to collect a fee if traffic volume exceeded a predetermined level. Enabling legislation would be passed by the legislature once a successful bidder was selected, and toll rates were to be included along with a schedule of estimated revenue. The franchise period, set at 30 years, was to commence when the tunnel opened for traffic in mid-

1997 or within 48 months after commencement of construction, whichever event occurs first.

Land Acquisition The government stated its intention to acquire and clear all lands as set forth in their Conforming Scheme project outline and would transfer these lands to the successful franchisee at no cost. The site for the toll plaza administration building would be made available to the franchisee at market value lease rates, and the term of the lease would be the same as the tunnel concession period. A work site and works area would be leased to the successful franchisee at $1 per year.

If a tenderer's proposal required additional land beyond that included in the Conforming Scheme, the proposer must take into account the additional time and cost to acquire these added parcels which might create unacceptable delays to the project.

A casting basin for construction of the immersed tube tunnel sections would be provided to the franchisee as part of the works area. All land and facilities made available to the franchisee were to be returned to the government in satisfactory condition, and if not the government could commence any required remedial work and backcharge the franchisee with one month's notice.

The Winning Proposal

In late summer of 1993, the Nishimatsu/Kumagai Gumi joint venture was awarded the franchise agreement for the BOT Western Harbour Tunnel project, and the Western Harbour Tunnel Company Ltd. was formed. Both construction firms, Nishimatsu and Kumagai Gumi, were to act as general contractors and subcontract a portion of the work to Gammon Construction and mechanical and electrical contracting firms.

The lead designer for the project is a Maunsell Acer/Parsons Brinckerhoff joint venture, and all the designers have considerable experience with other immersed tube tunnel projects. The construction period was designated as August 1993 to June 1997, a total of 47 months. Construction commenced on schedule; Kumagai Gumi has followed the same procedures used successfully in the construction of the Eastern Harbour Crossing. The tunnel casting yard for the construction of the 12 tube tunnel sections is located at Shek O Quarry. Each section is 372 feet (1,220 m) long, 28 feet (92 m) high, and 109 feet (358 m) wide.

The Western Harbour Crossing Ordinance

According to the invitation to tender, once the successful proposal had been selected, a new ordinance would be enacted based on the Tate's Cairn Tunnel Ordinance, which contains the rights and obligations of the franchisee and the government and the basic obligations to finance, design, commission, construct, maintain, and operate the new tunnel.

On July 31, 1993, Chapter 436, the Western Harbour Crossing ordinance, was enacted to grant a franchise for the construction and operation of a crossing between West Kowloon Reclamation and Sai Ying Pan, the terminus points for the new tunnel. The legislation recognized the following guarantors:

Adwood Company Ltd.
The Cross-Harbour Tunnel Company, Ltd.
China Merchants Holdings (Hong Kong) Co. Ltd.
China International Trust and Investment Corporation, Hong Kong (Holdings), Ltd.
Kerry Holdings Ltd.
China International Trust and Investment Corporation, Pacific, Ltd.

The franchise permits the company to design, construct, and complete the construction works; to operate and maintain the Western Harbour Crossing from the operating date until the expiration of the franchise period; and to collect tolls for the use of the tunnel by the public from the operating date until the expiration of the franchise agreement.

The Toll Stability Fund Chapter 436 creates the Western Harbour Crossing Toll Stability Fund, whose purpose is to "pay to the Company from the Fund an amount equal to the difference between the new revenue and the upper estimated net revenue for the year concerned." The Fund is to receive monies from excess revenue when the net revenue for a given year exceeds the upper estimated net revenue but does not exceed the maximum estimated net revenue, which is 50 percent of the amount in excess of the upper estimated net revenue per Schedule 5 (Table 11.1). If the revenue in a given year exceeds the maximum estimated net revenue the company must pay into the fund the amount in excess of the maximum estimated net revenue and an amount equal to 50 percent of the difference between the upper estimated net revenue and the maximum estimated net revenue. The definition of maximum and minimum estimated net revenue is established by referring to the schedules set forth in the ordinance: Column 4 of Schedule 5 is the maximum estimated revenue for the corresponding year, and Column 2 is the minimum estimated revenue for that year.

Toll Increases The government prefers that tolls be kept as low as possible and adjustment of tolls would take place by agreement of the franchisee and the governor in council. Depending on its net revenue for the year, the company may apply to the Secretary for Transport for a toll increase if the net revenue is less than the upper estimated toll revenue as set forth in Schedule 5. Schedule 4 of the ordinance anticipates that the company will be able to effect toll increases effective January 1 of the following years: 2001, 2005, 2009, 2013, 2017, 2021. However, should the net revenue of the company equal or exceed the upper estimated net revenue for that year, no toll increases will be allowed to take effect.

TABLE 11.1 Schedule 5 of Western Harbour Crossing Ordinance

Estimated Net Revenue (H.K.$ millions)			
Year Ending 31 July	Minimum Estimated Net Revenue	Upper Estimated Net Revenue	Maximum Estimated Net Revenue
1998	154	336	403
1999	201	399	471
2000	253	461	538
2001	506	768	865
2002	713	1,016	1,128
2003	794	1,106	1,221
2004	880	1,202	1,321
2005	1,190	1,570	1,711
2006	1,455	1,881	2,039
2007	1,549	1,983	2,143
2008	1,623	2,061	2,223
2009	1,876	2,369	2,551
2010	2,028	2,562	2,760
2011	1,892	2,405	2,594
2012	1,821	2,326	2,513
2013	2,212	2,815	3,038
2014	2,573	3,267	3,524
2015	2,733	3,474	3,749
2016	2,891	3,682	3,974
2017	3,507	4,449	4,797
2018	4,018	5,090	5,486
2019	4,220	5,355	5,775
2020	4,422	5,621	6,064
2021	5,192	6,583	7,098
2022	5,747	7,285	7,855
2023	5,726	7,286	7,864

Note: In this schedule "year" shall be construed having regard to the definition of "year" in section 42 of this Ordinance.

COMPETING BOT PROJECTS

The Western Harbour Tunnel project was advanced seven years earlier than the government's initial Information Memorandum had planned. And, although Kumagai Gumi has a stake in both Eastern and Western Harbour Tunnels, the company is concerned that these two harbor crossings will compete with each other and dilute the investors' return on investment. The situation could affect the cost of future BOT projects in Hong Kong, and could even make it more difficult to raise funds for future projects unless the host governments are willing to include a "noncompetition" clause in their contracts.

Table 11.2 from owner/operator NHKTC's 1994 annual report shows the profit trend for the Eastern Tunnel; however, Y. Matsumoto, chairman of NHKTC, antic-

TABLE 11.2 Profit and Loss Account for the Eastern Harbour Tunnel for the Period Ended December 31, 1994 (H.K.$)

	1989	1990	1991	1992	1993	1994
Turnover	2,459,944	135,410,342	201,225,175	299,213,528	348,426,997	385,674,535
Profit (Loss) before taxation	(3,588,766)	(126,289,682)	(59,565,930)	77,744,879	136,127,266	171,751,462
Taxation					(5,044,000)	(28,017,000)
Profit (Loss) for the year	(3,588,766)	(126,289,682)	(59,565,930)	77,744,879	131,083,266	143,734,462
Balance brought forward		(3,588,766)	(129,878,448)	(189,444,378)	(111,699,499)	19,383,767
Proposed dividend						162,750,000
Balance carried forward	(3,588,766)	(129,878,448)	(189,444,378)	(111,699,499)	19,383,767	368,229

Source: New Hong Kong Tunnel Company Ltd. Annual Report, December 31, 1994 (prepared by KPMG Peat Marwick, Hong Kong).

ipates that NHKTC will experience a 50 percent drop in revenue when the Western Harbour Tunnel crossing opens to the public in 1997. He is also concerned about the government's refusal to allow toll increases for the Eastern Harbour Tunnel:

> Since commencing operation in 1989, our toll schedule has remained the same, in effect a significant reduction when it is considered that Hong Kong's inflation rate has increased by 57% since we commenced operations. During this period, the Star Ferry has increased fares by over 90%, the MTR by 40% and cross harbour buses by over 100%.

The Matsumoto believes that if the current toll rates are not increased, the revenues will cause the return on investment to be substantially lower than that experienced by other public utility companies. NHKTC's ongoing discussions with the government of Hong Kong regarding toll increases have been to no avail, but it is continuing these talks in hopes of eventually obtaining a fair rate of return for its investors.

Matsumoto does assure stockholders that "The New Hong Kong Tunnel Company will continue to contribute to Hong Kong's development and growth." Hong Kong provides the Peoples Republic of China with more than 10 percent of that country's foreign investment, and more than 18,000 trucks per day cross the border from the PRC full of export goods—a container arrives or leaves a Hong Kong dock every five seconds or less. With this kind of traffic and prospects for even higher flow predicted, it appears that the Eastern and Western Harbour Tunnels will survive and even thrive.

THE PRC'S IMPACT ON HONG KONG

Gordon Wu, chairman of Hopewell Holdings and a major investor in Hong Kong, the PRC, and other Asian markets, was asked at a symposium in New York in the autumn of 1994 what would happen in Hong Kong after 1997. His answer: "1998!" However, the present government is planning ahead for its future when Great Britain turns the colony over to the Peoples Republic of China. Preparations for 1997 include major infrastructure projects totaling U.S.$20.3 billion.

Numerous development plans for growth have been created by the government as part of a master development plan.

- Metroplan is a development program specifically created to provide a framework where the public and private sectors can work together and make Hong Kong a better place in which to live and work. The Plan takes into consideration the relocation of the airport from Kai Tak to Chek Lap Kok.
- The Northwest New Territories are a specially targeted development area because of the proximity to the PRC. The Northwest New Territories Development Strategy Review was created to plan agricultural and industrial zones,

infrastructure requirements, environmental controls, and broad use of the region projected to 2011.

- The Southwest New Territories Development Strategy Review attempts to capitalize on this area, which is close to the developing western coast of the PRC.
- The Hong Kong Territorial Development Strategy encompasses a comprehensive transportation and industrial development plan for that island.
- The Kowloon Point Reclamation Planning and Urban Design Plan has attracted the attention of developer Wharf (Holdings) Ltd. The company has proposed a mixed land use development that fits within the government's Metroplan objectives.
- Central and Wanchai Reclamation will create 267 acres (108 hectares) of development sites for commercial residential and government office use and will set aside almost 60 acres (24 hectares) for open and leisure time usage.

As 1997 approaches, Hong Kong's potential as a major portal to China's 2 billion citizens is recognizable to everyone concerned, especially to BOT developers such as Kumagai Gumi.

12

FROM THE DARDANELLES TO DURBAN

The term BOT was supposedly coined by Turgut Ozal, Prime Minister of Turkey in the 1980s. Ozal, an engineer in a country where commerce, not engineering, was a national imperative, saw the value of the build, operate, transfer project as a way to propel his country into the twenty-first century. Ironically, the country where the term originated would take over a decade before actually initiating a BOT project.

Twice the size of California, with a predominantly Muslim population of 60 million, Turkey is one of the world's oldest civilizations. After the fall of the Roman Empire in the fifth century, the city of Constantinople remained the capital of the Byzantine Empire for 1,000 years. In 1453, the Ottoman Empire was formed and ruled the areas now known as Lebanon, Syria, Jordan, Israel, Saudi Arabia, Yemen, and Iraq.

During World War I Turkey joined forces with Germany, and after the armistice was signed Turkey was divested of much of its territory. A republic was declared several years later in 1924. Turkey joined the League of Nations in 1932 and remained neutral during most of World War II, finally declaring war on Germany in January 1945. Shortly thereafter the country became one of the founding members of the United Nations.

Because Turkey remained a democracy during the Cold War in a geographic area strongly influenced by the Soviet Union, it received a substantial amount of money from the United States to bolster its pro-Western status. The government of Turkey alternated between military and civilian rule in the late 1970s, and the latest military regime lasted from 1980 to 1983; martial law was finally lifted in 1984.

Under Turgut Ozal during the 1980s, Turkey's economy boomed, fueled by excessive government spending. However, the uncontrolled spending led to increased borrowing, which resulted in the accumulation of a national debt that would exceed U.S.$62 billion toward the latter part of 1993. The Turkish lira plummeted in value, and inflation rose to an annual rate of 150 percent. Gross domestic product fell by 6 percent in 1993.

President Ozal, who died on April 17, 1993, was succeeded briefly by Prime Minister Suleyman Demirez. In June of that year, Tansu Ciller was elected by the ruling party as its new leader. Prime Minister Ciller, the country's first female prime minister, is an economist who received her education at the University of Connecticut and Yale University. On taking office, Ciller's most immediate problem was how to deal with an economy in crisis.

Part of Prime Minister Ciller's program to prop up the country's economy included a $5 billion privatization program in an attempt to retire some of its national debt. In April 1995 the Office of Privatization Administration in Ankara indicated it would seek to sell off U.S.$60 billion of public assets.

FALSE STARTS

Although recent studies conducted by the World Bank indicated the need for, and the feasibility of, power plant BOT projects in Turkey, none have come to fruition to date. As of 1990, the German firm Siemens's attempts to complete a BOT deal in Turkey had spanned several years but never been finalized. Rumors concerning the start-up of BOT projects involving the Akkuyu Nuclear Power plant and the Ankara–Istanbul high-speed rail system have been subjects of engineering magazine articles for the past few years.

In 1991, a 1,700 MW cogeneration BOT project was announced by the Turkish government. A "protocol of intent" for this proposed U.S.$1 billion project was signed by the GAMA Group, one of Turkey's largest construction firms, Enron Corporation, and the Wing-Merrill Group of Houston, Texas. This BOT scheme remained in limbo for several years, but is now under construction as a non-BOT project. Enron Corporation will own 50 percent of the U.S.$545 million natural gas-burning power plant in Marmara and will also operate the plant when it comes on-line in 1997. Enron's partners in the project are Midland Electricity PLC of England (31 percent), GAMA (10 percent), and the Wing Group of Aspen, Colorado (9 percent).

In 1990, a working paper (entitled "The Build, Operate and Transfer ("BOT") Approach to Infrastructure Projects in Developing Countries," dated August 1990 (WPS 498)) prepared by Mark Augenblick and B. Scott Custer, Jr., for the World Bank revealed the current status of active or proposed BOT projects in Turkey:

1. Akkuyu Nuclear Power Plant	Abandoned
2. 1,000 MW coal-fired power plant	Contracts signed
3. Additional coal-fired power plants	Proposed
4. Hydro-electric power plants	Under construction (?)
5. Bosphorus Second Bridge	Under construction as non-BOT project

6.	Bosphorus Third Bridge	Abandoned
7.	Bosphorus Tunnel	Proposed
8.	High-Speed Rail Link, Istanbul–Ankara	Proposed
9.	Water Plant (Izmir)	Abandoned
10.	Ankara Metro	Proposed
11.	Toll roads	Proposed
12.	Port facilities and free trade zones	Proposed

As of 1995 only one of these proposed projects, the Ankara Metro, is under construction, but under a conventional owner-contractor arrangement, not as a BOT venture.

The Ankara Metro Project

The Ankara Metro Consortium is headed up by SNC-Lavalin, a large engineering and project management firm in Canada. Bombardier Inc.—a division of Canada's Bombardier Transportation Equipment Group, a world leader in the production of railroad rolling stock—is being called on to furnish the complete metro system for the Ankara Metro project. Turkey-based general contractor GMA has formed a joint venture with a mechanical and electrical equipment installer to perform the construction work. The first phase of this 33 mile (53.5 km) rapid transit system will consist of 12 stations along a 9 mile (14.6 km) route, and is scheduled to open in March 1997; the last railcar is scheduled for delivery in November 1997.

A Failed Power Plant Project

In their 1990 World Bank working paper, Augenblick and Custer detailed the tortured path one proposed BOT project took before it finally died a natural death from old age.

In 1984 the government of Turkey asked the Bechtel Group to investigate the feasibility of building a 600 to 1,000 MW power plant via the BOT method. After a year of investigation Bechtel concluded that a BOT project such as the one under consideration was possible at a cost of approximately U.S.$1 billion. Bechtel suggested that a consortium be formed consisting of Combustion Engineering from the United States and Kraftwerk, a German firm that would furnish the steam turbines for a 960 MW project. The other members of the team would be a Japanese construction company and TEK, the Turkish government-owned utility company.

The pro forma prepared by Bechtel was based on TEK providing payments to the consortium in currencies of the countries of the lenders and the equity investors to overcome any concerns about drastic swings in the currency conversion rates of the Turkish lira. The Turkish government agreed to guarantee TEK's monetary contributions and waived its requirement to collect a corporate income tax from the consortium. Although the government was willing to back TEK's obligations, it did not want to guarantee any debt repayment; the Export–Import Bank was agreeable to providing funding for the project, but required a sovereign guarantee.

The financial issue was resolved in January 1987 when the government announced its decision to issue an invitation to bid and selecting six bidders, who submitted their proposals in September of that year. The proposals were ranked according to the estimated power included in each proposer's bid. The successful tender was submitted by a consortium headed by Seapac Control Services Pty Ltd. of Australia and Westinghouse. The equity investors included: Seapac; the government of Queensland, Australia; TEC; and a group composed of Japanese contractor Chiyoda and several Japanese equipment manufacturers and trading companies. Debt financing was to be provided by the Export–Import Bank of the United States (USEXIM) and the Export–Import Bank of Japan (JEXIM), International Finance Corporation, a member of the World Bank Group, and number of commercial lending institutions.

The Turkish government favored Seapac's proposal over the earlier Bechtel deal because, among other reasons, the power generating plant would be larger than that proposed by Bechtel—1,050 MW instead of 960 MW. Another selling point was that the power plant would consist of three 350 MW units; therefore, it could be brought on-stream in stages.

The government of Queensland decided to withdraw from the group of equity investors in 1988. The Turkish government then embarked on a series of negotiations with each of the tenderers, trying to play one against the other to get the best deal. Bechtel withdraw from the competition, and in the end the government settled on the Westinghouse–Chiyoda group.

The Westinghouse–Chiyoda group announced that it had reached a final negotiated deal with the government in June 1989, but the Turkish government put the project on hold in July. The government then signed an agreement in principle with one of the other tenderers, a Japanese group that had proposed to construct a coal-fired plant at a cost of $1.3 billion. The agreement in principle was never converted into a contract, and the government's wheeling and dealing and vacillation finally killed the project.

THE NEW BOT MODEL IN TURKEY

In May 1994, the Turkish constitutional court declared null and void all powers previously granted to the Turkish government to privatize state-owned industry. It appeared that the existing law prohibited foreign concerns from owning land in Turkey; as one of the requirements of a BOT project is land acquisition, ostensibly it was illegal to form such a project in that country. Under the direction of Prime Minister Ciller in late 1994, new legislation was enacted permitting BOT projects.

Law 3465

Law 3465, "Granting Concession to Establishments Other Than General Directorate of Highways for Construction, Maintenance, and Operation of Highways (Motorways) with Controlled Access," was enacted in 1994, and became effective on

October 1, 1994. Having thereby cleared the way, the government announced its intention to solicit BOT proposals for the following specific types of projects:

Bridges
Tunnels
Dams
Irrigation channels
Water purification and sewage systems
Communications
Mining operations
Manufacturing plants
Environmental pollution protection systems
Highway and railway construction
Subway and highway garages
Seaport and airport construction

The solicitation of energy production, transmission and distribution BOT projects would be regulated by a law other than the 3465 BOT law.

The BOT Project Announcement

Announcements of BOT projects of interest to the government are advertised in the *Official Gazette,* a government publication, and in the two major newspapers with the highest circulation at least 30 days before the application deadline. Notifications of proposed BOT projects will also be advertised in foreign publications 45 days prior to the application deadline.

The BOT Investor

According to Turkish law, two types of companies can qualify as BOT investors:

1. Domestic corporations having the status of "Anonim Sirketi A.S."—companies already or to be established in Turkey. These domestic corporations do not have to be 100 percent privately owned; some of shareholders can be public institutions or corporations. Limited liability corporations (LLCs) do not meet the domestic corporation criteria.
2. Foreign companies that maintain a subsidiary or branch operating in Turkey in accordance with Law 6224, "Encouragement of Foreign Capital." In order for a foreign company to qualify,the potential investor must
   - prove that it has a financial structure strong enough to finance the project and submit audited and certified financial statements to that effect.

- have at least one shareholder of the potential team that can exhibit experience in the activity being proposed.
- submit documentation for evaluation including a feasibility report, work program, and cash flow statement.

Limited liability companies (LLCs) do not meet the qualifications for participation in BOT-type projects. BOT projects are governed by the High Planning Council of the government, and potential investors must obtain approval of the council before they can enter into any other agreements.

The BOT Agreement

In general, the government requires that all BOT agreements include the following:

Identification of all parties to the agreement
Subject and term of the agreement
General principles of the investment and services to be provided
Financing arrangements and loans to be instituted
Cost of the project
Terms and conditions of operation
Any delays and changes in cost during the investment period
Expropriation clauses
Guarantees
Requirement for monthly and annual operation reports
Audits to be submitted and dates for the audits
Security and environmental protection measures
Causes accepted as force majeure
Insurance as required by the project
Maintenance and repair procedures
Transfer of the assignment to another BOT project
Termination clauses
Terms of the transfer at the end of the period
Terms of the transfer before the end of the period
Responsibility, liability, and compensation requirements
Method for handling errors caused by the administration
Training to be provided
Applicable governing laws and resolution of conflicts
Expenses incurred and to be included in the agreement
Notification
Procedures for changes to the agreement
Other agreements, and other issues that form the basis of the primary agreement

Administration: The term "administration" is used by the Turkish government to denote a state economic enterprise, a public corporation, or a related ministry authorized by the High Planning Council to enter into an agreement with an investor or a BOT proposer.

Term: The term of the agreement or concession period may not exceed 49 years which is to include project completion, delays due to default of the administration and force majeure.

BOT Project Financing

The BOT project agreement must cover all issues relating to financing such as

- The total amount of investment to be financed
- Price escalation and the formulas used for the purpose
- The conditions that necessitate the use of subordinated loans
- Methods of financing tax payments and additional costs
- All the issues related to the loans that will be used to finance the project—the type, amount, terms, and covenants

Loan Types Three types of loans are acceptable instruments for BOT project financing in Turkey:

- **Senior Loan.** A loan obtained by the investor specifically to finance the project and related services. This loan will not carry the guarantee of the Turkish Treasure.
- **Subordinated Loan.** A short- to medium-term loan with partial or full guarantee by the Turkish Treasury obtained to finance unexpected costs of the BOT project resulting from conditions stipulated in the agreement, but not caused by the default of the investor. Subordinated loans can finance any losses caused by the administration, losses incurred by the investor resulting from force majeure, or any changes made by parties to the agreement.

Guarantees The investment team may be required to provide a guarantee of as much as 1 percent of the total project cost. This guarantee can take the form of cash in Turkish currency (lira); letters of guarantee with an unlimited time period issued by banks listed with the Ministry of Finance; securities such as government bonds, treasury bills, revenue sharing certificates, and similar liquid securities issued by the government.

Bonds Article 22 of Decree 94/5907 "Concerning the Method and Principles of Application Regarding the Realization of Some Investments and Services under the Build-Operate-Transfer Model," effective October 1994, includes the requirements for bonding:

Stand-By Loan: A loan obtained from the investor in order to finance any shortage of cash resulting from a default of the investor under conditions stipulated in the Agreement. Stand-by loans do not have the guarantee of the Turkish Treasury.

Article 22. The implementation contract includes provisions stating that at the beginning of the work, a bid bond and performance bond will be taken from the assigned company up to the value of one percent of the total investment cost that is the basis for the contract. Acceptable bonds are as follows:

A. Turkish money in circulation.
B. Bonds with no time limitation provided by banks indicated by the Ministry of Finance.
C. Government bonds, treasury bond that are acceptable as guarantee letters in tender, revenue sharing bonds that belong to the state and that are exchangeable to cash.

The Turkish Treasury may provide the investor with guarantees on behalf of the administration:

1. To secure payment to the investor for goods and services purchased by the administration.
2. To secure full or partial repayment of the subordinated loans obtained from international financial institutions provided that the provision of the loan is in accordance with the agreement.
3. To provide guarantee in favor of the Public Funds which may bear the financial obligations of the subordinated loans in case the Public Funds are assigned to obtain the subordinated loans for financing the BOT investment.
4. To secure repayment of the foreign loans other than subordinated loans if the shares of the investor are transferred to the administration before the end of the term in accordance with the related provisions of the agreement. However, this type of guarantee will not be available unless the registration related to the right to construct in favor of the investor is removed from the title deed.

No guarantee is provided by the Turkish Treasury as to the repayment of senior loans obtained from domestic financial institutions.

Termination of the Agreement

The administration may unilaterally terminate the agreement for any one of the following reasons:

Failure to perform on the part of the investor.

Breach of the conditions of the agreement by the investors.

Bankruptcy of the investors.

Declaration of reorganization of the investor under the Turkish Bankruptcy Code.

Default of the investor in payment (insolvency).

Determining the Price of the Project

The administration will deem any BOT proposal that offers the most competitive cost throughout the operating period to be the most acceptable bid. With the high rate of inflation in Turkey in recent years, price adjustment due to inflation is of paramount importance in any long-term financial deal. Price escalation must follow the prescribed formula devised to deal with inflation:

$$(a \times \text{CPI}) + (1 - a) \times M - X$$

where

a = coefficient between 0 and 1 depending on the type of investment and service.

CPI = Consumer Price Index, used depending on the type of goods or services produced.

M = the weighted average of increase in a major input cost.

X = a coefficient proposed by the BOT investor to account for a possible decrease in cost that would result from increase in productivity and market growth.

Alternately the price could be escalated by a rate equal to $\text{CPI} - X$.

Taxation and the BOT Investment

All transactions by the investors and the administration relating to BOT projects are exempt from a stamp tax and property purchase tax. Several transactions and related document preparation are also exempt from these taxes:

- Letters of guarantee given to the administration by investors
- The primary agreement and other agreements concluded with public institutions relating to the BOT project
- Expropriations made by the administration in favor of the investor
- Purchase of goods and services and the obtaining of loans to be used during the investment period
- The "operation" and "transfer" transactions between the administration and the investor
- Corporate tax issues

The corporate tax rate in Turkey as of the first half of 1995 was 25 percent. Withholding tax is based on profit after corporation tax which effectively increases the tax to 42.8 percent. This effective rate will be reduced by the government depending on the amount of investment incentive exempt from the corporation tax; however, the minimum tax cannot be lower than 20 percent of corporate income.

A 7 percent surcharge is then calculated on the minimum corporation tax. Several examples were provided by the government to illustrate these points:

Example 1

Corporate income	100.00
Corporation tax (25%)	25.00
Surcharges (25.0 × 7%)	1.75
Income tax base (100 − 25)	75.00
Income tax (20%)	15.00
Surcharge (15 × 7%)	1.05
Net profit after tax	57.20
Total tax burden	48.80

Example 2

Corporate income	100.00
Less: Investment Allowance	(100.00)
Corporation tax base	0.00
Corporation tax (25%)	0.00
Related compulsory funds (7% × 0)	0.00
Minimum corporation tax (20%)	20.00
Related compulsory funds (7% × 20)	1.40
Total tax burden	21.40

Withholding taxes for nonresident companies are required to be paid at the rates prescribed by the government:

Wages and salaries	26.75%
Professional services	21.40%
Rent	21.40%
Sale or transfer of intangible assets	27.75%
Royalties	21.40%
Interest payments on foreign loans	0.00%

THE IZMIT AND THE DARDANELLES STRAIT SUSPENSION BRIDGES

Two of Turkey's current BOT projects will add to one of the most heavily trafficked truck routes into Turkey, which passes slightly north of Istanbul before continuing due west and then southward into the heart of the country.

- The Izmit Bridge project is designed to handle heavy truck traffic and speed freight service into Ankara and directly south feeding into the country's existing arterial road system.

- The Dardanelles Straits Suspension bridge is to be designed to handle light truck and automobile traffic connecting Çanakkale and Eceabat, thereby allowing entry into the country without having to cross through Istanbul.

With the passage of legislation allowing the government to solicit BOT projects, the Ministry of Public Works and Improvements issued Requests for Proposals for both bridge projects. Their locations in the northeastern region of the country are shown in Fig. 12-1.

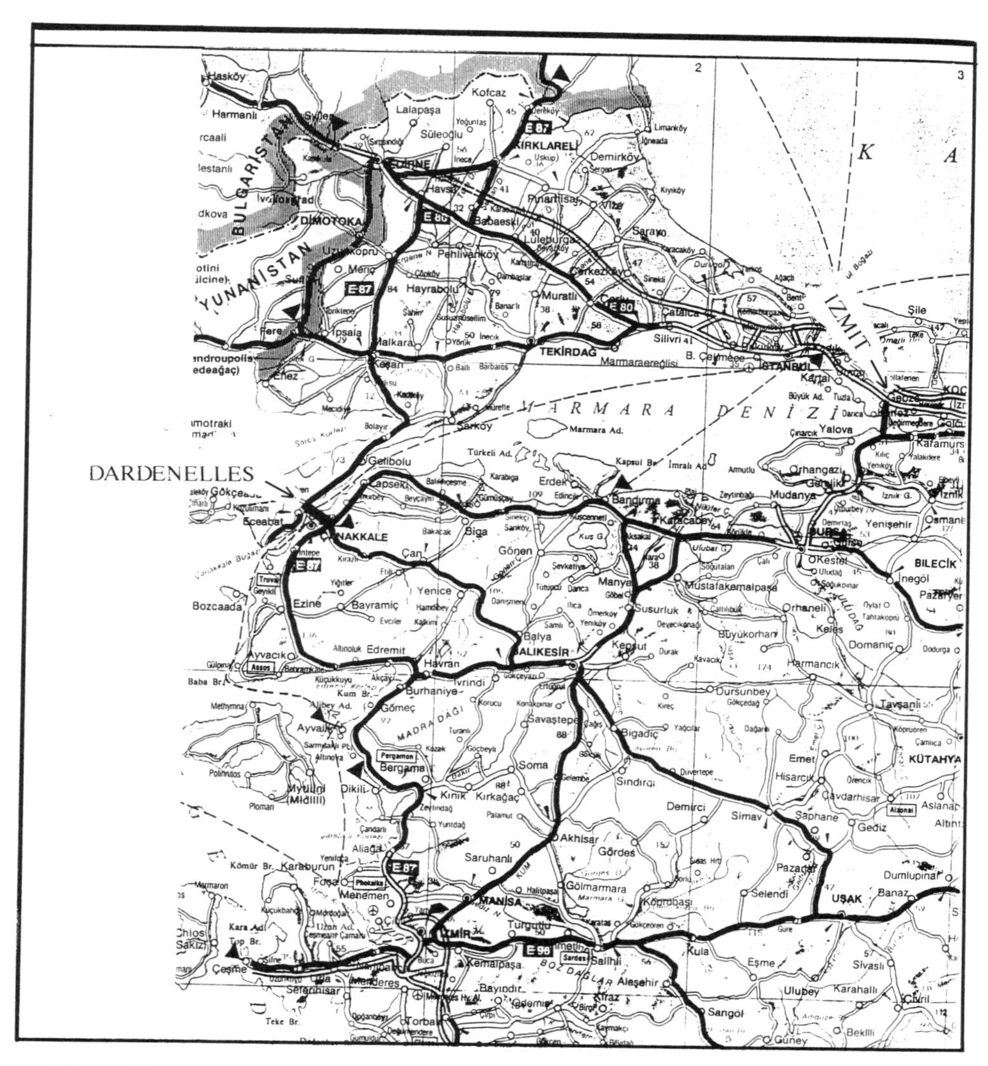

Figure 12.1. Northeastern portion of Turkey showing location of Izmit Bridge and Dardenelles Strait Bridge.

Bidder Profile: KiSKA Construction of Turkey

KiSKA Construction Corporation, headquartered in Ankara, Turkey, maintains branch offices in Istanbul, in Middlesex, England, and in Whitestone, New York. The company will celebrate its 30th anniversary in 1996; during the 30 years it has been in the construction industry, it has completed hundreds of noteworthy projects: water treatment projects in Turkey valued at U.S. $86 million and projects in the Middle East, Russia, and the United States totaling more than $2.1 billion. The company owns Lime and Brick Industrial Company, Inc., 3-A Aluminum Alloys Co., Inc., Aegean Industry Minerals Co., Kurtmak Kurtkoy Machinery and Industrial Co., and the Istanbul Tourism and Hotelling Co., all located in Turkey.

KiSKA Construction Corporation USA was founded in 1987 and established itself as a builder of prestigious high rise apartment complexes in the tough Manhattan, New York market. The company has a strong presence in civil engineering projects, particularly with shafts and tunnels for water systems, and transportation-related work including the rehabilitation of bridges and highway viaducts.

KiSKA participated in the construction of the Washington, D.C. Metro system and built the 14th Street/Park Road Tunnels with a total contract value of $90,983,000. The company obtained several contracts for the rehabilitation of expressways in New York City worth $65.4 million and completed Shafts 21B and 22B as part of New York City's $4 billion water supply expansion system. Rehabbing three bridges in Suffolk County, New York gained the company another $20 million in contract work.

Erben Arden, President of KiSKA Construction Corporation USA, was asked to assist in the preparation of tenders for both the Izmit and Dardanelles projects on behalf of his company. Arden worked with the Guy F. Atkinson Company of San Francisco and the engineering firm Sverdrup Corporation on the proposal for the Izmit Bridge project.

THE DARDANELLES PASSAGE BOT PROJECT

According to government directives, only domestic capital companies with "established technical and financial capabilities, who have been or will be founded, or who have guaranteed to be founded as set forth in Act 6224" would be permitted to submit bids on the Dardanelles Straits Suspension bridge project to connect Çanakkale and Eceabat. The Turkish government estimated that a company's financial resources should be in the U.S.$300 to U.S.$350 million range so as to ensure completion of the project.

The KiSKA, Atkinson, and Sverdrup joint venture was to be one of eight groups bidding on the Dardanelles project and the only one that would contain American entries. Subsequently, all three firms decided to withdraw from the competition on both projects. One consortium consisted of Tekfen Insaat (Turkey), Impregilo (Italy), Bilfinger Berger (Germany), Campenon Bernard (France), and Transroute (France). Other groups were composed of Enka Insaat (Turkey), Ishawajima–

Harima and Marubeni Corporation (both of Japan), and Trafalgar House (England).

Contract details were being offered by the Directorate of Motorways of the Department of Highways for 10 Turkish liras (which equates to U.S.$270 (February 1995 rates). Plans and location maps scaled to 1/5000 and 1/1000 scale were to be made available to the successful contractor. The Ministry of Public Works and Highways Dardanelles Straits bridge information included a preliminary design of the cable-stayed bridge, the span, navigational clearance specifications, deck cross section, and other "outline" specifications data in the form of a design schematic.

No traffic count was provided with the bid information, however; each bidder was to base its revenue stream for the pro forma on its own traffic study. KiSKA's Arden commented that the detailed design information available would ensure that all Dardanelles Bridge project bidders based their construction estimates on the same basic design, but without a traffic study baseline, each tenderer would probably assume different traffic flow rates, resulting in different revenue streams and differing return on investment scenarios.

The closing date for the Dardanelles Strait project was April 25, 1995.

THE IZMIT BAY CROSSING PROJECT

The Izmit Bridge BOT project, for heavy truck traffic and speed freight service into Ankara and directly south into the country's existing arterial road system, was announced in June 1994. The General Directorate of Highways (KGM) issued a prequalification document that set forth the project scope and the qualifications necessary to be considered for the KGM's tender short list. The short-listed companies or consortiums would then be invited to submit detailed proposals for the BOT project. Those included in the short list had to have met the following qualifications:

- Proven experience in the design and construction of major infrastructure projects, particularly on long-span bridges in seismic zones similar to Izmit Bay
- Proven experience and managerial capability in managing major transportation projects
- Necessary financial strength and ability to secure a sound financial package
- Experience and managerial capability in traffic management, operation, and maintenance of tolled highways and bridges

KGM reserved the absolute right to short list and select any organization it deemed qualified for the project. The format of the prequalification proposal had three parts:

Organizational Details: A description of the company's history and a list of the directors and principal shareholders were required along with profiles of key personnel. Details of participation in similar infrastructure projects within the last 10 years must also be provided.

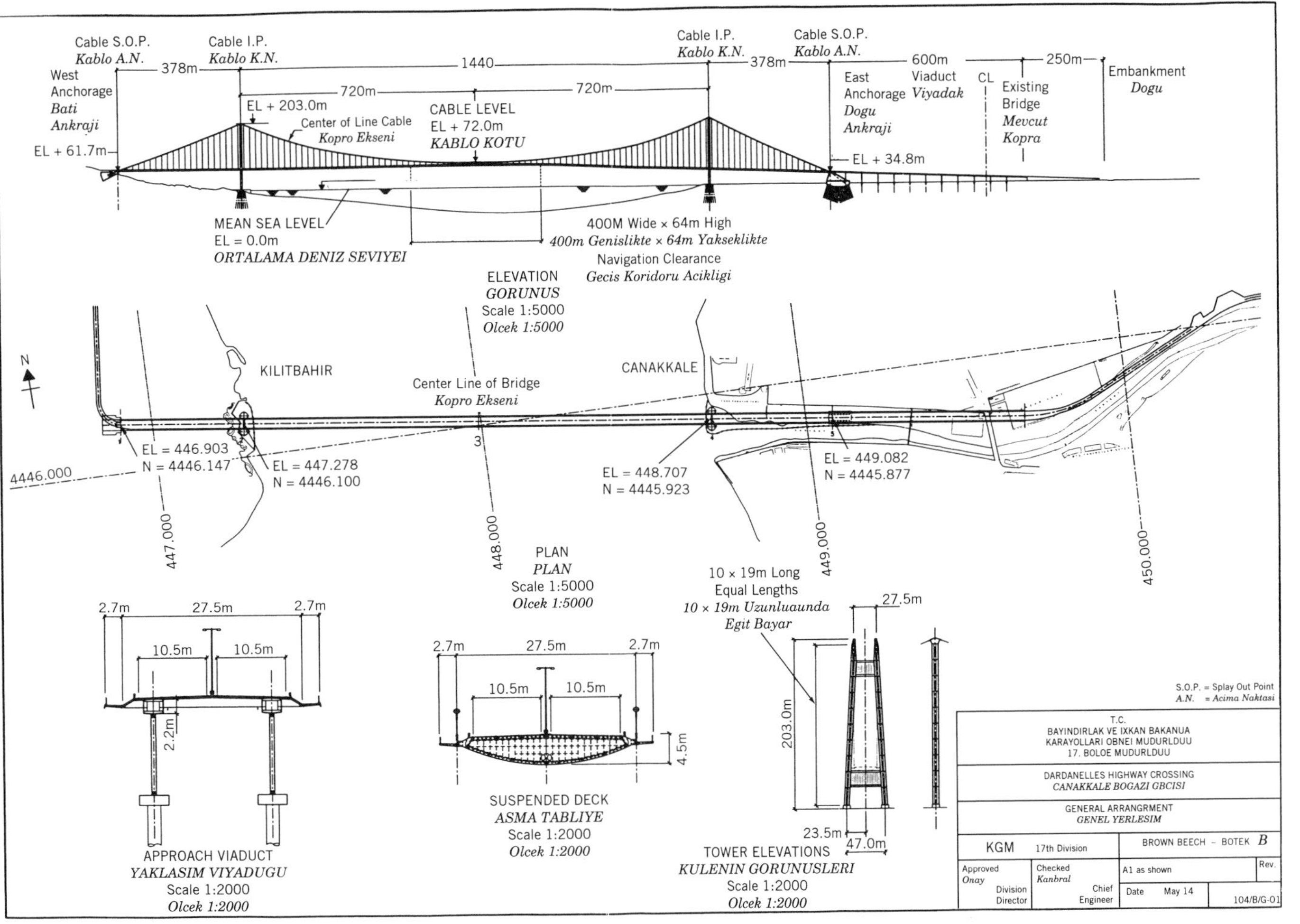

Figure 12.2. Schematic of the Dardanelles Strait Crossing Bridge.

Financial Information: A statement of the authorized capital of the proposer and certified copies of financial statements for the past five years plus the current year's statement, even if uncertified, were required. These statements could be prepared in either English or Turkish. The financial setup of the proposed consortium or joint venture was to contain the names and registered addresses of all shareholders, the proposed capital structure, and the percentage to be held by each shareholder. Copies of shareholder agreements, if available, and the types of guarantees which the consortium or joint venture would provide were to be submitted.

Engineering and Construction Information: The applicant must provide evidence of the experience of the consortium in similar civil engineering projects, the experience and track record of the contractor, the details of major holdings in plant and equipment required for the project. The applicant's specific and relevant engineering experience must include, but is not limited to, geotechnical/seismic engineering, wind evaluation testing, long-span bridge construction, and deep-sea foundations.

The government's RFP documents provided anticipated traffic flow data, but design information was the responsibility of the proposer. The applicants were to provide an outline of their proposed method of attacking the project. The applicants' design approach was to be fully explained and submitted with a design milestone schedule and a detailed management structure.

Erben Arden, President of KiSKA Construction Corporation USA, noted that the Izmit Bridge project created a difficult evaluation situation for the government. A sophisticated design could result in higher construction costs that would have to be supported by higher toll rates whereas a low bid might result in lower toll rates but produce a lower quality structure.

The Outline Implementation Proposal

The construction of the proposed six-lane Izmit Bay crossing would not be an easy task. Not only is the bay beneath the bridge an active commercial and naval vessel passageway, but the area in which the bridge is to be built is an active seismic zone. The Sea of Marmara was declared a tsunamigenic region—susceptible to tsunami waves. The soil strata under the seabed was determined to be not conducive to conventional foundation structures.

The Izmit Bay Crossing prequalification document from the General Directorate of Highways provided prospective proposers with a fairly definitive scope.

Bridge Structure and Related Motorways

The code requirements for the Izmit Bridge were similar to those used for the design of the existing Fatih (Second Bosphorus) Bridge; the International Technical Specifications for Highway Bridges of the General Directorate of Highways of Turkey were to be incorporated as well. But there were a number of constraints that the government would impose on each proposer making the design and construction of this bridge an even more formidable task. KGM included in their request for proposal the "preferred bridge configurations by which the design criteria should be determined":

- The design must allow for a total service life of 100 years
- Structural flexibility should accommodate any potential seismic movements, including asynchronous motions
- A minimum number of piers within the sea channel with adequate horizontal navigational clearance for both commercial and naval vessels must be provided
- Overall bridge length, including viaduct spans on the south shore, would be about 4,000 meters (approximately 2.48 mi)
- The overall width will be 35 meters (114.8 ft) including pedestrian walkways on each side

A long-span cable-supported bridge (suspension, cable-stayed, or hybrid) was deemed by KGM to meet the above criteria, although this was not to be construed as placing a limit on other types of bridge design.

The alignment of the motorway and bridge had several existing conditions to deal with. The motorway would consist of a 3 kilometer (1.86 mi) section from an interchange near Dilovasi, the bridge itself, and a 43 kilometer (26.66 mi) section running southerly to Orhangazi. The six-lane bridge would cross the bay in a straight alignment between Kara Burun on the north side and Dil Burnu on the south shore.

The bridge alignment was to connect to a limestone rock promontory at Kara Burn:

- The village of Dilovasi, directly north of the planned bridge location, had to be considered in the approach road design.
- An active quarry is in operation at the axis of the bridge alignment on the north coastline.
- A dual-track railway tunnel for the line that serves as the primary connection between Ankara and Istanbul is located on the north shore.
- An abandoned single-track railway is also located close to the north coastline.

On the south shore the bridge would connect with the tip of the peninsula:

- The bridge aligns with the west side of a currently operational naval communications center.
- Several small residential buildings are also located on the peninsula.
- The planned bridge alignment coincides with the west side of a lagoonlike lake within the peninsula.

Marine and Meteorological Conditions

The maximum water depth along with planned bridge alignment varies from 60 meters (197 ft) near the south end of the crossing to 30 meters (98 ft) to 50 meters (164 ft) along the north half of the crossing.

Very high winds prevail in Izmit Bay and suitable testing for wind effects was to be identified in the proposal along with evidence of an advanced level of investigation required for a structure of the magnitude planned.

Geotechnical and Seismic Considerations

The 1,500 meter (4,920 ft) long North Anatolian seismic fault zone extends from eastern Turkey to mainland Greece and is divided into three strands: southern, middle, and northern. The northern strand passes beneath the northern portion of the Sea of Marmara, and Izmit Bay is a part of this seismic zone.

In Turkey, the most severe seismic zone is labeled Zone 1, and the least severe, Zone 4. Izmit Bay falls within the Zone 1 area, and horizontal ground acceleration coefficients are greater than 40 percent of acceleration due to gravity.

Izmit Bay has been the site of several seismic events with intensities of 8 on the Richter Scale and above, including the Great Istanbul Earthquake of 1894. The 1963 Yalova–Çinarcik earthquake, which measured 6.4 on the Richter Scale, occurred within 12 miles (19 km) to 25 miles (40 km) from the site of the proposed bridge.

KGM advised applicants to take into account the asynchronous nature of the ground motion when foundation design is considered for structures on both sides of the fault trace, and to submit a geotechnical evaluation of liquefaction potential with their proposals. KGM advised applicants that the soil strata under the seabed on the northern portion of the Dil Burnu south shore consisted largely of deep alluvial soils with no indication at what level bedrock could be found. Thus, deep geotechnical bore-hole explorations would be necessary for the proposer to determine the depth of alluvial soils and their susceptibility to liquefaction under earthquake effects.

Since the Sea of Marmara is a tsunamigenic region, a tsunami wave propagation study was also to be included for the design of off-shore and/or low-altitude onshore structures.

The RFP Consortiums

KiSKA Construction, although it had spent considerable time and work on bid preparations, withdrew from the competition when one of the lead partners in the group lost interest in the project. Thus, the Izmit Bay Crossing short list, as of April 1995, was as follows:

1. Anglo–Japanese–Turkish Consortium

Enka Ins. San A.S.	Turkey
Ishikawajima–Harima Heavy Industries	Japan

Itochu Corporation	Japan
Marubeni Corporation	Japan
Mitsubishi Heavy Industries Ltd.	Japan
NKK Corporation	Japan
Trafalgar House Corp. Development Ltd.	Great Britain
2. I.B.K.O. Consortium	
Impregilo SpA	Italy
Campenon Bernard SGE	France
Bilfinger + Berger	Germany
Dragados Construcces	Spain
Tekfen Insaat TES S.A.	Turkey
Transroute International S.A.	France
3. Izmit Korfez Gecisi Ve Otoyolu	
Dyckerhoff & Widmann AG	Germany
STFA Ins. A.S.	Turkey
Doğus Ins. Tic, A.S.	Turkey
Spie Batignolles	France
Cubiertas Myzov S.A.	Spain
Salini Constructtor SpA	Italy
4. Izmit Bay Crossing Consortium	
Hochtief AG	Germany
Balfour Beatty	Great Britain
Garanti Koza Ins.	Turkey
Botek A.S.	Turkey
5. Bouygues Vinsan	
Bouygues S.A.	France
Vinsan A.S.	Turkey
6. Izmit Bay Crossing Group	
Yapi Merkezi Ins. San A.S.	Turkey
Wayss & Greytagag	Holland
Interbeton BV	Holland
Freyssinet Yapi Sis San A.S.	Turkey
Freyssinet International Inc.	France
Parsons Brinkerhoff International	U.S.A.
ICF Kaiser Engineers Inc.	U.S.A.
Norcensult Group	Norway

The closing dates for bid proposals for this complex project was set by the Turkish government at 5:00 P.M., July 18, 1996.

THE FUTURE OF BOT IN TURKEY

The economic situation in Turkey in mid-1995 was still precarious and not the most fertile ground for risk-laden BOT projects. Comparing the Turkish financial situation with recent events in Mexico, one local expert said "Mexico's situation is nothing new, we invented it." Can Yesilada, chairman of the Privatization Administration says that "nobody understands that high risk means high returns"—but investors in BOT project certainly do understand. An economically shaky host government may present a risk too high for all but the most daring BOT investors. However, Turkey's triple-digit inflation rate is dropping along with its interest rates, and foreign investors are beginning to come back into the country.

TRANSPORT INFRASTRUCTURE IN SOUTH AMERICA

The government of South Africa, much like that of Turkey, is exploring methods by which BOT projects can be made attractive to potential investors. But in Pretoria, the administrative capital of the country, legislators have not been able to mount an effective campaign to enact the necessary laws that would permit the creation of BOT projects. Instead, several investor groups have agreed to proceed to design, finance, construct, operate, and maintain several private toll roads based on a "handshake" agreement with the government.

Motorists using Route N2 heading north from Durban to the Mtunzini Toll Plaza and Route N3 leading east from Durban in a northern direction to Pretoria will be stopping at three toll stations along the way. A major portion of this highway system has been financed, designed, constructed, operated, and maintained by the private corporation Tollcon (Toll Road Concessionaires Pty Ltd.).

THE ORIGINS OF THE PRIVATIZATION INITIATIVE

The World Bank's 1994 World Development Report revealed that Sub-Saharan Africa's roads, with a total value of $13 billion representing one-third of all of those built in the past 20 years, had eroded because of lack of proper maintenance. Once roadways begin to deteriorate, lack of regular routine maintenance will only hasten the deterioration. Road maintenance in Equatorial countries appears to be a high cost item, perhaps due to extremes in temperature.

The highway system in South Africa was superior to most other Sub-Saharan countries, but maintenance was clearly the funding priority. In the 1980s the state road authorities in South Africa were unable to commit any substantial amount of funding to new highway construction because all their limited resources were required for the maintenance of existing highways. The government realized that with its resources thus committed, private-sector involvement must be considered if any significant highway expansion program was to take place.

Agent Agreements

In 1984 a toll road privatization movement arose after seven groups of contractors and bankers responded to an Expression of Interest issued by the then Minister of

Transport Affairs. Of the 7 groups of contractors responding, 4 groups were selected and allowed to submit bids for two potential toll road agreements, Toll Road Concessionaires Pty Ltd. (Tollcon), and Toll Highway Development Pty Ltd. (THDC).

Interim agreements were issued to both of the private consortiums in 1986 with the intention of concluding negotiations within a short period of time and awarding full concession agreements within months after the construction began. However, final negotiations never materialized, and the roads continued to be operated on the "agent" arrangement.

There was precedent for the agent concept. The National Roads Amendment Act of 1983 allowed the South African Roads Board to designate certain sections of the national road system as toll roads and appoint agents to collect the toll charges, and, in effect, to operate these toll roads. No provision existed then or today to permit full privatization of any government-owned highways, nor is there legislation on the books to permit the awarding of BOT-type concessions to developers to design, build, finance, construct, and operate toll roads.

Since 1986 these two companies have been acting as agents of the state with respect to the financing, design, construction, operation, and maintenance of certain sections of the national highway system. Tollcon was involved in the N1 and N3 program, and THDC was awarded the N17 roadway construction. Work on these projects began in 1988, and by 1990 various "private" toll road sections were open to the public.

The Legislation Roadblock

During the early 1980s the general public was apathetic toward the construction of toll roads in general and privatized toll roads in particular. But the levying of tolls on existing freeways brought forth very strong resistance from motorists. To neutralize some of the public reaction against the concept of new tolls, the Minister of Transport announced wide-ranging toll reductions. Tolls on existing roads would be changed so that the revenue stream produced would be at a level that would cover the cost of maintenance and rehabilitation. This change in policy was sufficient to turn away any developer looking for a return on investment; this program thus effectively scuttled the highway privatization movement.

When legislation to create privatized toll roads was introduced into the two houses of the South African legislature in September 1988, it was summarily rejected. Legislators who attempted to enact the necessary laws to launch a privatization program found so many stumbling blocks along the way that the movement languished and finally died in August 1990.

Privately Operated Toll Roads

Although they had no concession agreements, Tollcon and THDC had continued with their plans to construct toll roads by acting as agents of the state. Pursuing this plan they were able to attract R1 billion (rands; U.S.$497.000) in private-sector

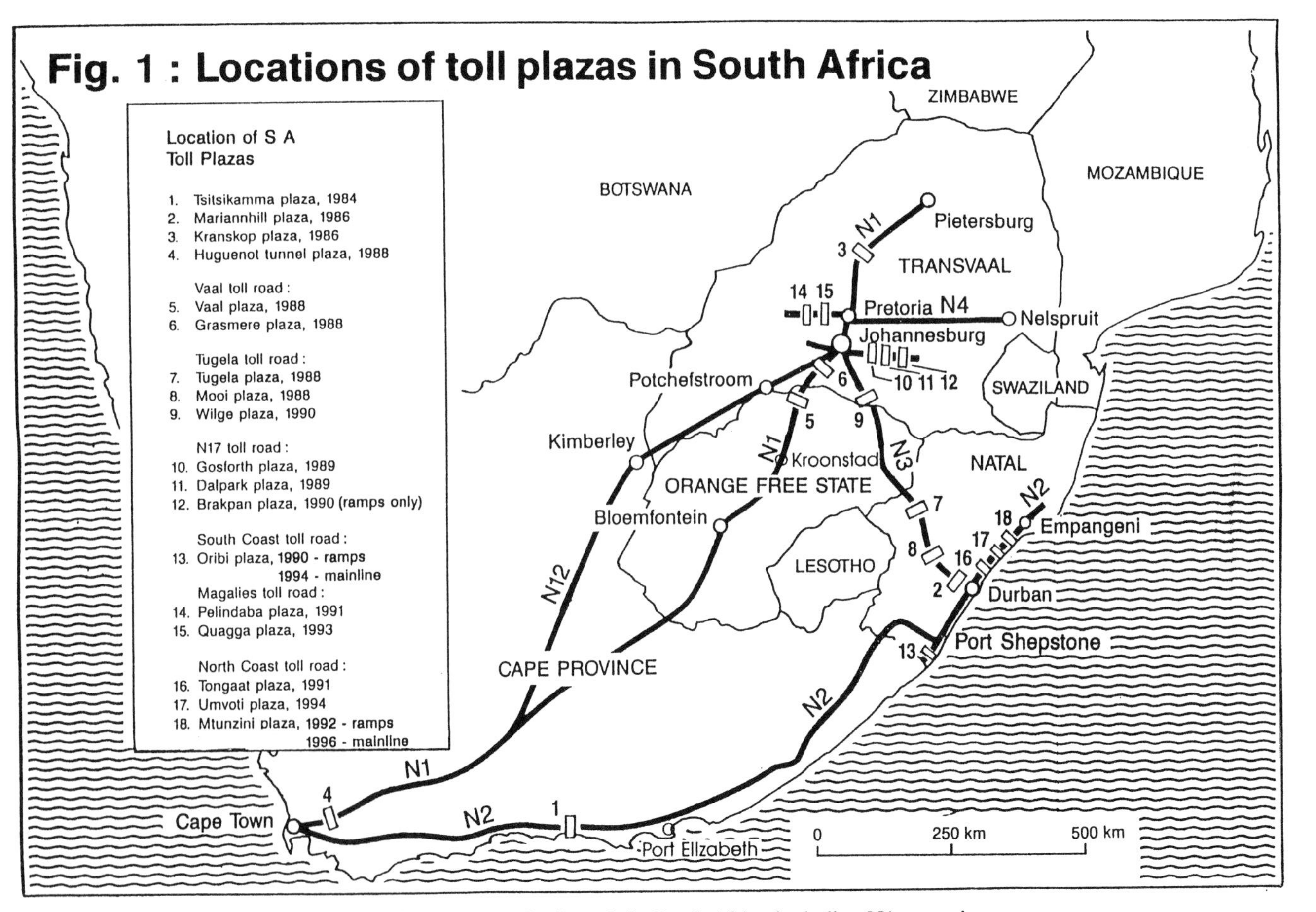

Figure 12.3. Network of toll roads in South Africa including N1 extension.

financing by 1987, which allowed them to continue the privatization of certain specific highways. When final concession agreements were reached in 1989, the agreements were conditional on the passage of the enabling legislation. But the legislation never took place because of the concerns of the communities were these roads were to have been built.

Without public support, the new toll roads would have no chance to succeed. An anti-toll road campaign mounted by the media at that time generated the threat of "toll busting"—toll evasions and low road traffic counts that no private toll road operation could survive for long. Thus, some of the toll routes selected for the private concession agent program did not appear to be financially viable without some form of government participation. The government, instead of providing cash contributions, transferred some of the recently constructed national road sections financed by the National Road Fund to the private companies at no cost. Thus, the private developers were able to install tolls and operate and maintain these networks, and then apply the revenue stream toward the short-fall from the newly constructed toll roads.

The government went a step farther when the South African Roads Board assumed full financial responsibility for the projects and loans taken out by the private companies on April 1, 1991. That left the operators with only the responsibility to operate and maintain the highways and eliminated the burden of amortizing the debt incurred to construct the facility.

Since the 1983 amendment to the National Roads Act in which toll road legalization was introduced, 12 South African toll roads covering a distance of 660 kilometers (407 mi) have been established as of 1994. Figure 12.3 depicts the network of these toll roads and the location of their toll plazas.

The Reconstruction and Development Program

The first South African toll road pilot program produced a gross revenue stream of R689,666 (approximately U.S.$190,000) in the 1984/1985 fiscal year. By the end of the 1993/1994 fiscal year this income had increased to R198.2 million (about U.S.$55 million). A major portion of the increase in traffic, and hence revenue, may have occurred because of the Reconstruction and Development Program (RDP) instituted by Nelson Mandela's African National Congress to spur faster economic growth and increase infrastructure development.

Jay Naidoo, the minister in charge of RDP, stated that much of the growth being experienced in the South African economy in mid-1995 was a result of the RDP and the increase in manufacturing was linked to RDP expenditures. He said, "We are beginning to see a kickstart in the construction industry. We are confident that infrastructural programs can be the vehicle for a growth strategy."

If the privatization of toll roads gets a portion of the RDP's kickstart, Tollcon will probably be the recipient of some of the concession agreements. Tollcon's major shareholders are three prominent South African construction companies listed on the Johannesburg Stock Exchange. Their minority shareholders include some of the country's major financial institutions. Tollcon is, at present, the largest operator of toll roads in South Africa, and manages on behalf of the South Africa Roads Board about 400 kilometers (247 mi) of toll roads encompassing the N1 and N3 highway networks. The average daily volume of traffic passing through their five toll plazas is about 36,000 vehicles, producing a revenue stream of R150 million (U.S.$45 million) per annum.

THE N1 PROJECT

The N1 project was conceived with the intention of extending a highway from the northern end of the existing Kranskop toll road at Middelfontein about 120 kilometers (74 mi) to Pietersburg, the capital of the Northern Transvaal province. The present location of the N1 is shown in Figure 12.4, and Figure 12.3 shows the proposed extension. The need for this road had existed since 1975, but unavailability of funds delayed the program until 1990 when some degree of urgency was attached to the commencement of construction. The government began to investigate the BOT concept for the road and decided to implement a competitive bidding process.

The Request for Proposals

The N1 tender requirements indicated the South African Roads Board's intention to contract with a private entity to design, construct, operate, and maintain a toll road for a period not to exceed 10 years. Because the Minister of Transport was vested with the authority to set toll charges, the proposers could not be held responsible for risks associated with establishing a revenue stream. So the government devised a schedule of monthly payments based on net toll revenue assumptions arrived at by conservative estimates of traffic flow and traffic growth.

The contractor in its tender would specify the period for which the monthly payments must continue and these payments were to constitute full compensation for the financing, construction, operation, and maintenance of the toll road over the tendered period. The Roads Board would then award a concession agreement based on the most competitive terms of the submitted proposal and would remit payments to the contractor escalated for cost of living increases over the term of the agreement. These payments would not be effected by levels of actual toll revenue. This arrangement was somewhat similar to the shadow tolling used in Great Britain.

Risks Associated with the N1 Project

In the case of N1, the contractors were required to procure all necessary funding at their own risk, not an uncommon occurrence in a conventional BOT project where the developer assumes risks associated with revenue and formulates the pro forma based on projected traffic flow and the cost to construct and fund the project. The

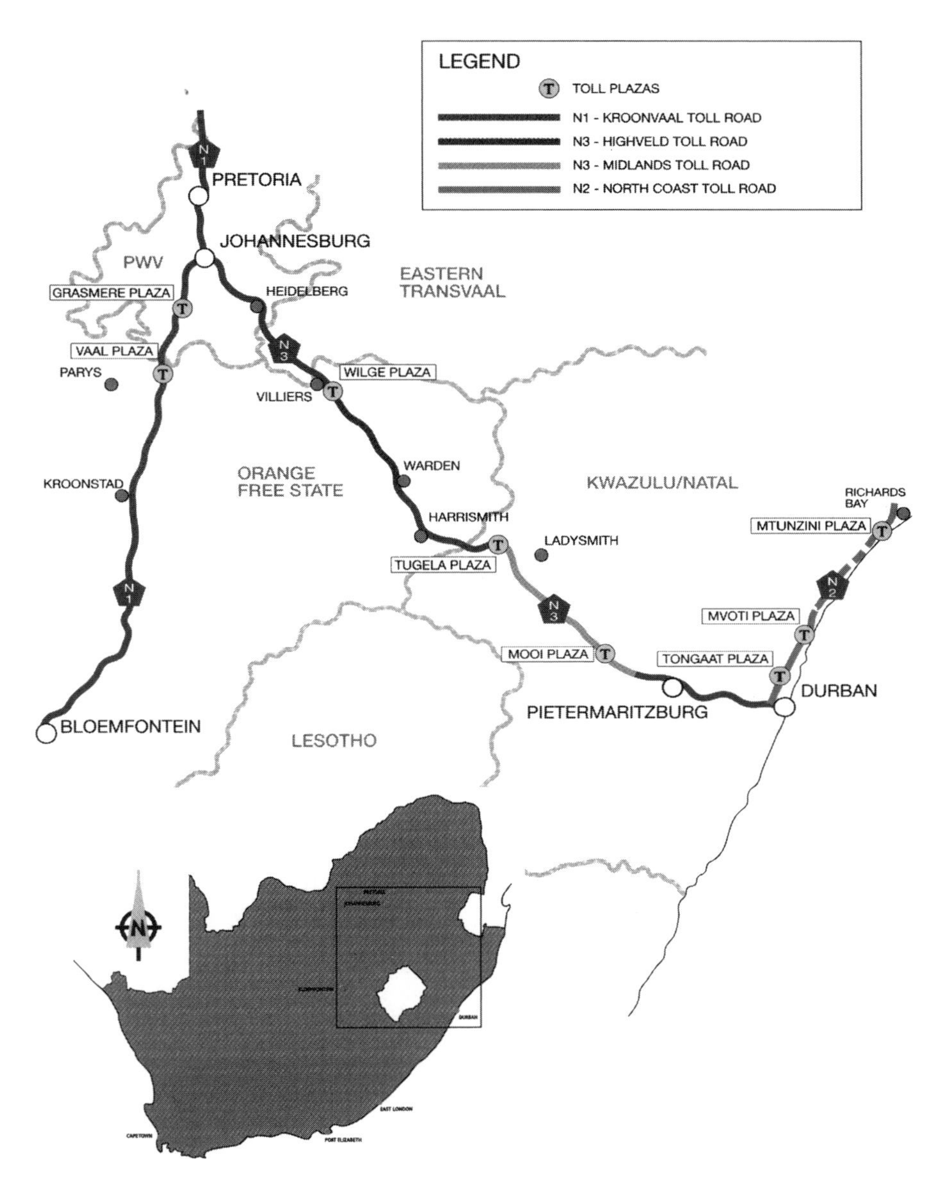

Figure 12.4. The existing N1 highway without the proposed extension and the N3 leading out of Durban.

developer accepts the risk that traffic may fall below projected estimates or, conversely, exceed estimates. But the risks involved with the N1 were not those of most conventional BOT programs:

1. The project held the normal construction-related risks of delays and cost overruns.
2. The usual risks relating to life-cycle costs, maintenance and rehabilitation costs over the term of the concession period existed.
3. The cost of funding was higher than normal. Foreign investors were reluctant to become involved in South Africa during this period while government changes were in progress. Local lenders were demanding a rate of return comparable to that of the stock market before they would consider such an investment.
4. The financial viability of the project was contingent on meeting certain traffic flows as the government would be bound by a stipulated monthly payment.
5. Political risks could arise from changes in government policy. South Africa was in the process of having its first truly democratic election.

The government recognized the risk potential associated with this program, and the developer's compensation would be based on the predetermined monthly payment from the government, not from the flow of traffic. The tender that was formulated by the government included the following provisions:

1. Construction risk to the contractors would be in the form of a lump sum, all-risk contract.
2. Life-cycle costs related to performance of the toll road pavement would rest with the contractor.
3. All risks associated with financing, save inflation, would be borne by the contractor.
4. As far as the revenue stream was concerned, the Roads Board would make scheduled payments to the developer irrespective of the actual levels of net toll revenue.
5. Political risks rested with the contractor.

Prequalification took place in December 1993, five qualified tenders were selected in January 1994, and four bids were received by the Board on March 25, 1994. Three of the bidders were South African firms and the fourth was from Kuwait.

The four tenderers who responded to the conditions set by the Board exhibited extreme reluctance to assume any of the following risks:

1. They did not wish to accept responsibility for toll road pavement maintenance over the full duration of the concession period.

2. They were unwilling to accept responsibility for fluctuations in the real cost of capital over the concession period.
3. Political risk was a major concern and a key issue for discussion.

Although the scheduled payments to the successful concessionaire were to have been guaranteed by the South African Roads Board, the four tenderers wanted the full faith of the South African government behind these guarantees, not just a guarantee from a statutory body of that government. It was evident to the Board that these conditions were being foisted on them by the proposer's lenders who were concerned about the Board's ability to meet these payments in a time of political uncertainty—a feeling that was widespread among the financial community at that time.

After considering all of these objections, the Board made a conditional award to the firm of Murray & Roberts and LTA and agreed to negotiate some of the terms and conditions that were found most objectionable by the proposers. An agreement was signed in January 1995 for a contract period of 285 months (23.75 years). Other than the requirement for federal government guarantees, all other goals of the tender were met.

THE FUTURE OF BOT IN SOUTH AFRICA

The future of BOT in South Africa may rest with the Reconstruction and Development Program. According to Colin Campbell, managing director of LTA, one of the partners in the N1 program, "The private sector cannot become meaningfully involved in filling the enormous requirements of the RDP without the willing and active partnership of government." He believes that for most new infrastructure development, government participation may well have to take the form of a subsidy to afford the private sector an adequate return on investment. These subsidies would not have to be in the form of a transfer of money but could consist of the transfer of an existing asset such as a highway, airport, or power distribution system to the private developer.

The Reconstruction and Development Plan seems to be working well and within it lie the seeds of a true public/private partnership and possibly the future of BOT in South Africa.

13

THE PHILIPPINES AND ITS BOT VISION OF THE FUTURE

In early 1995, Fidel V. Ramos, President of the Philippine Republic, pushed the button that fired up a 330 MW combined-cycle power plant at Barangay Kitang on the Bataan Peninsula. This power plant project was built by a consortium headed by ABB Power Generation Ltd., Kawasaki Industries, and Marubeni Corporation, the Japanese trading company. This U.S.$40.3 million project was only the latest of 19 power plant projects that had come online since Ramos took office. The sum total of foreign investment he has attracted for these types of projects has exceeded U.S.$17 billion—and far more is needed.

Ramos is clearly a man with a vision of what his country should aspire to be in the twenty-first century. A graduate of the U.S. Military Academy at West Point, he won a hotly contested presidential election against Corazon Aquino in 1992, and one of his campaign promises to the electorate was to provide the country with an adequate supply of electrical power by mid-1994.

Electric power in the Philippines has been substandard for years; brownouts were an everyday occurrence in 1992 and 1993—even the National Power Corporation, the generator of 95 percent of the country's electrical requirements, was often without electricity. The deficiency in delivering adequate power had lost the Philippines an estimated 400,000 jobs and over U.S.$3 billion in revenue per year.

To fulfill Ramos's vision and allow the Philippines to meet the social, political, and economic challenges that it must face in the next century, a vast array of infrastructure projects had to be put in place with power generation at the top of the list. Billions of dollars must be committed to infrastructure construction, funds that the Philippine government may not be able to deliver. President Ramos saw clearly that alternate sources of financing would be required to advance the development his nation's infrastructure, and he called on the private sector to provide that alternate source of capital. The vehicle through which it would be injected into the economy would be through the BOT concept and its many variations.

THE FILIPINO BOT PROGRAM

The BOT program envisioned by the Ramos government included many variations of the theme: build, own, operate (BOO); build, transfer (BT); build, transfer, operate (BTO); build, lease, transfer (BLT); rehabilitate, operate, transfer (ROT); rehabilitate, own, operate (ROO); develop, operate, transfer (DOT); and contract, add, operate (CAO) projects.

Republic Act 6957

The Philippines became the first country in Asia to enact a law specifically for the BOT process. Its first such law was enacted in 1987, but it was on July 9, 1990, that Section 110 of the Republic Act 6957 was passed. This law, known as the Build-Transfer-Operate Law, was "An act authorizing the financing, construction, operation and maintenance of infrastructure projects by the private sector and for other purposes." Section 1 of the Act defines BOT as follows:

> This is a contractual arrangement whereby the contractor undertakes the construction, including financing, of a given infrastructure facility, and its turnover after completion to the government agency of LGU (local government unit) concerned, which shall pay the contractor its total investment expended on the project, plus a reasonable rate of return thereon.
>
> The BOT scheme shall be adopted for financially viable infrastructure projects, i.e., the project cost should be recovered from tolls, fees, rentals and other charges on the facility users, as well as other related income of the Agency/LGU concerned. The agency/LGU shall pay the contractor based on an agreed amortization schedule.

The list of designated projects that could employ BOT were as follows:

1. Highways, road, bridges, interchanges, tunnels
2. Rail-based projects packaged with commercial development opportunities
3. Non-rail-based mass transit facilities, and navigable inland waterways
4. Port infrastructure, piers, wharves, quays, storage, handling, and ferry services
5. Airports and air navigation facilities
6. Power generation, and the distribution of electrification
7. Telecommunications, backbone network, and terrestrial and satellite facilities
8. Irrigation
9. Water supply, sewage, and drainage

10. Land reclamation and dredging
11. Education and health infrastructure
12. Industrial estates, including infrastructure facilities and utilities
13. Markets and slaughterhouses
14. Warehouses and postharvest facilities
15. Public fish ports and fish ponds, including storage and processing facilities
16. Environmental and solid wastes management facilities, such as collection equipment, composting plants, incinerators, landfills, and tidal barriers

And to make certain that prospective developers fully grasped the broad scope of projects available for BOT, each type of project listed by the government included the catchall phrase "and related facilities."

R.A. 6957 contains a comprehensive compilation of the bidding procedures and bid content; the bid evaluation criteria, the provisions for negotiating a contract; typical contract terms and conditions; the assurance of compliance by the contractor; the procedures for adjustment of tolls, fees, rentals, and other charges; and contract termination provisions.

Republic Act 7718 On July 26, 1993, R.A. 7718 was enacted as an amendment to R.A. 6957. Several provisions were added to the BOT law:

- In the case of foreign contractors, Filipino labor shall be employed or hired in the different phases of construction where Filipino skills are available.
- When the project requires a public utility franchise, the facility operator must be Filipino; or if a corporation, it must be duly registered with the Security and Exchange Commission and have at least 60 percent Filipino ownership.
- If the project has difficulty obtaining financing, direct government appropriations can be obtained. The Official Development Assistant of a foreign government may be tapped up to 50 percent of the project cost. The balance of the financing must be provided by the project consortium.
- Rehabilitation of an existing facility or facilities is also eligible for BOT consideration.
- Repayment can be effected not only by the collection of tolls, fees, and rentals, but by nonmonetary payments such as the grant of a portion or percentage of reclaimed land or in the form of a share in the revenue of the project.

The BOT project development process envisioned by the government is shown in Figure 13.1. The flow of a BOT project through various government agencies is shown in Figure 13.2.

In May 1994 the government allowed foreign firms to submit unsolicited proposals for projects involving high-tech industrial ventures and set a goal of obtaining at least U.S.$400 million investment in this area.

The BOT Project Development Process

Step One: Project Advertisement
➤Agency develops projects with CCPAP assistance, issues, request for proposals (RFP), Advertises project, etc.
➤CCPAP or agency holds informational conference(s)

Step Two: Prequalification
➤Agency calls a prequalification conference
➤Proponents submit prequalification documents and are prequalified

Step Three: Bidding, Evaluation, Award
➤Prequalified proponents submit financial and technical proposals
➤Agency evaluates proposals and award is made

Step Four: Negotiation and GOP Approvals
➤Proponent negotiates and executes project agreement with agency
➤Proponent secures approval of contract from the Investment Coordination Committee (ICC), Monetary Board, and Department of Finance
➤Department of Justice reviews contract and issues protocol letter

Step Five: Project Proponent Implements
➤Proponent undertakes Environmental Impact Assessment (EIA)
➤Proponent secures Environmental Clearance Certificate (ECC) from the Department of Environment and Natural Resources
➤Proponent registers project company with SEC
➤Proponent negotiates financial incentives with Board of Investments
➤Proponent secures financing
➤Proponent secures Monetary Board and Department of Finance approval of financing secured

Figure 13.1. The Philippine Government's BOT Project Development Process.

THE AGRO-INDUSTRIAL ZONE

The Philippine government developed a plan for the systematic economic growth of the country with focus on certain strategic geographic areas and with an eye to developing key industries within these strategic zones. The government called these designated growth areas "Agro-Industrial Zones" (Figure 13.3), and the planned strategy was to build a substantial infrastructure base in each zone. The development of specific industrial types for each zone would be aided through the granting of special government assistance and incentives. The Agro-Industrial Zones were identified with these industries:

Calabarzon. This area would specialize in textile and garment manufacture, and consumer electronics production.

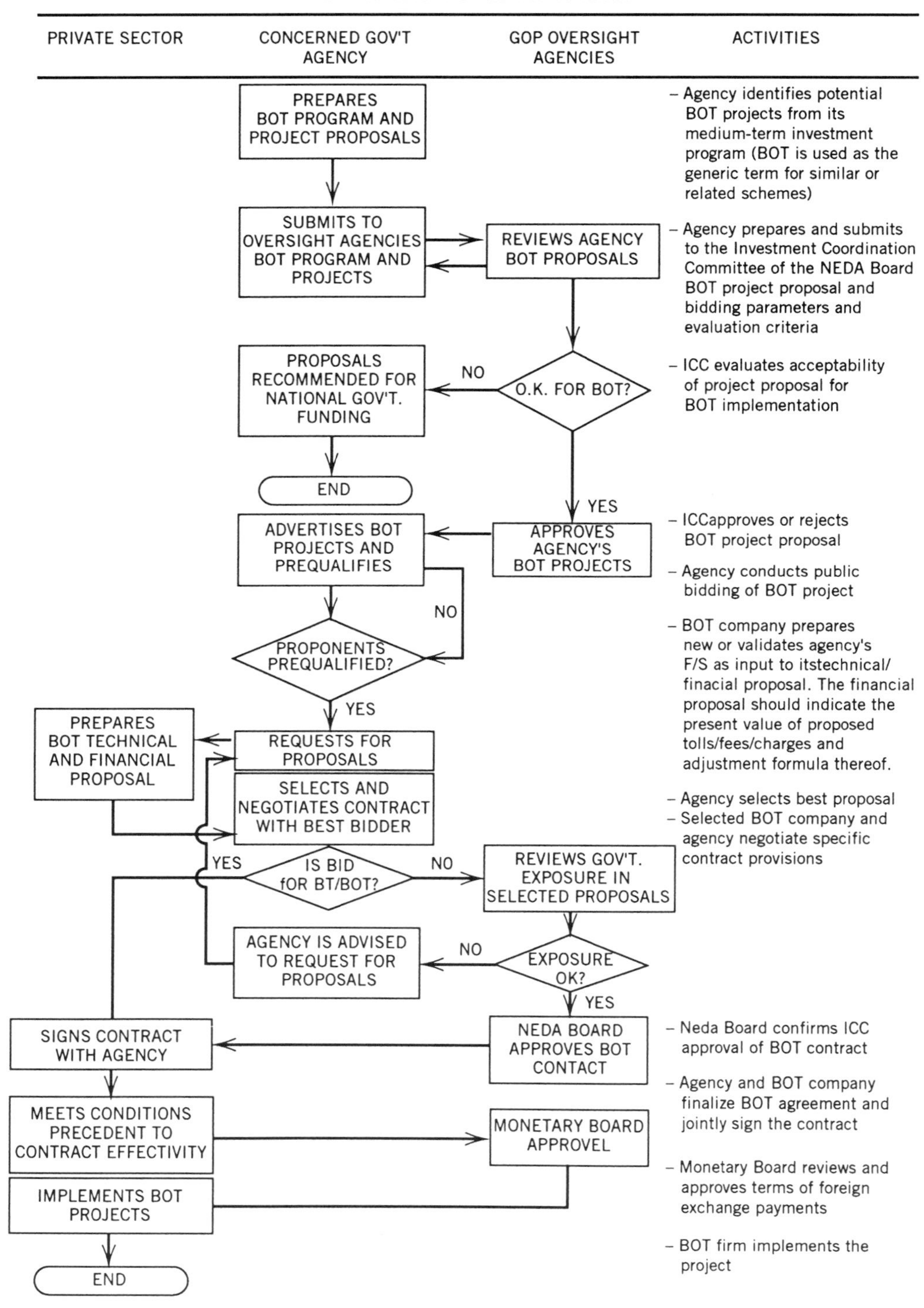

Figure 13.2. Flow chart with routing of BOT projects.

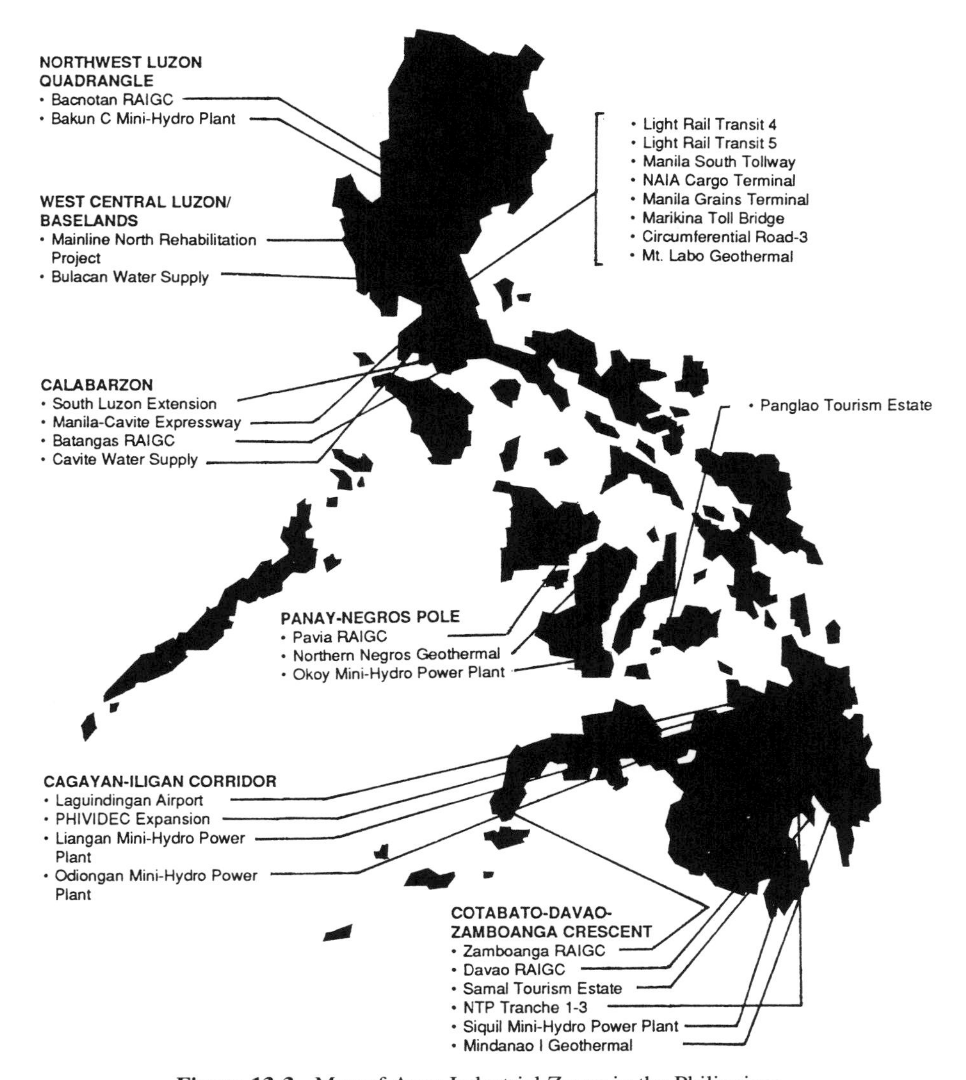

Figure 13.3. Map of Agro-Industrial Zones in the Philippines.

Cagayan–Iligan Corridor. This area would be developed as a heavy industrial center because of its access to hydroelectric power sources. It would also focus on agricultural industries and become a trade center.

Northwest Luzon Quadrangle. Due to its proximity to Taiwan, Hong Kong, Japan, and Korea, this area would become a financial center.

Cotabato–Davao–Zamboanga Crescent. The Crescent was scheduled to become the agriculture center of the country and a major export center for fresh and processed foods.

Panay–Negros Pole. This region was designated for the development of light manufacturing, specializing in furniture, jewelry, toys, handicrafts, and kitchen and housewares products.

Leyte–Samar Zone. Because of its access to geothermal and hydroelectric power, this area would develop as another heavy industrial center.

West Central Luzon Baselands Area. This area would specialize in ship and aircraft manufacture, regional warehousing, and trans shipment, and defense industry manufacturing.

BOT projects were designated to play a critical role in the government development plan. To streamline the BOT proposal review and approval process, the government created a special BOT Center. Ensuring "transparent bidding, clear negotiation process, attractive contractual incentives and investor friendly regulatory framework," are the BOT Center's goals, according to Roberto F. de Ocampo, the Center's Presidential Action Officer.

The *Invitation to Invest in Infrastructure Development,* a government packet for distribution to prospective investors, contained the short list of BOT projects shown in Table 13.1.

ELECTRIC POWER, THE PRIME MOVER IN INDUSTRIAL DEVELOPMENT

Power generation is the driving force behind any planned long- or short-term development of infrastructure. The Philippine government has put a high priority on the creation of hydroelectric, geothermal, and fossil fuel/gas–powered electrification plants. President Ramos announced his intention to have a completed energy plan in operation by the year 2005, a plan that included the major islands of Negros, Leyte, and Mindanao.

Gordon Wu of Hopewell Holdings, Hong Kong, was an early participant in the program. In 1989, the National Power Corporation, the predominant power supplier in the Philippines issued a tender for a 200 MW gas turbine plant in southern Luzon. Hopewell was the successful bidder on the project with a winning bid of U.S.$40 million that incorporated second-hand gas turbines. Wu obtained financing from a series of banks for this gas turbine plant; however, his next project in the Philippines, a 100 MW gas turbine contract worth U.S.$40 million, his company financed by itself.

THE PAGBILAO POWER PLANT PROJECT

In the late 1980s the Philippine government sought a proposal for a 770 MW coal-fired power generating plant to be constructed as a BOT project. Fourteen qualified bidders were selected from an international competitive bidding process, and Hopewell Holdings, one of the companies of Gordon Wu's Hopewell Group, was selected as the lowest qualified bidder.

The agreement with the Philippine government did not dictate how the construction contract should be structured, but Hopewell wanted to lock in a lump sum price for the construction portion of the project through a turnkey agreement. Hopewell invited five selected international construction companies to submit turnkey bids on the construction portion of this BOT project.

Hopewell selected Mitsubishi Corporation of Japan and Slipform Engineering Ltd. of Hong Kong, based on Hopewell's experience with this team on similar projects in the Peoples Republic of China. Mitsubishi is one of Japan's best known *zaibatsu* (giant conglomerates) and could call on technical assistance from any of its various manufacturing affiliates.

Hopewell Holdings Ltd. (HHL) became the project's sponsor; the Philippines' Hopewell Power Corporation (HPPC) was designated as the project company; and Hopewell Energy International Ltd. (HEIL) was created to provide equity in the project company. The Mitsubishi–Slipform consortium became the contractor, and their bid formed the basis for the turnkey contract. The National Power Corporation would act ostensibly as the customer; as part of the deal the NPC would construct support facilities such as transmission lines, and some ancillary facilities, and would provide temporary construction power, access roads to the site, and site land. The project was to start on April 1, 1993, and the targeted completion date was set as July 31, 1996.

The Importance of the ECA The Energy Conversion Agreement is one of the fundamental documents that a developer must have before any attempts are made to obtain debt financing in the Philippines. This document is essentially the project agreement, spelling out the actual buy–sell agreement between Hopewell and the National Power Corporation. The key provisions of Hopewell's ECA included the following commitments from Hopewell, the National Power Corporation, and the Philippine government:

- The company will build, own, and operate the project for 25 years, after which time it will transfer title to NPC at no cost.
- NPC will provide the support facilities required for the operation of the Pagbilao Power Plant at no cost to the company.
- Force majeure consists of events not under government control and events under government control.
- If a specified rate of return is not achieved and therefore adversely affects the project, certain buy-out provisions will be triggered.
- The company is subject to bonus/penalty provisions spelled out in the contract.
- The company is liable for damages due to delays, abandonment of the project or in the unlikelihood that it is not completed.
- The company must post a U.S.$16 million bond to insure that it will pay any penalties to NPC due to delays.

TABLE 13.1 Short List of Pipeline BOT Projects, Philippine Infrastructure Privatization Program

Project Name	Location (Region/Growth Zone)‡	Agency§	Cost 1994 Est. (U.S.$ in millions)
Transportation			
1. NAIA Cargo Terminal	NCR	MIAA	84.80
2. Manila Grains Terminal	NCR	PPA	95.48
3. Light Rail Transit 4	NCR	DOTC	678.40
4. Light Rail Transit 5	NCR	DOTC	279.84
5. Mainline North Rehabilitation	West Central Luzon/Baselands	PNR	236.68
6. Laguindingan Airport	Cagayan–Iligan Corridor	DOTC	32.86
Telecommunications*			
7. NTP Tranche 1–3	Mindanao	DOTC	182.32
Tourism			
8. Panglao Tourism Estate	Bohol	DOT	42.69
9. Samal Tourism Estate	CDZC	DOT	44.74
Highways			
10. Manila South Tollway	NCR	DPWH	204.40
11. South Luzon Extension	CALABARZON	DPWH	78.60
12. North Luzon Tollway	West Central Luzon/Baselands/NCR	DPWH	186.16
13. Manila–Cavite Expressway	CALABARZON/NCR	DPWH	33.24
14. Marikina Toll Bridge	NCR	DPWH	6.77
15. Circumferential Road 3	NCR	DPWH	340.05

Water Systems			
16. Bulacan Water Supply	West Central Luzon Baselands	LWUA	37.10
17. Cavite Water Supply	CALABARZON	LWUA	78.18
Power			
18. Small Hydro Program	Nationwide	NPC	645.55
19. Northern Negros Geothermal	Panay/Negros Pole	PNOC	108.44
20. Mt. Labo Geothermal	Bicol Region	PNOC	303.48
21. Mindanao I Geothermal	CDZC	PNOC	303.62
Industrial Estate†			
22. PHIVIDEC Expansion	Cagayan-Iligan Corridor	DTI	6.12
23. Batangas RAIGC	CALABARZON	DTI	77.93
24. Pavia RAIGC	Panay/Negros Pole	DTI	17.36
25. Bacnotan RAIGC	NWLGQ	DTI	41.0
26. Davao City RAIGC	CDZC	DTI	25.61
27. Zamboanga RAIGC	CDZC	DTI	13.73

*NTP: National Telephone Program
†PHIVIDEC: Philippine Villanueva Development and Economic Commission
RAIGC: Regional Agro-Industrial Growth Center
‡CALABARZON: Cavite/Laguna/Batangas/Rizal/Quezon Zone
CDZC: Cotabato/Davao/Zamboanga Crescent
NCR: National Capital Region
NWLGQ: NW Luzon Growth Quadrangle
§DOTC: Department of Transportation and Communication
DPWH: Department of Public Works and Highways
DTI: Department of Trade and Industry
LWUA: Local Water Utilities Administration
MIAA: Manilla International Airport Authority
PNOC: Philippine National Oil Company
PNR: Philippine National Railway

- NPC's payment schedules will be calculated on the net agreed generation capacity available. These fees will constitute 95 percent of the project company's revenue consisting of capital recovery fees, fixed operating fees, and service fees.
- The government guarantees performance of NPC's obligations.
- The ECA is governed by the laws of the Republic of the Philippines.
- Approvals from the appropriate government agencies must be obtained by the company.

Financing the Pagbilao Plant

The total cost of the Pagbilao Power project was set at U.S.$973 million, of which U.S.$933 million was to be provided by Hopewell and U.S.$40 million by the NPC. Hopewell was to provide equity financing amounting to U.S.$235 million.

Hopewell invested U.S.$205 million and obtained U.S.$10 million each from three lenders: Commonwealth Development Corporation (CDC), Asia Development Bank (ADB), and the International Finance Corporation (IFC).

Debt financing of U.S.$698 was the next requirement. Armed with a signed ECA and approvals from the Philippine Securities and Exchange Commission, the Board of Investment, and the NEDA, Hopewell dispatched a senior executive of the International Finance Corporation to the Asia Development Bank with a written request to consider the possibility of financing the Pagbilao Power BOT Project. The company ultimately obtained financial commitments from the following group of lenders:

Export–Import Bank of Japan (JEXIM)	U.S.$367.3 million
Export–Import Bank of the United States (USEXIM)	U.S.$172.4 million
International Finance Corporation	
"A" Loan	U.S.$ 60.0 million
"B" Loan	U.S.$ 23.3 million
Asian Development Bank	U.S.$ 40.0 million
Commonwealth Development Corporation	U.S.$ 35.0 million
Total	U.S.$698.0 million

ASSESSING THE FINANCIAL RISKS OF BOT PROJECTS

In a September 1993 staff presentation, Latif M. Chaudry, senior investment officer for the Asian Development Bank (ADB), set forth the risks to be analyzed in the decision to finance BOT projects. The guidelines he discussed have universal applicability to all such projects:

The Sponsors What is the sponsors' financial capability? Are they financially able to take on the project in question, and do they have experience with projects of a similar nature? Does the lender feel comfortable with the sponsors, and does a feeling of trust prevail? Chaudry cautioned his staff that if they didn't trust the sponsors, then they shouldn't make a loan.

The Turnkey Construction Contractor The same criteria that is used to evaluate the sponsor can be applied to the contractor—with the added proviso that each of the major subcontractors and suppliers must also be scrutinized.

Potential Delays If project revenues are delayed, how will the debt be serviced?

Cost Overruns When costs overruns occur, who will pay for them?

Design Parameters What if the design parameters are not met? What will be the consequences?

Insurance Does the contractor carry all-risk insurance? For Hopewell's Pagbilao project, for example, critical material and equipment were to be shipped by sea—the contractor had to have marine cargo insurance so that any equipment or materials damaged or lost at sea would be adequately insured for replacement. Also, lenders must require loss of revenue and business interruption insurance.

Post-Commission What happens if something goes wrong after commissioning? Are the project operators experienced enough to handle these kinds of problem? Are there adequate warranties/guarantees in place to cover these situations?

Force Majeure If unanticipated or uncontrollable events occur, what affect will they have on the project? Are any contingency plans at hand?

Buyer Obligations What if the buyer (in Hopewell's Pagbilao project, the NPC) cannot meet its obligations?

Chaudry presented a list of problems associated with BOOT/BOT projects and posed the question, Why are these kinds of projects so complicated to develop? Again a banker's review of the potential hazards facing a BOT project is worth of consideration:

1. *A poorly developed feasibility report and the absence of a realistic power purchase agreement and/or project agreement.* Some project sponsors are lured by the possibility of earning very high rates of return on their investment and may not prepare a comprehensive proposal, anticipating that the host country will not scrutinize it thoroughly. Many of these BOT projects are multimillion or billion dollar endeavors, so a complete, thorough, and realistic feasibility report is essential to define all terms and conditions clearly for both sponsor and host country.
2. *Private parties demanding unreasonable profits.* Expectations of high profits by some entrepreneurs may be handily dismissed by the host country.
3. *Countries with no legal or institutional framework.* Some countries have no legal or institutional framework to deal with BOT projects; therefore, passage

TABLE 13.2 Priority Projects Available for Bidding, 1995–1997

Year	Sector/Project	Cost (U.S.$ millions)
1995	**Power**	
	50MW Zamboanga Diesel	50.0
	80MW Mt. Apo II Geothermal	215.34
	Small Hydro	625.20
	Bataan GT Conversion to CCGT	40.0
	Malaya GT Conversion to CCGT	30.0
	11MW Bohol Diesel	11.0
	Calaca I Coal ROT	Concept
	Malaya Thermal ROT	Concept
	Manila Thermal ROT	Concept
	Leyte–Cebu Connection	Concept
	Transportation	
	LRT IV	678.40
	Metro Manila Skyway	596.9
	Industrial Estates	
	PHIVIDEC Expansion	6.55
	Resorts	
	Samal Island	44.74
		Total 2,613.13
1996	**Power**	
	Tongonan Geothermal	550.0
	6MW Bohol Diesel	6.0
	Small Hydro	136.51
	Small Hydro	17.68
	Leyte–Mindanao Connection	Concept
	Transportation	
	LRT V	279.84
	South Luzon Extension	69.72
	North Expressway	256.69
	Manila–Cavite	108.44
	Circumferential Road	122.88
	Water	
	Bulacan Water Supply	37.10
	Cavite Water Supply	78.18
	Industrial Estates	
	Pavia RAIGC	18.60
	Bacnotan RAIGC	37.72
	Davao RAIGC	22.86
	Resorts	
	Panglo Island	42.69
	Waste Management	
	Metro Manila Solid Waste Management Program	Concept
		Total 1,777.31

TABLE 13.2 *(Continued)*

Year	Sector/Project	Cost (U.S.$ millions)
1977	**Power**	
	11MW Bohol Diesel	11.0
	Kalayaan Pumped-Storage 3 & 4	159.0
	200MW Luzon Gas Turbine	200.0
	Masinion I	O&M
	Transportation	
	Mainline North Rehabilitation	76.80
	Industrial Estates	
	Zamboangra RAIGC	12.25
	Batangas RAIGC	83.49
		Total 342.54

of needed legislation may either be not of the proper type or take an inordinately long period of time to enact.

4. *Resistance by civil service.* If civil service bureaucrats resist the privatization of traditionally public projects, there will be difficulties and delays in the implementation of the BOT project.
5. *Insufficient reward for risk.* The host country must recognize the risks associated with a BOT project and be willing to adequately reward a sponsor.
6. *Less-than-thorough scrutiny.* When a bank is being requested to fund a project in excess of U.S.$1 billion, nothing can be taken for granted. The lender must thoroughly scrutinize to every aspect of the project.
7. *Protracted negotiations.* An enormous amount of legal work is required for these types of projects, and loan agreements with many lending institutions will usually be formulated. The technical issues that surface must be resolved quickly and professionally. Budgetary problems will most certainly arise, and all of these weighty issues will consume days, weeks, months, and even years.
8. *Lack of government approvals.* Dozens of complicated permitting procedures must transpire, requiring review by dozens of government agencies—a time—consuming and frustrating experience, especially in developing countries.

UNSUCCESSFUL PHILIPPINE POWER BOTs

To the experienced sponsor/developer, the rewards of BOT projects outweigh the risks, and the Hopewell Group has developed an enviable track record in completing BOT projects in the Far East. Gordon Wu continues to be bullish on Philippine

TABLE 13.3 Twenty-Three Power Generating Projects in Operation in the Philippines.

Project Name	Proponent	Type	Capacity (MW)	Fuel	Comm'l. Operation
1. Navotas Gas Turbine 1–3	Hopewell Holdings Ltd. (Hong Kong)	BOT	210	Diesel	1/15/91
2. Benguet Province Mini–Hydro	Hydro Electric Dev't. Corp. (Phils.)	BOO	22	Hydro	6/92
3. Subic, Zambales Diesel Power Plant I	Enron Power Corp. (USA)	ROL	28	Bunker C	1/18/93
4. Toledo Cebu Coal Thermal Plant	Atlas Consolidate Mining & Dev't. Corp. (Phils.)	ECA	55	Coal	2/93
5. Navotas Gas Turbine 4	Hopewell Energy Int'l. Ltd. (Hong Kong)	BOT	100	Diesel	3/30/93
6. Limay, Bataan Combined-Cycle Gas Turbine Power Plant "Block A"	ABB/Marubeni/Kawasaki Consortium (Swiss/Japan)	BTO	210 90	Diesel	SC 4/93 CC 10/94
7. Gas Turbine Power Barges	Hopewell Tileman Ltd. (Hong Kong)	ROM	270	Diesel	1st qtr. 93
8. Clark Air Diesel Plant	ELECTROBUS (Phils.)	ROM	50	Diesel	1st sem. 93
9. Pinamucan, Batangas Diesel Power Plant	Enron Power Corp. (USA)	BOT	105	Bunker C	7/93
10. Iligan City Diesel Plant I	Alsons/Tomen (Phil./Japan)	BOT	58	Bunker C	7/31/93
11. Binga Hydro Power Plant	Chiang Jiang Energy Corp. (China)	ROL	100	Hydro	8/93

12. Calaca, Batangas Diesel Power Barges	Far East Livingston (Singapore)	OL	90	Bunker C	9/93
13. Limay, Bataan Combined-Cycle Gas Turbine Power Plant "Block B" (Simple-Cycle)	ABB/Marubeni/Kawasaski Consortium (Swiss/Japan)	BTO	210 90	Diesel	SC 11/93 CC 1/95
14. Iligan City Diesel Plant II	Alsons/Tomen (Phil./Japan)	BOT	40	Bunker C	12/93
15. Mak-Ban Binary Geo. Plant	Ormat Inc. (USA)	BTO	15.73	Geo. Steam	3/94
16. Subic, Zambales Diesel Plant	Enron Power Corp. (USA)	BOT	108	Bunker C	3/94
17. Naga Thermal Plant Complex	Salcon (Phils.)	ROM	203	Coal/Diesel	4/94
18. Mindanao Diesel Power Barges	Mitsui/BWES (Japan/Denmark)	BTO	2×100	Bunker C	Unit I 4/94 Unit II 7/94
19. North Harbor Diesel Barges	Far East Livingston (Singapore)	OL	90	Bunker C	9/1/94
20. Navotas Diesel Power Barge	Van Der Horst Ltd. (Singapore)	OL	120	Bunker C	10/94
21. Engineering Island Power Barge	Sabah shipyard SDN, BHD (Malaysia)	OL	100	Naptha	11/94
22. Bauang, La Union Diesel Power Plant	First Private Power Corp. (Phils.)	BOT	215	Bunker C	2/95
23. Malaya Thermal Power Plant 1&2	KEPCO (South Korea)	ROM	650	Bunker C	09/15/95

TABLE 13.4 Eight Power Generating Plants under Implementation and Fourteen Power Plant Projects Being Readied for Bidding

Projects under Implementation					
Project Name	Proponent	Type	Capacity (MW)	Fuel	Comm'l. Operation
1. Bataan EPZA Diesel Plant	Edison Global (Hong Kong)	BOO	58	Bunker C	6/30/94 Partial (Delayed)
2. Bac-Man Binary Geo. Plant	Ormat Inc. (USA)	BTO	15.73	Geo. Steam	3/94 (Delayed)
3. Cavite EPZA Diesel Plant	Magellan Cogeneration Utilities (Phils.)	BOO	63	Bunker C	8/14/94 Partial (Delayed)
4. Ambuklao Hydro Power Plant	Miescor (Phils.)	ROL	75	Hydro	10/30/95
5. Pagbilao, Quezon Coal-Fired Thermal Power Plant	Hopewell Energy Int'l. Ltd. (Hong Kong)	BOT	700	Coal	Unit 1 6/95 Unit 2 1/96
6. Tongonan Leyte Geothermal Power Plant	PNOC–EDC (Phils.)	BOO	200 400	Geo Steam	1996 1997
7. Sual. Pangasinan Coal-Fired Thermal Power Plant	Hopewell Holdings Ltd. (Hong Kong)	BOT	1000	Coal	1999
8. Casecnan Hydro Electric Plant	National Irrigation Administration (NIA)	PPA	140	Hydro	2004

Projects for Private Power Participation

1. Gen. Santos Diesel Power Plant	Contract negotiation with Dragon Oil on-going	BOO	50	Bunker C	July 1996
2. Bataan Thermal Power Plant	Preparation of bidding documents ongoing. Bidding date to be announced	ROM	225	Hi-Vis	1996
3. Zamboanga Diesel Power Plant	Bid evaluation completed and for presentaion to CAC and NP Board	BOO	100	Bunker C	01/97
4. Small Hydro Projects (First Priority)	Re-bidding completed 12/1/94; Bid proposals under evaluation	BOT	Various (296)	Hydro	1998
5. Masinioc Coal-Fired Thermal Power Plant	For Bidding: to be announced	BOT	300	Coal	1998
6. Mindanao Coal-Fired Thermal Power Plant	Bid proposals under evaluation. For presentation to CAC and NP Board	BOT	200	Coal	12/99
7. Panay Coal-Fired Power Plant	Differed indefinitely	BOO	100	Coal	3/2000
8. Caltex/Texaco Cogeneration	Under negotiation (subject to competing proposals per BOT Law)	BOO	300	Bunker C	2000
9. Natural Gas Projects	For re-bidding: Tentative Schedule: Oct. 26, 1995	BOT	1200	Natural Gas	2000
10. Mindanao Coal II Power Plant	For Bidding: to be announced	BOO	150	Coal	2000
11. Bulanog–Batang HE Plant	For Bidding: to be announced	BOT	150	Hydro	2003
12. San Roque HE Plant	For Bidding: to be announced	BOT	390	Hydro	2005
13. Calacal Coal Thermal Plant	For Bidding: to be announced	ROM	300	Coal	Not available
14. Aplaya Diesel Power Plant	For Bidding: to be announced	ROM	100	Bunker C	Not available

power generating plants; at a symposium held in New York on November 19, 1994, he told the audience that he was working on a huge 1,320 MW power generating plant in the Philippines—his fifth such power plant project in that country.

But not all Philippine power plant projects have progressed so smoothly. In 1985 a 620 MW nuclear power plant project was completed at Morong near Manila. However, this U.S.$2.1 billion project never became operational because of government concerns over safety. Because Westinghouse Electric Corporation was involved in this Bataan nuclear power plant, the government banned Westinghouse and its licensed technology from any further participation in Filipino power plant projects during ongoing legal disputes. The Philippine government may have more than one bone to pick with Westinghouse Electric, according to *Engineering News Record*. A 1992 article alleged that the company had paid bribes to a friend of Ex-President Ferdinand Marcos in order to receive the contract for the power plant.

On October 20, 1995, the Philippine National Power Corporation issued a request for tender to either convert the moribund 620 MW Bataan nuclear station into a 1,200 MW gas-fired combined-cycle power plant or build a greenfield 1,500 MW gas-fired, combined-cycle power plant under a BOT agreement. This was actually the second time this particular call for proposals had been issued. A previous request for tender had produced proposals that were scrapped by the NPC when it was determined that the low bidder—Consolidated Electric Power Asia Ltd., a unit of Hopewell Holding and Mitsubishi Corporation—was planning to utilize a gas turbine developed by Westinghouse Electric Corporation. The second bidder at that time was Enron Development Corporation from Houston, and when the government threw out all bids, howls could be heard all the way from Texas.

THE PHILIPPINE BOT PROCESS FORGES AHEAD

Since 1987, the government in the Philippines has embarked on a major campaign to divest itself of public assets as part of the country's privatization program. So far more than U.S.$3.6 billion in sales of these assets has occurred. The BOT process is going full steam ahead; projects with a total value of U.S.$4.733 billion (Table 13.2) have been announced for the period 1995 to 1997, and the government is offering private investors who participate in the program a host of sweeteners. The incentives range from four- to six-year tax holidays and tax credits for purchasing domestic capital equipment or hiring local labor to arranging financing packages with state-owned banks.

The Philippine private power initiative is going strong. The Department of Energy in Metro Manila in October 1995 listed 23 projects with a total generating capacity of 3,393.73 MW in operation (Table 13.3). Eight projects were under implementation that would add another 2,651.73 MW, and 14 projects were being negotiated or being prepared for bid that would increase generating capacity by another 3,861 MW (Table 13.4).

Alan T. Oritz, executive director of the Coordinating Council for Philippine Assistance Program (CCPAP) stated, "To put the investment climate we have created here in perspective, it took the United States five years after enabling legislation was enacted before the first privately owned power plant was completed. We have been able to do that in just two years." President Ramos, speaking at a October 25, 1995 New York dinner marking the 75th anniversary of the Philippine–American Chamber of Commerce, announced that Manila had shed its image as the "brownout capital of the world": "We started out in the 1960s as the Asian country most likely to succeed after Japan. In the next quarter century we became a basket case. We are a nation that has had to reinvent itself." And the reinvention of the Republic of the Philippines is well underway.

14

FROM INDONESIA TO THAILAND—AND IN BETWEEN

The World Bank identifies the East and South Asian markets, exclusive of Japan, as areas where U.S.$1.2 trillion to 1.5 trillion will be required to meet the needs of infrastructure growth over the next decade. From an annual requirement of U.S.$110 billion in 1995, regional infrastructure needs are projected to grow to U.S.$203.6 billion by the year 2004 (Figure 14.1). This represents a commitment of 7 percent of each country's Gross Domestic Product (GDP), and funds of this magnitude will not be available from public coffers. The gap is slowly being filled by the public sector as BOT toll roads, and several water and waste treatment plants are underway in Malaysia. The Indonesian and Thai governments have embarked upon campaigns to attract private investors, and a number of BOT projects are underway and/or in the proposal development stage in both countries. In China, Gordon Wu of Hong Kong's Hopewell Holdings built two power plants and a six-lane, 304-kilometer (188 mi), U.S.$1.5 billion toll road in two phases; Phase 1 between Hong Kong and Guangzhou was completed in 1994, and the second phase, between Macau and Zuhai, is due for completion in 1996. As shown in Figure 14.2, China offers the greatest potential of any Asian country for public/private infrastructure partnerships, attaining a level of U.S.$90 billion per annum by the turn of the century.

Four elements are responsible for this surge in interest in infrastructure development in this part of the world.

1. Accelerated growth in urbanization in the East and Southeast Asian countries, bringing with it the need for adequate power, potable water, waste treatment plants, telecommunications, and transportation systems.

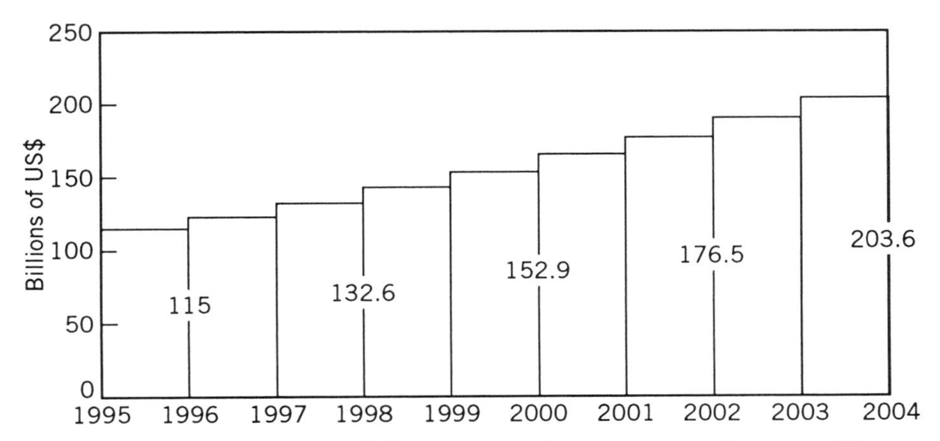

Figure 14.1. Projections of Asian infrastructure requirements over next ten years. *[Source: World Bank.]*

2. Underinvestment in past decades coupled with neglect of existing infrastructure.
3. Economic growth in the region as manufacturing industries chase low-cost, efficient labor. This growth can be sustained only by building infrastructure to keep pace with that industrialization. It has been estimated that for every 1 percent growth in per capita GDP in Asia, investment in infrastructure should also increase by 1 percent.

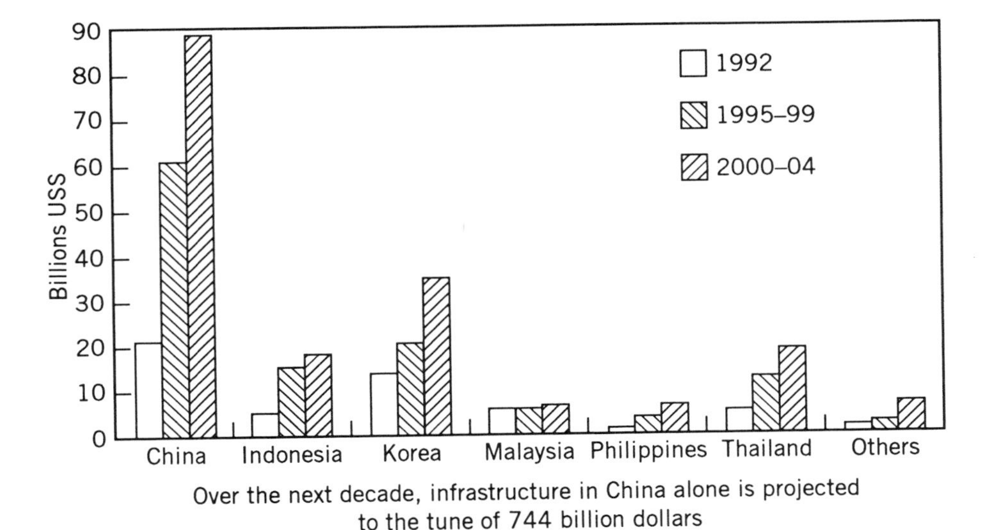

Figure 14.2. Average annual investment in infrastructure in Asia with China far and away the largest potential investor. *[Source: World Bank.]*

4. Need for infrastructure capabilities so that a country can enter the world of globalization and multinational corporations.

However, a number of elements must be in place before developers will be willing to enter these markets. Government objectives must be defined, enabling legislation must be in place, political and public opinion must be convinced that these new public/private partnerships are necessary for internal growth, and risk taking must be assessed and the rewards determined accordingly.

Until the Asian countries fully develop their own bond market infrastructure, they must rely upon the region's more mature stock market, the syndicated loan network, and off-shore financing (referred to by some as "suitcase bankers") for their financing needs and industrial growth. Most Asian governments don't issue long-term government debt. This is an inhibiting factor in the development of local bond issues. Without mature government bond markets, corporate bonds lack the benchmarks on which to base their rates and liquidity, and only the more liquid secondary markets will allow potential investors to feel more comfortable with long-term risk and fixed income return. Short maturity bonds such as South Korea's three-year limit is built upon expensive bank guarantees, and Taiwan's corporate bond market has even shorter maturity rates. Malaysia has fixed income paper with maturity dates of not more than five years. These short redemption periods impede the development of the longer-term securities that are needed to finance the area's infrastructure.

Francis Yeoh Sock Ping, Managing Director of Malaysian YTL Corporation, a dynamic construction-finance-power plant conglomerate, was quoted as saying that Asian infrastructure can be financed by Asian currencies and that the region recognizes the need for an Asian bond market for its long-term funding needs. In the meantime, foreign investment is being sought by Asian governments eager to accelerate their infrastructure placement programs. In fact, even governments that are actually experiencing surpluses want to attract foreign funding for infrastructure financing.

THE INFRASTRUCTURE CONFERENCE

A common meeting ground for government officials, investors, contractors, and designers has been in a conference environment. Here a private organization such as Singapore-based AIC Conferences provides the setting in a regional resort hotel and invites host government officials and prominent private and public infrastructure managers as speakers, thereby attracting attendees anxious to learn about the opportunities in the region. One such conference took place in the Grand Hyatt Hotel in Jakarta, Indonesia, in late September 1995. Several high officials in the Indonesian government welcomed bankers, engineers, developers, and builders to apprise them of the government's desire to attract investors for infrastructure development.

INDONESIA

The archipelago that comprises the country of Indonesia includes more than 17,000 islands, including Java, Sumatra, Kalimantan (most of Borneo), Sulawesi (Celebes), and West Irian. The country's population is reaching 190 million, and its Gross National Product (GNP) is increasing at a annual rate of about 7.3 percent. The economy of the country had been based upon oil production, but President Suharto was the force behind his government's attempt to develop a manufacturing-based economy. As a result, Indonesia experienced a 24 percent increase in manufacturing over the past few years. When the demand upon Indonesia's existing infrastructure outpaced its ability to meet these new economic demands, the government began a systematic attempt to interest private investors in participating in the growth in their country's power, telecommunications, and transportation expansion and development program.

Indonesia's U.S.$23.7 billion investment potential for infrastructure projects has not gone unnoticed by the global business community. As of the first quarter of 1995, more than U.S.$12.2 billion in foreign investment has flowed into the country. Their current five-year development plan REPELITA V requires U.S.$14.5 billion for transportation projects alone, of which 44.1 percent will be offered to the private sector. Demand for toll roads is expected to increase by 15 percent per year, and demand for all infrastructural services has been estimated by the government to be 6 percent of GDP per year.

What Happens When President Suharto Leaves Power?

One uncertainty in the foreign investment future in Indonesia is President Suharto. He has been the prime mover behind the country's development. In the 1960s, the country had a per capita income of less than U.S.$70, and 60 percent of Indonesia's population lived below the poverty level. In 1965 the army crushed an attempted coup by the Indonesian Communist Party (PKI), and President Sukarno was shunted aside by General Suharto, who became president in 1968. The PKI was banned after several hundred thousand suspected Communists were slaughtered. At that time, government policy swung sharply to the West and non-Communist Asian countries.

Suharto is now in his sixth term as president and has derived most of his popularity from the way in which he has guided his country's economic growth. Indonesia's economy was based upon exports of oil and gas. When oil prices collapsed in the 1980s after OPEC withered on the vine, the country's economic base eroded. Suharto encouraged the growth of low-cost manufacturing to take up the slack and gradually replace dependence on fossil fuel exports. Now in 1995 gas and oil exports make up approximately U.S.$9.7 billion of the country's total U.S.$40 billion annual exports, a reduction of 50 percent from the dollar amount represented in 1972.

The 74-year-old President Suharto told advisers that he may not seek a seventh term in 1998, but potential successors have not yet come forward. All signs indicate that Indonesia plans to follow Suharto's lead. When Suharto spoke at a World Infrastructure Forum held in Jakarta in October 1994, he reinforced his country's intentions to invest U.S.$53 billion in infrastructure by the end of the century, stating that private investors would be given a U.S.$18 billion share in this market. Speaking at the same conference, Coordinating Minister for Industry and Trade Hartarto reinforced his message and enticed potential investors by stating, "We offer an attractive 22 percent return on investments." If the Suharto engine fueling the economy stops along the way, will there be someone else around to fire it up again? That is the question facing potential investors.

Telcommunications and Power

The 1989 Republic Indonesia Law No. 3 on Telecommunication opened the door for private investment in this field. As a result, an agency designated TELKOM was created to deal with domestic telecommunications service, while INDOSAT was established to handle international services. Because of the complexity of Indonesia's geographic and demographic distribution, massive amounts of money are required to create an effective and cohesive telecommunications system. As of 1995, less than 5 percent of Indonesian homes have telephone service, and about 60 percent have no access to electricity. On the other hand, private investment in telecommunications in Indonesia is occurring in the area of cellular phone operations and paging facilities.

Data communications is another sector designated for growth by the government and one in which private developers are being invited to participate. The sixth REPELITA will require a total investment of RP (rupian) 19.4 trillion (U.S.$8 billion) for the installation of an additional five million telephone lines and accommodations for 600,000 mobile phones. The government plans to supply RP11 trillion (U.S.$5 billion), and the remaining RP8.4 trillion (U.S.$3 billion) must come from the private sector.

The power requirements in Indonesia are increasing by about 15 percent per annum and will require massive doses of investment for the construction of power plants and electrical distribution lines. Environmental issues come into play when power generation plans are discussed; however, Indonesia has its ANDAL (environmental impact assessment) program in effect to deal with these concerns. Their program requires an assessment to be organized in the following stages:

Formulation of the design concept development plan.

Identification of key environmental issues concurrent with the design development.

Collection of baseline surveys and evaluation of that information.

Evaluation of designs to predict the impact and determine quantifiable measurements where possible.

Development of prescriptions for the mitigation of any identified impacts on the program.

Specification of the monitoring of mitigation and protective works during construction and operation of the facility.

Implementation of the monitoring and enforcement program.

The Indonesian Infrastructure Market

Sunaryo Sumadji, Director of Urban Road Development, Directorate General of Highways, spoke on the feasibility of investing in Indonesian toll road projects. He said that with the implementation of REPELITA V, the growth rate of cities such as Jakarta, Surabaya, Semarang, and Medan has been very high, and the nation's highway system must be greatly improved and expanded to keep pace with this growth. The national policy affecting highway construction allows for the application of tolls to expedite the development of these roadway systems, which previously had been designated as freeways. This national highway policy stipulates that

- Toll roads are national assets and are to be controlled by the government.
- Toll roads are to be built as an alternative to the existing roads, not the contrary.
- There must be no pressure of any kind on society to use them. In other words, the development of toll roads has to be integrated to the existing road network.

Republic Indonesia Law No. 13 enacted in 1980 contained Article 14, which included the requirement that the construction of toll roads was to be carried out without placing a burden on the state budget. In 1986 a presidential decree was more specific and invited the private sector to invest in the construction of toll roads, and in 1987 the Capital Investment Coordinating Board (BKPM) declared that capital investment in roads was open to both local and foreign investors. The 1980 Republic Indonesia Law No. 13 on Roads, Government Regulation No. 26 of 1985 on Roads and Government Regulation, and Regulation No. 8 of 1990 on Toll Roads established the groundwork for public/private highway partnerships.

Under Indonesian law the government would invite private investors to participate in infrastructure development on a BOT basis or BTO (Build, Transfer, Operate). In the case of a BOT project, the concession agreements would stipulate that upon expiration of these agreements, ownership would pass to Jasa Marga, a state-owned company. When the BTO vehicle was to be implemented, the toll road would be handed over to Jasa Marga for operation during the concession period after construction had been completed. The investor in that case would receive a portion of the toll revenue.

The government provided potential investors with the following guidelines:

- The private enterprise must cooperate with Jasa Marga as the sole company appointed to undertake toll road development in Indonesia.
- The toll roads being proposed must become an integral part of the national highway network and represent an alternate road, which means that another road must exist so that the user can choose between the two.
- The highway must conform to the technical specifications stipulated by Bina Marga, the Directorate General of Highways.
- The duration of the concession period must allow investors to recover their investment, which is to include profit.
- The government considers toll tariffs to be far reaching and of social interest; therefore, the authorization to establish the level of toll tariffs must remain with the government.
- Tariff levels must be calculated on "vehicle operation costs and time value where expenses in addition to toll tariffs may not exceed 70% of the expense incurred where non toll roads are used."

Allocation of Risks The Indonesian government recognized the risks faced by both public and private entities. Risk faced by the government include

- Land acquisition delays caused by the government.
- Changes in monetary laws and regulations that would influence the investment.
- Engineering design changes requested by the government.
- Force majeur.

If such events did take place, the government would compensate the investor by granting an extension to the concession period.

Risks faced by the investor include

- Engineering changes that must be made because of government regulations.
- Investment cost increases due to any number of unforeseen conditions. (Inflation in Indonesia as of 1995 was averaging 10 percent per year.)
- The cost of money.
- Delays in completing the project.
- Changes in toll tariffs.
- Traffic flow that is considerably less than projected.

Djoko Ramiadji, a former Directorate General of Highways official and now executive and commissioner of six infrastructure-related private businesses in Indonesia, indicated that, in his opinion, there were ways in which the government could share risks and responsibilities during the design, finance, construction, operation, and maintenance of these revenue-producing projects. Ramiadji suggested that the government could provide

outright grants,
loans,
purchase of bonds,
payment of shadow tolls,
various guarantees,
offers of land at no cost to the investor or at concessionary rates,
offers to construct complementary infrastructure projects, and
permission to capture value in other ways.

Ramiadji said that private sector risks included the miscalculation of traffic projections and the resultant reduction in revenue streams. This risk could be eliminated, he offered, if the government enacted legislation guaranteeing revenue stream payments.

Opinions of a Private Developer—Trafalgar House Michael G. W. Adams, Business Development Manager, Trafalgar House Corporate Development Limited, has been the company's point man for the Cikampek-Padalarang Toll Road Project in Java and spoke at the Infrastructure Indonesia 1995 Conference on a number of topics for which he has first-hand knowledge. Referring to risks associated with BOT projects, he said, "Whilst formal risk assessment is a very necessary process in deciding how to structure a project, one must not lose sight of the fact that a major project, particularly a privately financed one, in reality never quite turns out in the same way as expected at its outset. This is hardly surprising when one considers that the typical life of a concession-financed project is around 20 to 30 years."

Adams said that inflation in certain areas of the world may have a considerable impact on project costs. Some countries recognize this fact. For example, Trafalgar Houses' contract for the Queen Elizabeth Bridge in England was indexed for inflation; however, their contract with the government of Hong Kong for the Tate's Cairn Tunnel project was not. Adams went on to state that inflation is a risk that concession companies and, more particularly, their lending institutions are not prepared to accept in Asia. He questioned who is best suited to assess this risk and deal with it most competitively.

Projected traffic flows are another risk, and this is generally placed at the feet of the concessionaire. Even though tolls are regulated by the government, which is the normal practice, it is also common practice for the concessionaire to look to the government to index these toll rates on an automatic basis to protect the concessionaire from the effect of inflation. Nevertheless, political considerations enter into this equation, and many governments are reluctant to institute this automatic, indexed, adjusted toll schedule.

According to Adams, the financial risk associated with BOT projects revolves around interest rate risk, particularly if funds are raised through off-shore financing. The concessionaire will attempt to guard against these two types of risk (Interest Rates & Exchange Rates) by negotiating some form of automatic indexing of the toll rate to compensate for both interest and exchange rate fluctuations; however, their success in doing so depends upon the overall deal with the government.

Land acquistion risks are somewhat lessened in countries with strong planning agencies and directives in place. Even so, experience confirms that the risks are real. When developing the Birmingham Northern Relief Road, Trafalgar House underestimated the cost of permitting and has committed significantly more resources to the project than originally anticipated in order to obtain a fully approved plan. When land acquisition responsibility rests with the government, the concessionaire still faces one of two risks.

1. The government will not be able to acquire the land that was included in the concessionaire's proposal.
2. The concessionaire will have to pay more for the land than originally anticipated.

Political risks abound, and changes in government, so states Michael Adams, can lead to punitive legislation, restrictions on repatriation of earnings, or appropriation of assets. Operating risks can somehow be mitigated by the concessionaire by contracting out these services to companies with successful operational experience.

Indonesia's Transportation, Telecommunications and Power Development Program

Since the operation of toll roads was initiated in Indonesia in 1978, approximately 500 kilometers (308.6 mi) of main and access roads have been built. The Second Long-Term Development Plan (PJP II), established in 1994, intends to create an additional 2600 kilometers (1605 mi) of toll roads as follows:

Under construction or completed (Figure 14.3)	113.70 kilometers (70 mi)
Under investment process (Figure 14.4)	539.96 kilometers (333 mi)
Offered to investors (Figure 14.5)	770.00 kilometers (475 mi)
Further development (Figure 14.6)	1194.40 kilometers (737 mi)

As late as 1994, 19 public/private partnership projects, with a value of approximately U.S.$11.8 billion, were in the pipeline in Indonesia, including 9 power plants with a combined generating capacity of 7182 MW. A contract to install 442,000 telephone lines was underway, and four tolled highway projects were in various stages in the pipeline. Seven international consortiums had presented proposals for a four-phase, 82-kilometer (50 mi) light rail system between Jakarta, Tanggerang, and Bekasi, and a review in late 1995 was being conducted by the Ministry of Transport. The projected cost of this light rail project was estimated at U.S.$3.1 billion and would require 10 years to complete.

Political Uncertainty and Nepotism Make Strange Bedfellows

The coal-fired Paiton Power complex, a scheduled Build, Own, Operate project sponsored by the Indonesian Ministry of Mines is an example of how things can

TOLL ROADS IN OPERATION

No.	Road Link	Main Road (km)	Access (km)	Starting Year of Operation	Jasa Marga/Investor
1.	Jagorawi	47.00	9.00	1978	Jasa Marga
2.	Jembatan Tol Citarum	0.22	0.69	1979	Jasa Marga
3.	Jembatan Tol Tallo Lama	0.21	0.67	1981	Jasa Marga
4.	Jembatan Tol Mojokerto	0.23	0.60	1982	Jasa Marga
5.	Semarang Artery (Seksi A&B)	15.00	—	1983	Jasa Marga
6.	Pintas Serang (Serang By Pass)	8.4	—	1984	Jasa Marga
7.	Jakarta—Tangerang	27.00	6.00	1984	Jasa Marga
8.	Prof. Dr. Ir. Sedyatmo	14.00	—	1985	Jasa Marga
9.	Belmera	34.00	9.00	1986	Jasa Marga
10.	Surabaya—Gempol	43.00	6.00	1986	Jasa Marga
11.	Jakarta—Cikampek	73.00	11.00	1988	Jasa Marga
12.	Cawang—Tomang	17.00	—	1989	Jasa Marga
13.	Ir. Wiyoto Wiyono, MSc.	17.00	—	1990	PT. Citra Marga Nusaphala Persada
14.	Cakung—Cikunir	9.00	—	1990	Jasa Marga
15.	Padalarang—Cileunyi	35.6	28.77	1991	Jasa Marga
16.	Tangerang—Ciujung	34.2	—	1993	PT. Marga Mandala Sakti
17.	Surabaya—Gresik (Dupak—Tandes)	3.5	—	1993	PT. Marga Bumi Matraraya
	Subtotal PJP I	378.16	71.73		
18.	Ciujung—Serang Timur	13.5	—	1994	PT. Marga Mandala Sakti
19.	Surabaya—Gresik (Tandes-Kebomas)	12.20	—	1994	PT. Marga Bumi Matraraya
20.	Serang Timur—Cilegon Timur	13.80	—	1995	PT. Marga Mandala Sakti
21.	JORR Sections S (P. Pinang—L. Agung)	8.80	—	1995	PT. Marga Nurindo Bhakti
22.	Harbour Road (T. Priok—Ancol Timur)	4.30	—	1995	PT. Citra Marga Nusaphala Persada
	Subtotal PJP II	52.60	—		
	TOTAL	430.76	71,73		
	GRAND TOTAL	502.49			

Figure 14.3. Toll roads under construction or completed.

TOLL ROAD UNDER INVESTMENT PROCESS

No.	Road Link	Length (km)	Remarks
I	JAKARTA & Sekitarnya		
1.	JORR Seksi E2, E3 & N	10.00	—preconstruction
2.	JORR Seksi W1	9.70	—preconstruction
3.	JORR Seksi W2	11.17	—under negotiation
4.	Jakarta—Serpong	13.04	—land acquisition
II.	JAWA BARAT		
1.	Cirebon—Palimanan	26.30	—land acquisition
2.	Cikampek—Padalarang	59.00	—preconstruction
3.	Akses Citatah	6.00	—under negotiation
4.	Cileunyi—Nagrek	24.00	—under negotiation
5.	Cilegon—Bojonagara	10.00	—prenegotiation
6.	Cilegon—Labuhan	44.00	—prenegotiation
III.	JAWA TENGAH		
1.	Semarang—Batang	75.00	—under negotiation
2.	Semarang Seksi C	9.75	—under negotiation
3.	Semarang—Demak	25.00	—prenegotiation
4.	Solo—Yogya	45.00	—prenegotiation
IV.	JAWA TIMUR		
1.	Surabaya—Mojokerto	38.50	—preconstruction
2.	Pandaan—Pasuruan	31.50	—under negotiation
3.	Gempol—Pandaan	13.00	—under negotiation
4.	Jembatan Surabaya—Madura	5.00	—under negotiation
5.	Gresik—Tuban	75.00	—prenegotiation
	TOTAL	539.96	

Figure 14.4. Toll roads under negotiation or prenegotiation.

change in a political climate that can swing from a controlled democracy to one that resembles anarchy. Initially this Paiton coal-fired power generating plant was to be built by state-owned Perusahaan Umum Listrik Negara (PLN) and was to consist of four 400-MW plants and four 600-MW plants resulting in a facility that would have a total output of 4000 MW.

In 1990 the government, desirous of accelerating the construction of this much-needed facility, decided to invite the private sector to bid on the four 600-MW plant phase. That same year, Intercontinental Energy, Inc., of Hingham, Massachusetts, was awarded a contract to build two of the four 600-MW units. In the meantime, the government also wished to accommodate one of President Suharto's sons, Bambang Trihatmojo, who owned Bimantara Baya Nusa. Bimantara had initially expressed interest in becoming part of a consortium to build some of Paiton's power plants ever since it had teamed up with Gordon Wu's Hopewell Holdings in 1991. When

TOLL ROAD OFFERED TO INVESTOR
(Launched on April 4, 1995)

Road Link	Length (km)	Remarks
I. Jalan Tol Sadang—Palimanan		
1. Sadang—Subang	37.0	FS
2. Subang—Dawuan	52.5	FS
3. Dawuan—Palimanan	24.5	FS
II. Jalan Tol Ciawi—Sukabumi—Citatah		
4. Ciawi—Sukabumi	53.5	FS
5. Sukabumi—Cianjur	31.0	FS
6. Cianjur—Citatah	30.0	FS
III. Tol Kanci (Cirebon)—Batang		
7. Kanci (Cirebon)—Pejagan (Tegal)	43.0	PFS
8. Pejagan (Tegal)—Pemalang	52.0	PFS
9. Pemalang—Batang	40.0	PFS
IV. Jalan Tol Semarang—Solo		
10. Semarang—Solo	80.0	PFS
V. Jalan Tol Solo—Mojokerto		
11. Solo—Mantingan	40.0	PFS
12. Mantingan—Ngawi	35.0	PFS
13. Ngawi—Caruban	34.0	PFS
14. Caruban—Kertosono	53.0	PFS
15. Kertosono—Mojokerto	47.0	PFS
VI. Jalan Tol Lingkar Luar Timur Surabaya		
16. Simpang Susan Waru—Tanjung Perak	24.0	FS
VII. Jalan Tol Pandaan—Malang		
17. Pandaan—Malang	29.5	FS
VIII. Jalan Tol Pasuruan—Probolinggo		
18. Pasuruan—Probolinggo	40.0	PFS
IX. Jalan Tol Medan—Binjai		
19. Medan—Binjai	24.0	PFS
Jumlah	770.0	

Note: FS = Feasibility Study
PFS = Pre Feasibility Study

Figure 14.5. Opportunities for investors.

Hopewell Holdings bowed out, Bimantara formed a joint venture with Intercontinental Energy.

In October 1991 the government awarded two units at Paiton to this company. In order to avoid any impression that favoritism entered into the award, two other 600-MW units were awarded to the BMMG consortium, a group made up of PT Batus Hitam Perkasa, Mitsui & Company of Japan, Mission Energy U.S., and General Electric.

In June 1992 the government's negotiating team decided that Intercontinental

INDICATION FOR TOLL ROAD DEVELOPMENT

No.	Road Link	Length (km)
1	Jakarta North—South Coridor	17.00
2	Bandung Inner Ring Road & Akses Gede Bage	37.00
3	Bandung—Lembang	10.00
4	Cileunyi—Tj. Sari—Dawuan	50.00
5	Bogor Ring Road	10.50
6	Nagreg—Ciamis—Cilacap	150.00
7	Demak—Kudus—Tuban	195.00
8	Cilacap—Pejagan	100.00
9	Probolinggo—Banyuwangi	160.00
10	Ngawi—Babat	115.00
11	Poros Utara—Selatan Surabaya	20.00
12	Medan—Serdang	10.00
13	Medan—Ring Road	15.00
14	Tj. Morawa—T. Tinggi—Parapat	125.00
15	Bandar Lampung—Bakauhuni	60.00
16	B. Lampung—Bandara Beranti	20.00
17	Pel. Panjang—B. Lampung	12.00
18	Palembang—Bitung	70.00
19	Uj. Pandang—Mandai	15.00
20	Selat Bali	2.90
	TOTAL	1,194.40

Figure 14.6. Toll roads under construction for further development.

Energy, Inc., would not be able to attract enough funding to satisy their requirements. As a result, President Suharto dropped them and began to negotiate the deal with BMMG two months later. With Intercontinental Energy out of the picture, B. J. Habibie, Minister of Research and Technology, with President Suharto's approval, invited Germany's Seimens AG and PowerGen, a United Kingdom corporation to collaborate with Bimantara on a power purchase agreement for Paiton II.

This maneuver should have come as no surprise since President Suharto's family is involved in a number of businesses in Indonesia. A study completed by several international consulting firms over the past decade lists Indonesia as having the most corrupt business atmosphere of any country in the world.

As of mid 1995, BMMG has put almost all its U.S.$2 billion funding requirement in place. Chase Asia and IBJ Asia will be the coordinating banks for the total project, and U.S. Exim and Japan Exim may provide U.S.$1.4 billion support.

Another problem looms on the horizon. In August 1994 PLN was converted from a state-owned agency to a limited corporation and will lose the government subsidies it once enjoyed. As a result, PLN formed subsidiary companies and issued stock. And because PLN lost its status as a monopoly and the government assistance that went with it, it expanded its capacity and now has a surplus of electricity to sell in the Java-Bali integrated grid. If all the Paiton power-generating plants are completed on schedule, a surplus of power could exist into the first decade of the next century. And if these new projects produce power at a cost higher than PLN when completed in 1998 and Suharto is out of power, a political boondoggle of immense proportions could be in the offing, affecting other public/private ventures either in the pipeline or coming on stream.

THAILAND

The rapid economic expansion taking place in Thailand brought with it many of the ills an emerging nation must face. Traffic jams are legendary in Bangkok. With little or no mass transit available, private vehicles of all shapes and sizes pack the street of this capital city. Traffic congestion and the ensuing air pollution problems in Bangkok have gone from bad to worse, demanding immediate attention by the government.

The Express and Rapid Transit Authority of Thailand built the First Stage Expressway (FSE), joining transportation systems from the north, east, and south and allowing vehicles to traverse the city of Bangkok without entering its bumper-to-bumper traffic jams in the central city. The government reasoned that by building the FES and providing access around Bangkok, the expected traffic flow of 230,000 vehicles per day would save travelers approximately 39 minutes in travel time and substantially reduce fuel consumption. They reasoned that it would also lower the high air pollution levels produced by cars idling in long lines of traffic.

With the completion of the First Stage Expressway in the 1980s, the government looked to accelerate the construction of the second stage of the Master Transportation Plan. The Second Stage Expressway (SES) would connect to the First Stage Expressway from Bangkhlo to Changwattana and Phyathai to Sri Nakarin and thereby create a beltway around central Bangkok. Budgetary pressures and the need to implement the Second Stage Expressway as rapidly as possible led to a government decision to invite private sector participation in the project and request tenders for a BOT proposal.

The government stated that the following goals were to be met upon completion of the SES.

- Alleviate traffic congestion in the central Bangkok metropolitan area by creating a ring road formed between the SES and the FES and a new expressway to the north and east.
- Save 20 percent in travel time during peak periods. This increase should produce fuel savings and lower vehicle maintenance expenses.
- Encourage commercial and residential development along the SES route, resulting in increased economic development.
- Create new investment and employment opportunities that might lead to the development of entire new communities.

- Provide a psychological lift to motorists accustomed to recurrent traffic jams and travel delays.

The method of construction (i.e., BOT) would present the private sector with an opportunity to participate in the economic growth of Thailand, thereby easing the citizen's tax burden and allowing the project to progress much more rapidly than the normal public project.

Japan's JICA Always Ready to Assist Their Businessmen

The Japanese government is always on the alert for ways in which to provide economic aid to a neighboring country, anticipating that this assistance may result in more business for one or more of their nationals. The Japan International Cooperation Agency (JICA) saw the Bangkok traffic dilemma as an opportunity to assist the Thai government and offered to conduct an economic and social feasibility study with respect to the planning of this Second Stage Expressway. JICA completed two studies—one on a north-south route (Bangkhlo-Changwattana), which would be approximately 25 kilometers (15.4 mi) in length, and an eastern route (Phyathai-Sri Nakarin), which would be about 13.5 kilometers (8.3 mi) long. After an 18-month study, JICA submitted its report to the Thai government in January 1984. Along with the report, JICA submitted surveys, detailed design drawings, and an environmental study, all performed by International Consultants Company Limited, a firm hired by the Japan International Cooperation Agency.

On June 30, 1987, the government of Thailand passed a resolution to permit The Expressway and Rapid Transit Authority of Thailand to grant a concession to the private sector to design, build, finance, and manage the Second Stage Expressway for a period of 30 years. When the tender was issued, only two bidders responded—The Bangkok Expressway Consortium and the Thai Expressway Development Joint Venture. On September 20, 1988, the Cabinet gave its approval for the Expressway and Rapid Authority of Thailand (ETA) to enter into an agreement with the Bangkok Expressway Consortium, conditional upon approval by the Public Prosecution Department. This approval was granted on December 22, 1998, and the terms and conditions follow:

1. The investor(s) was to have a registered capital of approximately 20 percent of the project. Since the project was valued at slightly in excess of U.S.$1 billion, this represented a rather significant requirement.
2. All costs of construction and related financing risk were to be assumed by the concessionaire without any loan guarantees by the Thai government.
3. The expressway was to be built according to the required standards of the ETA, construction was to commence in 1990, and the 38.5-kilometer (23.7 mi) portion was to be completed by 1995.
4. The concessionaire was to refund the cost of land acquistion, with interest to the government.

5. Toll rates were to be set initially at 30 baht (U.S.$1.39 in 1995 dollars) for the urban section of the expressway and 15 baht (U.S.$0.70 in 1995 dollars) for suburban portions of the expressway. A discount of 5 baht (U.S.$0.23) would be offered if both sections of the expressway were traveled at the same time.
6. The toll revenue collected in the urban area was to be shared with ETA and the concessionaire at the rate of 40/60 for the first year, 50/50 for the second year, and 60/40 for the third year through the end of the concession period.
7. Title to the expressway and all ancillary structures was to revert to ETA after the 30-year concession period had expired.

The Bangkok Expressway Company Limited (BECL) was composed of 8 shareholders, 11 on-shore banks, and 30 off-shore lending institutions:

Shareholders:	Kumagai Gumi
	Ch. Karnchang Company Limited
	The Thai Military Bank
	Bangkok Bank Limited
	The Bureau of the Crown Property Limited
	Krung Thai Bank Limited
	The Bank of Asia Limited
	The Siam Commercial Bank Limited
On-Shore Lenders:	Bangkok Bank Limited
	The Thai Military Bank
	The Bank of Asia Limited
	Krung Thai Bank Limited
	The Siam City Bank Limited
	The Union Bank of Bangkok Limited
	The Siam Commercial Bank Limited
	Bangkok Metropolitan Bank Limited
	First Bangkok City Bank Limited
	Nakornthon Bank
	Bangkok Bank of Commerce Limited
Off-Shore Lenders:	Lead managers of the 30 banks including Credit Lyonnaise, DKB Asia Limited, LTCB Asia Limited, National Westminster Bank Plc.
Project Manager	Kumagai Gumi Company Limited

For the Expressway and Rapid Transit Authority of Thailand:

Engineers:	Daniel, Mann, Johnson & Mendenhall
	Asian Engineering
	Consultants Corporation Limited
	DeLeuw, Cather International Limited
	Thia DCI Company Limited
Independent Certified Engineer (ICE)	Louis Berger International

and Independent Design Checker (IDC):	Thai Professional Engineering Consultant Company Limited

Kumagai Gumi the major shareholder, was awarded the construction contract, and organized BECL along the lines of a holding company.

Funding the SES Project

Bangkok Expressway Company Limited shareholder capital was U.S.$226 million, and BECL executed the On-shore Credit Facilities Agreement on March 21, 1989, which provided U.S.$924 million in construction and development financing. Bangkok Bank Limited was the lead bank in this 11-bank on-shore financial group, and maturity of the on-shore loan was set at March 2009, with repayment commencing in March 1996. Repayment of the principal and interest for the Second Stage Expressway was to be serviced by BECL's portion of the toll revenue.

Work began in March 1990, and Kumagai Gumi contracted for the construction as a construction manager rather than a general contractor, bringing a staff of 300 on board. Most of the expressway designed by Freeman Fox Intercon was match-cast prestressed concrete segments to create spans up to 47.8 meters (157 ft). A consortium composed of German contractor Bilfinger + Berger and Ch. Karnchang Company Limited of Bangkok operated the precast concrete plant, which was to produce a total of 14,780 segments, each of which weighed about 40 tons. The city of Bangkok is built almost entirely on piles as is the expressway. A U.S.$60 million contract was awarded to Kin Sun Kier Beazer (KSKB) to furnish and install hollow prestressed concrete piles driven to a design depth of about 40 meters (131 ft).

Although construction proceeded smoothly, the completion of the project ended in disaster for Kumagai Gumi. The expressway assets were seized by the government of Thailand just prior to its being opened to the public. One source indicated that this expropriation occurred because of a problem between the Thai government and the Thai ETA. Another authoritative source said that the dispute was between Kumagai Gumi and the government of Thailand over the amount of toll charge to be levied. The concession agreement called for a toll charge of 30 baht-US$1.38, but the government insisted on a rate half that amount. When Kumagai Gumi refused to open the expressway until the government honored its original commitment, the highway was taken over by the government. The Expressway did open in August 1993, and the toll was 30 baht-US$1.38 not the 15 baht-US.69¢ requested by the government. Negotiations between Kumagai Gumi and the government had been ongoing to arrive at a permanent solution.

Although a public stock offering of the Bangkok Expressway was planned for 1994, the Thailand Security and Exchange Commission ordered it postponed until the project lenders and shareholders could resolve "certain issues," intimating that BECL was overleveraged. Morgan Stanley was to have managed the stock offering along with Goldman Sachs and S. G. Warburg. BECL had intended to issue 130 million shares of stock and was advised that a price of U.S.$1.36 per share could be expected. This price would have yielded U.S.$176.8 million. Whether this price represented a real stockholder issue or whether it was an action taken at the request of the Thai government to tighten the screws a little more may never be known.

An adversarial host government can create a great deal of stress on a concessionaire, as was the case in this instance. Kumagai Gumi sold its 65 percent share in the project to Ch. Karnchang, a military contractor, in March 1994 in order to settle the dispute over the toll charge issue. Project management was assumed by a consortium of Ch. Karnchang, Bilfinger + Berger (the precast concrete suppliers), and Tokyo Construction Company. These three companies were also involved in assembling a team to extend the SES another 6.17 kilometers (10 mi).

As a result of the government action that took place upon completion of the Second Bangkok Expressway, you might think that other foreign consortiums would be wary of making any deals with the Thai government, but this was not so. The Thai government, still concerned about moving people quickly through Bangkok, were looking for developers to build an elevated railway system in Bangkok on a Build, Operate, Transfer basis. International bankers were skeptical after the Kumagai Gumi, Second Expressway System fiasco. One banker was overheard to have said, "If they can't persuade rich car owners to pay a realistic toll, how are they going to persuade poorer commuters."

Prospective developers for elevated rail system would require sources for financing. Bankers would be reluctant to offer financing inasmuch as gov't might renege on their deal with developers as they had done to Kumagai Gumi on the SES.

The government arm twisting on the SES project does not seem to have deterred other existing or potential investors. Several years after Kumagai Gumi had signed its concession agreement for the SES, a subsidiary of Hopewell Holdings (Hong Kong) signed an agreement to build a mass transit community train (CT) and provide Bangkok with an integrated light rail system. Under the agreement that became effective on December 6, 1991, Hopewell (Thailand) Limited [H(T)L] entered into a project that would extend almost to the end of the twentieth century.

The Bangkok Elevated Road and Train System

Hopewell (Thailand) Limited will build and operate the CT and tollway system for 30 years, after which all their assets will be transferred to the State Railway of Thailand (SRT). The project consists of the construction of an integrated railway system, an urban toll road, and associated local roadways. This network will be built mostly on SRT right-of-way land and will be divided into two main lines:

The north-south line with a length of 34.2 kilometers (21 mi) from Rangsit in the north to Don Muang Airport and on to Hualumpong, where it crosses the Chao Phraya River and continues on to Ponimit in the south.

The east-west line, 25.9 kilometers (16 mi) long, commencing at Huamak in the east, running through Makkasan, intersecting the north-south line at Yommaraj, where it crosses the Chao Phraya River and ends at Thonburi and Taling. At Makkasan there will be a 3.3-kilometer (2 mi) branch line to Mae-

Nam and Bangkok Port, consisting of only an elevated railway and community train.

The elevated system will be constructed in five stages:

Stage	Connection	Start	Complete
1	Don Muang-Yommaraj	December 1991	December 1995
2	Yommaraj-Hualumpong	December 1992	December 1996
	Yommaraj-Hua Mak	December 1992	December 1996
	Makkasan-Mae Nam	December 1992	December 1996
3	Rangsit-Don Muang	December 1993	December 1997
4	Hualumpong-Wong Wian Yai	December 1994	December 1998
	Thonburi-Yommaraj	December 1994	December 1998
5	Wong Wian Yai-Ponimit	December 1995	December 1999
	Taling Chan-Thonburi	December 1995	December 1999

In addition to the elevated system, H(T)L will carry out various real estate development activities on SRT property. These development rights will remain in effect during the 30-year transit system concession period. By this agreement, H(T)L is granted authority to establish and collect tolls, community train fares, rent from their developments, and other related revenue.

Construction work began in May 1993 and as of February 1995, the piling operation for Stages 1, 2, and 3 was almost 25 percent complete after 4162 piles had been driven. Substructure construction was underway as of June 1995, according to Suri Khuanmon, Director, Project Development Bureau, and trackwork at Bangsue were under construction. The project was behind schedule as of mid 1995, and H(T)L submitted an updated schedule for Stages 4 and 5. It was under review at that time by the State Railway of Thailand.

As of late 1994, several BOT projects were in the pipeline in Thailand, with a total value of U.S.$13 billion. They included a 25-year concession for a large sports complex in Bangkok, with an estimated value of U.S.$1 billion to be built by a consortium led by Japan's Mitsui & Company.

VIETNAM

When the United States announced in 1995 that diplomatic relations with Vietnam would be opened, a new era began in the reconstruction and industrialization of this country that had been ravaged by war and civil strife. The potential that Vietnam, Laos, and Cambodia have with respect to the availability of low-cost labor and as vacation and recreational areas is not lost on the rest of the world.

Market research analysts at Shimizu Corporation, one of Japan's Big Six contractors and developers, have been eyeing the Vietnam market for years. Yo Hisatomi, Research Director at Shimizu Corporation, stated that Vietnam has considerable oil reserves and a climate that lends itself to development as an Asian tourist attraction. He noted, however, that the absence of a strong infrastructure base would hinder any attempts to attract industry and tourism. In 1993, when Hisatomi visited Vietnam, local electrical power sources were so unreliable that hotels needed a 100 percent backup emergency generator system.

The State Planning Committee (SPC) in Vietnam is charged with the responsibility to develop BOT projects. Whether they are willing to assume risks related to exchange rates and indexing remains to be seen. Nevertheless, they seem to be moving forward. In 1995 the government was seeking proposals for a BOT toll road between Saigon and Vung Tau, a southern coastal city. At that time, they estimated that U.S.$20 billion in private investment capital would be required to construct the necessary infrastructure, transportation, and telecommunication projects in their country. Feasibility studies are underway for a 25-year BOT concession for a tunnel, cargo complex, and control tower at the Noi Bai Airport in Hanoi, and conceptual proposals are being developed for a 360-MW powerplant in Ban Mai. Also under study is a BOT port redevelopment scheme for the area south of Ho Chi Minh City.

CAMBODIA

Cambodia is the least developed of these three former French colonies. Because its civil war ended just a few years ago, it is a country that needs nearly total infrastructure upgrading, replacement, and expansion. Cambodia's infrastructure requirements would, more than likely, start with power-generating plants and end with port construction. If the country is to prepare to receive visitors and businessmen, it will need an airport that can handle the influx of passengers and freight. Government officials would like to build a new airport at Siem Reap, which is estimated to cost U.S.$450 million. They would also like to see a new jetway added to the existing Phnom Penh Pochentong International Airport at a projected cost of U.S.$100 million.

LAOS

The crying need in Laos is for power-generating plants. Given the geography of the country; it has been estimated that 18,000 MW could be generated from the hydroelectric resources indigenous to the hilly terrain of the country. If the resources were available to harness and use all this hydroelectric power, Laos would be a net provider of electrical power to adjacent countries.

THE NEED FOR PUBLIC/PRIVATE PARTNERSHIPS

The governments of these three countries want to invite private investors to work with them, and there certainly is a need for all forms of infrastructure construction. Despite the strides taken so far to invite private investment into the Southeast Asian market and the number of significant infrastructure projects that have been completed or are under way, much more work needs to be done to ensure the future of public/private partnerships in Asia. What is lacking is a sophisticated government program and the necessary assurances that private investment groups will find a cooperative, stable government, willing to assume some risks in exchange for firm guarantees and commitments that developers can take to the bank.

15

MEXICO AND ITS LATIN NEIGHBORS

In contrast to a projected East Asian infrastructure investment of U.S.$1.5 trillion during the next decade, the Latin American demand for U.S.$800 billion pales in comparison. Nevertheless, it is still a huge amount of money in any language. Mexico, Brazil, Argentina, Columbia, and Chile have begun the "privatization" process in the classic fashion—selling off government-owned companies such as airlines, airports, telecommunication, petrochemical, and power generation plants. These moves are being undertaken to improve the efficiency of operation of these facilities and to obtain much needed funds to close gaps in government budgets or to reduce foreign debt.

For a time, economic growth in Latin America was ebbing, and foreign investment shied away from the area, but as some South and Central American economies began to display growth in the 1980s and 1990s, private investors looked southward. Coupled with each government's desire to upgrade and expand its own infrastructure, public and private interests are beginning to converge south of the border. For example, as of the third quarter of 1995, the toll road program in Argentina, Brazil, Chile, Columbia, Mexico, and Panama calls for a total of 11,297 kilometers (6973 mi) of new highway construction for which U.S.$4.731 billion in private investment would be sought.

CHILE

Chilean Transportation Ministry spokesman Pablo Orozco in early 1995 indicated that his government would provide concessionaries with traffic projections and compensate the concessionaires for the shortfall if minimum traffic levels were not achieved. Orozco indicated that funds obtained from toll revenues collected on existing government-owned toll roads would be used for that purpose. Maximum toll rates would be based upon user benefits, but the concessionaires would be required to try to keep actual toll charges below these limits. For example, Orozco said that the maximum toll charge on the 100-kilometer (62 mi) Santiago to San Antonio road is U.S.$1.95 ($0.03 per mile), an amount far below the maximum level set at U.S.$5.26.

Chile's program is a most ambitious one, having as its goal the construction of a total of 3700 kilometers (2283 mi) of toll roads slated to be built between 1995 and 1997. Approximately 80 percent of this highway system is scheduled to be constructed with concession agreements if the government is able to attract U.S.$1.2 billion in private investment funding. However, as of October 1995, only 200 kilometers (123 mi) of concession-type toll roads have been built in the country.

Chile's program to attract foreign investment in power-generating plants is unique. Whichever power-generating plant has the lowest rate at a given time during the day is allowed to plug into the distribution grid first; the second lowest price gets plugged in next and so on down the line. These electricity generators are allowed to use the break-even rate of the last company plugged into the grid to establish their selling price. This program encourages efficiency and should attract foreign investors with new and efficient equipment and trained operation management teams.

ARGENTINA

Robert Cruz, president of the Buenos Aires Access Road Concession Oversight Board, reported that accurate government traffic projections are essential to the success of his country's BOT program, and long-term concession agreements are required to ensure that the most competitive toll rates are in effect. The Build, Operate, Transfer transportation program in Argentina is fueled by 80 percent foreign financing and concession bidding by Argentine-foreign corporation joint ventures. The average concession periods are for 20 years, and so far toll rates have been maintained at levels approximating U.S.$0.01 per kilometer.

BRAZIL

Brazil wanted to build only 850 kilometers (524 mi) of concession-type toll roads in 1995, but the cost of this program in its final bidding stages required the government to seek private investors willing to provide U.S.$1.1 billion in order to participate in this new highway system.

One of Brazil's critical path infrastructure projects is power generation. The state holding company, Eletrobras, owns both state-of-the-art hydroelectric dams and big city distribution systems along with many less attractive holdings. Eletrobras piled up a mountain of debt during its construction of conventional and nuclear power plants during the 1980s, and any sale of assets must deal with this issue. Several situations in Brazil, unless curbed, will provide roadblocks to foreign investment in power generation and distribution systems. Government-controlled electricity rates are well below those required to turn a profit. These low rates were established as one attempt to curb inflation that was running as high as 2500 percent in 1993 but that was brought down to a more manageable 30 percent in 1995. The government does allow electricity rates to be adjusted somewhat for inflation, but this is done in an arbitrary way and with little policy consistency.

COLUMBIA

In November 1995 Columbia announced a massive three-year highway expansion program to construct more that 800 kilometers (494 mi) of tolls roads, using the concessionaire approach. The first portion of this U.S.$2.2 billion program was activated in July 1995 with the commencement of construction of the first phase of the 61-kilometer (37-mi) Bogota-Villavicencio Highway. Carlo Julia Romero, Director General of Roads and Infrastructure in Columbia, noted of one of the problems that plagued Mexico's BOT program—that of cost overruns. He has initiated several procedures that can be brought into play should cost increases occur.

When change orders that could increase the contract value of the toll road are presented by the contractor and approved by the government, Romero indicated that his department will offer to extend the developer's concession period or allow an increase in toll rates as long as the increase does not exceed the rate of inflation by 7 percent.

MEXICO

Mexico's highway program in the late 1980s was more ambitious than its South American neighbors, partly because of the impending North American Free Trade Alliance (NAFTA), which was expected to create increased traffic to and from the United States, and partly because newly elected President Salinas was seeking increased national status among the community of nations. In 1989 the Mexican government under President Carlos Salinas de Gortari announced its five-year, U.S.$10 billion highway expansion program (Figure 15.1), one that promised to construct 5000 kilometers (3086 mi) of Build, Operate, Transfer concession-type toll roads with private sector funding. By the end of 1991, more than 50 percent of these toll roads were under contract to private investors who had funded U.S.$3.4 billion for a total of 32 road and bridge projects.

Funding the Highway Program

President Salinas planned to initiate this private toll road program by auctioning off about 3000 kilometers (1852 mi) of construction work to private sector concessionaires. The basis of the auction would be low bid in terms of time and costs based

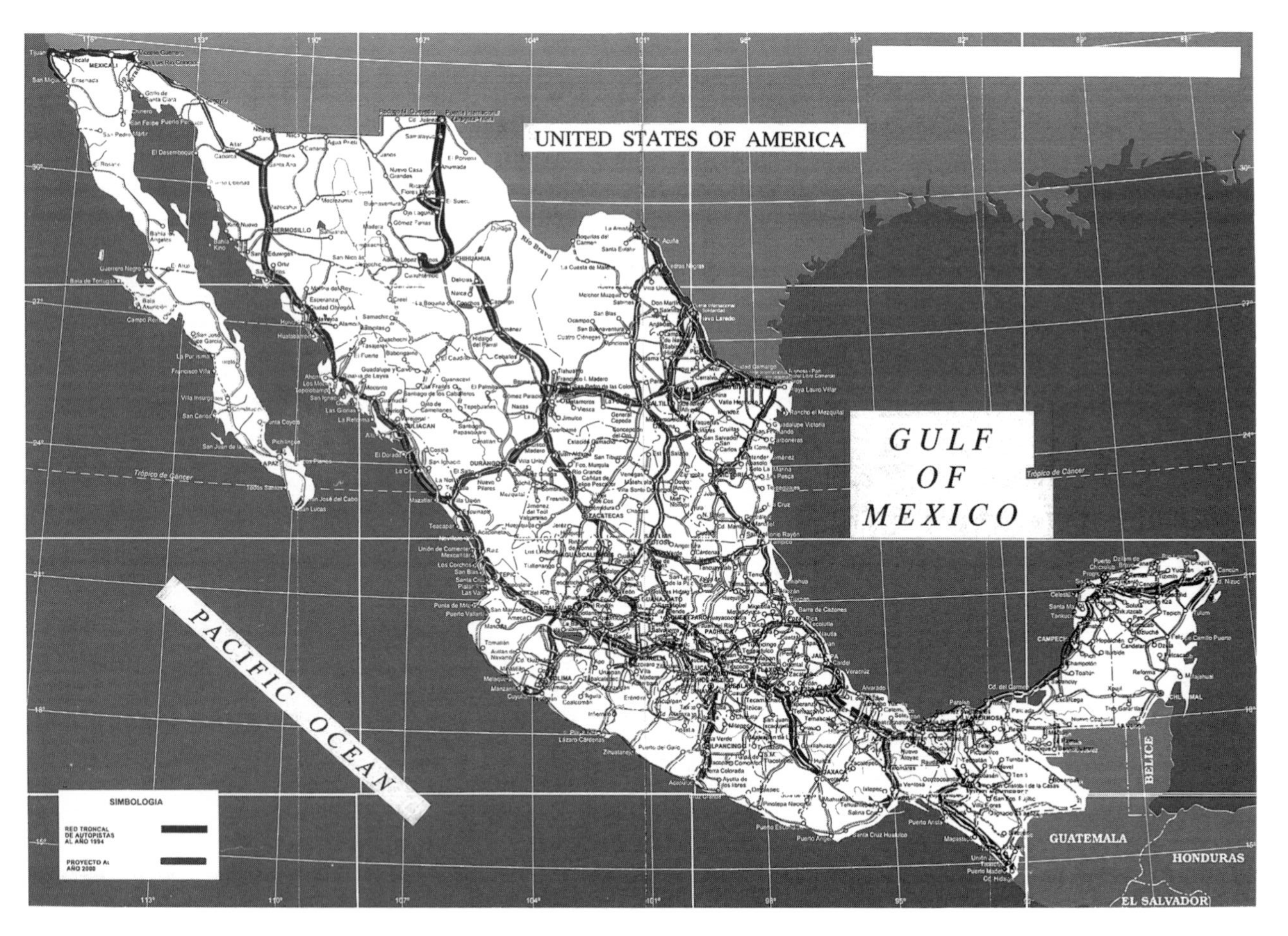

Figure 15.1. Mexico's national highway toll road master plan.

upon data supplied by the government relating to projected traffic flows and anticipated toll charges. This program caught the attention of many Mexican contracting companies eager to obtain work in a somewhat depressed market. They saw this new approach to highway construction as a way to obtain a profit not only from the construction contract but also from long-term revenue from the operation of the toll roads.

The government expected the concessionaire to put up 15 percent equity, which was considered "sweat equity," since it would be assigned to the concessionaire in exchange for partial payment on the construction contract. Contractors who were not shareholders in the joint venture could receive direct equity as partial payment for their construction-related services. When the toll road was completed and open to the public, the contractors were expected to recoup their investment through receipt of reasonable dividends as the toll road produced a revenue stream. The contractors would have the option of reducing their share of equity by selling off all or a portion of their ownership at a profit. Upon termination of the concession period, as with other BOT projects, the toll road facility would be turned over to the government.

In order to ensure maximum transparency, which implies open bidding or arms length transactions, the concessionaire would be required to establish with a local bank a trust account that would administer the financing arrangements during the concession period on behalf of the shareholders, the lenders, and any government agencies involved in the project. The government could elect to participate in the equity of these trusts by contributing public funds, creating a "tax holiday" for the shareholders, or employing any other means it chose. Nonequity funding would be provided by lending institutions in the normal fashion; however, some of these loans could receive the backing of the Mexican government.

Although the program started out in this manner and several large Mexican construction companies saw opportunities to obtain what looked to be lucrative construction contracts and additional revenue over the concession period, the private toll road concession scheme quickly became a quagmire.

The Economic Plight of the 1990s

Mexico's financial problems and economic slump were well publicized in mid 1994, particularly in light of President Clinton's firm stand on providing U.S. financial assistance to shore up the economy of its southern neighbor. In December 1994 the peso was devalued, unemployment reached record heights, and inflation exceeded 45 percent. Adding to the dilemma of those construction companies engaged in private infrastructure projects, banks were so busy restructuring outstanding loans that they had little time to negotiate new ones. Most commercial projects were put on "hold" when bank loans dried up. The federal government, looking to bring cash into their coffers, began to "privatize" public corporations.

Government statements were made early in 1995 that state-owned Petroleos Mexicanos S.A. (Pemex), one of Mexico's most powerful national symbols, would never be placed upon the auction block. But 10 months later in November 1995, a large chunk of Pemex was offered for sale. The government estimated that it could obtain anywhere from U.S.$1 billion to as much as U.S.$4 billion for 60 Pemex petrochemical plants that would be offered for sale to private concerns. The Mexican people were upset by their government's change in policy and became very vocal about the government's abrupt about-face. On October 15, 1995, a referendum taken in the coastal town of Minatitlan revealed just how unpopular this decision to split up Pemex was. Out of 42,067 voters in this town, only 350 voted for the sale. The cry of "Gringos out of Mexico" resounded throughout the southeastern part of the country where most of the Pemex refineries and petrochemical plants were located. Divestiture of government-owned facilities got off to a bad start.

The Toll Road Concession Program

Within this framework, the Mexican private toll road concession program gestated, was born, and experienced traumatic growing pains. Three of Mexico's major construction companies participated in the program, anticipating construction profits and reasonable rates of return on their investments, but many of these projects are swimming in a sea of red ink. The companies that bought into this concession program are looking to keep their heads above water until the economy turns around. Triturados Basalticos y Derivados, S.A. de C.V. (Tribasa); Empress ICA Sociedad Controladora, S.A. de C.V. (ICA—pronounced *Eek-a*); and Grupo Mexicano de Desarrollo, S.A. (GMD) built most of the toll roads included in President Salinas' 1989 national toll road highway plan, and not too many of their ventures have been profitable ones for reasons we shall soon examine.

Grupo Mexicano de Dessarrollo, S.A. Founded in 1975, GMD completed more than 350 major construction projects in Mexico and in 1993 became a publicly held company with listings on the Mexican and New York Stock Exchanges. In 1994 the company's gross revenue was U.S.$607,960,000, but due in part to the depressed nature of the Mexican economy and revenue shortfalls on some of their concession-type toll roads, they reported a net loss of U.S.$20,427,000.

GMD's investment in toll road concessions is substantial. As of December 1994, their investment included the following highways:

Toll Road Entity	Investment in New Pesos (in thousands)
Autopista del Occidente, S.A. de C.V.	197,946
Promotora de Autopistas del Golfo, S.A. de C.V.	25,100
Grupo Concesionario del Oeste, S.A.	17,168
Concesionaria Mexicana de Vias Terrestres, S.A. de C.V.	11,437
Autopistas de Guerrero, S.A. de V.C.	4,205
GMD, ICA, Tribasa Asociacion en Participation	412
Compania Nacional de Concesiones, S.A. de C.V.	50

The total investment converted to U.S.$ at a currency exchange rate of 7.5 pesos equates to an investment of U.S.$34,175,733.

GMD has the dubious distinction of becoming the first concessionaire to be awarded a BOT toll road project—the 245-kilometer (151 mi) highway from Cuernavaca to Acapulco in July 1989. They formed a joint venture with ICA and Tribasa, maintaining a 34 percent interest while the other two participants in the project each had a 33 percent interest. This highway, initially named Autopista del Sol (Highway of the Sun), would later be called Headache Highway and ultimately prove to be one of the many bitter pills these three companies would swallow as a result of the Mexican government's headlong rush into a program that was hastily conceived and executed.

The Terms of the Highway of the Sun Agreement The initial contract sum for this project was U.S.$780 million when construction began in 1981, but upon completion in July 1993 cost overruns had increased the final cost to U.S.$1.3 billion. This, of course, was a major factor in the financial problems faced by investors, but there were others.

The Mexican government provided all bidders with traffic studies and directed each bidder to prepare a proposal based upon the projected traffic flow and a toll rate established at 138 pesos (U.S.$46 based upon the exchange rate of 3 pesos to the dollar). The cost per mile would equate to U.S.$0.30, which was quite high by comparison with other countries (i.e., $0.04 to $0.07 per mile for many U.S. toll roads). With Mexico experiencing economic problems, 40 percent inflation and high unemployment during the late 1980s and early 1990s, in retrospect, it would have appeared obvious to developers that the average citizen would avoid these toll roads and seek other free, alternative routes. The initial concession period for the Highway of the Sun was 14 years and 8 months commencing January 1989. However, in November 1994 the concession period was extended to 30 years, ostensibly to allow GMD a longer time in which to recoup their investment. Although tolls were inflation indexed to a point, increasing the already expensive toll charges simply produced less traffic. So it appeared that all GMD could do was to wait it out, hope for a reduction in inflation and interest rates, a decrease in unemployment, and a general turn around in the nation's economy.

Victor Hardy Mondragon, Technical Director of GMD's Infrastructure Group, said that the government's Request For Proposal for these concession toll roads contained three basic objectives.

1. An award would be made to the bidder submitting the lowest construction cost.
2. An award would be made to the bidder requiring the shortest concession period.
3. An award would be made to the bidder requiring the least amount of government participation.

Although all bidders were provided with highway construction specifications, Mondragon noted that the absence of federal gross vehicle weight limits made it difficult to project maintenance costs. If heavily loaded trucks frequented the highway, premature failures could be expected, but at the time of bidding, based upon government-furnished traffic projections, it was difficult to assess the cost of future repairs accurately.

Higher construction costs could ultimately result in lower maintainance costs, but bidders were reluctant to add costs to their estimates for fear of producing noncompetitive proposals. Private developers were prohibited by law to own the land upon which these roads were to be built, and the Mexican government purchased all the land and associated rights-of-way for the project. After establishing a price for the land, the government directed all bidders to include that cost in their proposal.

The GMD group was the successful bidder on the Autopista del Sol project. It would be responsible for the design, construction financing, operation, and maintainance of the toll road during the concession period after which title would pass to the government. GMD was to provide a surety bond in the amount of U.S.$2.13 million, which was to remain in force until the end of the concession period, and was obliged to maintain a minimum capital stock of U.S.$2.6 million during the entire concession period.

A syndicate of Mexican banks led by Banca Serfin, S.A., provided construction financing for the highway project up to 40 percent of the total amount of the investment but not in excess of $U.S.373 million. Any construction costs above this amount would be funded by the joint venture partners, the GMD group.

The Cost of Doing Business in Mexico The magnitude of construction financing becomes more apparent when we look at the situation in Mexico since the time of devaluation in 1994. The final cost of the Autopista del Sol, which was about U.S.$1.3 billion, meant that approximately U.S.$927 million would have to be financed by partners GMD, ICA, and Tribasa, hence the name Headache Highway.

When asked "What went wrong with the toll road concession program," Mondragon responded, "Everything went too fast. We found ourselves building the highways before everything was in place. Banks were unable to convert their return on investment to present value. The government is going to have to change some laws—for instance there are no provisions in the law for arbitration. The government was interested in low initial construction costs, and based upon the initial short term concession periods, these low capital costs would most certainly translate into high maintenance costs. When we built the Autopista del Sol, the government specifications were prepared in such a way so as to produce the lowest possible bids, there was no regard for future maintenance costs."

Regarding the low traffic count on this highway and other similar toll roads, Mondragon said, "BOT projects, and not just toll road projects, will gain acceptance only if they provide quality service. If a concessionaire provides a quality product, one that improves the quality of life, they will be able to charge more for that product."

In addition, the financial demons at work in Mexico's economic crisis had produced commercial financing rates of 47 to 52 percent. The rate of inflation from December 1994 to October 1995 has been officially established as 43.63 percent. Mondragon said that any commercial developer daring enough to consider construction on, say a U.S.$5 million project, would be able to borrow 80 percent of the cost of construction from the local banks but at interest rates ranging from 40 to 50 percent. Imagine what revenue levels would have to be produced merely to cover the U.S.$1.6 million to 2 million per annum interest charges!

Mondragon said that these extraordinary interest rates do not truly reflect the actual state of the Mexican economy, which seems to limp along hoping for better conditions. He said that businessmen and contractors alike remain optimistic that better days lie ahead. He recalled that, when the Autopista del Sol construction began, GMD was financing the project with relatively short-term money, 7 to 10 years. This was the only source of funding available in the country at that time, and because they were short-term loans, the funds bore interest rates approximating 34 percent.

The net result of all these factors—extraordinary cost overruns, substantially reduced traffic flow with a resultant low revenue stream, and the high cost of financing—resulted in the obvious, a combined loss of approximately $U.S.90 million for the operation of the Cuernavaca-Acapulco toll road for the years 1992, 1993, and 1994.

Undaunted by the state of the economy and the lost revenue on their toll road concessions, GMD is forging ahead. As of December 1995, GMD was negotiating a light rail project in Mexico City with the government with partner Tribasa. Mondragon said that when they submitted their bid in 1994, the currency exchange was 3 pesos to the dollar. Now that the exchange rate is 7 pesos to the dollar, negotiations with the government revolve around who will be responsible to absorb the 4-peso deficit. The project must go ahead, because the traffic congestion in and around the city is intense at all hours of the day. Because the greater Mexico City area with its population of 23 million is ringed by mountains, it is subjected to frequent air inversion problems and cars at idle or those barely moving increase air pollution to unacceptable health levels. The government is undertaking immediate measures to reduce traffic and, hence, pollution.

The government has already instituted a program to limit traffic coming into the city. Depending upon an odd or even number on the vehicle's license plate, the driver will be able to enter the city only on certain designated days; however, everyone is free to drive in Mexico City on Saturday to do their week's shopping.

Other Markets and Sources of Income GMD is actively seeking other markets. It is involved in a hydroelectric plant project in Columbia known as Mille Uno. This BOT project has what Mondragon referred to as a "Take or Pay" clause. The power purchase portion of their U.S.$500 million design, build, finance, operate, maintain contract is the most important part of that agreement in that minimum and maximum power generation production levels have been established. If the minimum power output of the plant is not sold, the government will pay the company the difference between actual revenue and minimum revenue. If sales of the generating plant's output is greater than the maximum level, GMD and the government will split those excess profits on a 50–50 basis.

Joint ventures with other countries is another option. GMD is negotiating several joint venture agreements with U.S. and European contractors to bolster sales and revenue. Mondragon said that GMD will continue to explore global markets even when the Mexican economy returns to a firm economic footing.

Like so many other Mexican companies during the economic downturn, GMD tapped the international debt market in search of capital. On February 17, 1994, the company obtained U.S.$250 million in medium-term Eurobonds due February 2001. Interest of 8.25 percent per annum was to be paid semiannually in U.S. dollars but when Mexico's withholding tax on interest payments is added, the effective interest rate becomes 8.675 percent, which is still a bargain when compared to local interest rates. Certain restrictions were placed upon the issuance of these Eurobonds:

- The consolidated net worth of GMD must be maintained at or above 130 percent of the aggregate principal amount of the notes outstanding at all times.
- The combined net worth of the guarantors (GMD affiliate companies) must exceed 130 percent of the outstanding principal amount of the notes. Combined net worth is calculated to exclude certain intercompany assets and to include certain off-balance-sheet liabilities.
- Subject to certain customary exceptions, GMD may not permit the incurrence of liens on any of its assets.
- GMD must maintain direct or indirect control over 51 percent of each of the guarantors at all times. The net amount received in connection with this issuance (U.S.$246,875,000) of which approximately U.S.$45 million was applied to the payments of debt with Banco Mexicano, S.A. The remaining proceeds were to be retained by GMD as working capital.

Although this infusion of capital helped to ease the cash flow problems of the company; the year 2001 is looming closer. Unless the economy picks up substantially in Mexico and/or GMD is able to obtain profitable work outside the country, GMD will be faced with a huge repayment obligation just six years down the road.

Tribasa's Mexico City-Toluca Toll Road Project Although GMD obtained the first Mexican toll road concession award, Tribasa is credited with completing the first such highway project in Mexico, the 22-kilometer (13.5 mi) highway between Mexico City and the industrial city of Toluca in 1990. This project was refinanced, and the highway was expanded in 1992 when Tribasa became the first company to secure a Mexican concession-type toll road successfully in the international capital markets.

This U.S.$313 million deal, representing a refinancing of an existing highway, married Mexican fixed-income securities to Eurodollar securities. Lehman Brothers underwrote U.S.$207.5 million of the Eurobond issue, and International Finance Corporation (IFC) underwrote U.S.$10 million and purchased a like amount for their own account. This 10-year issue consisted of global despository units (GDUs) issued by Nacional Financiera, the trustee for the toll road project. These GDUs were backed by amortizable toll revenue participation certificates called TRIPS (Toll Road Indexed Participation Securities) and carried an 11 percent coupon. During the course of the bond offering, the government extended Tribasa's concession period from 2.3 years to 10 years, thereby allowing them to raise the targeted U.S.$313 million. However, not much of this funding remained with Tribasa; U.S.$105.1 million of the proceeds went toward payment of construction costs, and U.S.$102 million was turned over to the Mexican government in exchange for the lengthening of the concession period. Another portion of these funds were used by Tribasa to plow back into the Cuernavaca-Acapulco project.

Lehman Brother's success in obtaining capital led to their participation in another BOT highway project, the Mexico City-Guadalajara toll road being built by a consortium consisting of Tribasa, GMD, and ICA.

Mexico City-Guadalajara Toll Road Certainly the most ambitious highway project attempted in Mexico, this 538-kilometer (332 mi) toll road connected Mexico's largest city with the country's second largest metropolitan area, Guadalajara, which has a population of 4 million. The first stage of this project beginning at the city of Maravatio ended about 16 kilometers (10 miles) from the center of Guadalajara, and construction on the second phase began in 1992. It added 343 kilometers (212 miles) to complete this toll road, which will connect with existing highways that link Mexico City with Toluca. With the completion of this road in September 1994, travel time between Mexico City and Guadalajara has been cut in half to just about 4.5 hours. A one-way trip is U.S.$59, which equates to U.S.$0.27 per mile, expensive by other country's standards. In fact the average cost to travel on Mexican toll roads is $.15–$.18 per mile, exceeded only in Japan where the cost per mile is about U.S.$.205 per mile. This compares to an average of U.S.$0.04 per mile in the United States and U.S.$0.13 per mile in Europe.

Tribasa has been involved in a number of other concession toll road projects in Mexico: the 22.5-kilometer (13.9 mi) Ecatepec-Piramides toll road and the 47-kilometer (29 mi) Aeremeria-Manzanillo toll road. The Ecatapec-Piramides toll road (Figure 15.2) with its 12-peso toll charge is a winner for Tribasa since it is so heavily traveled by tourists and Mexican nationals who wish to visit one of the country's top attractions, the Aztec ruins at Teotihuacan where the Pyramid of the Sun and The Pyramid of the Moon date back to 250 to 600 A.D. (Figure 15.3). The Ameria-Manzanillo, by contrast, is sparsely traveled.

Grupo Tribasa created the Tribasa Toll Road Trust 1 to attract investors to participate in the funding of these two roads in order to have them upgraded, operated, and maintained at a profit. The Trust raised U.S.$200 million from the sale of notes in November 1993. These notes made Mexican history because they were the first successfully syndicated securitization in the international capital market with U.S.-dollar-denominated securities supported by Mexican-peso-denominated cash flow.

(*a*)

(*b*)

Figure 15.2. Tribasa's Ecatapec-Piramides toll road.

(*a*)

Figure 15.3. (a) The Pyramid of the Moon at Teotihuacan.

(*b*)

Figure 15.3. (b) The Aztec ruins dating back to 500 A.D. at Teotihuacan.

A debt service fund was established when the Tribasa note sale took place. This fund was set up to ensure that U.S. dollars would be available to repay debt obligations on a timely basis. The reserve fund was funded at first from proceeds obtained from the initial note offering. Subsequent funding was to come out of preamortization cash flow and would be maintained at certain levels through the concession period. Although this scheme appeared to have all the safeguards required by the international investment community, the downturn of the Mexican economy may yet be the downfall of this financing scheme.

With inflation rates of 35 to 45 percent snapping at the heels of the Mexican worker, discretionary income quickly disappears; with unemployment at 10 percent, workers will travel less. And coupled with the rising cost of gasoline and auto repairs, all these events point to drastically reduced toll road travel. But Tribasa continues to build highway systems in Mexico. In 1995 Tribasa completed the Colegio Militar highway project and was awarded a 25-year concession contract. But this time the Mexican government employed an independent consultant to prepare the traffic studies and traffic projections, and the government allowed Tribasa to construct the project under a design-build contract. Since concessionaire Tribasa completed the Penon-Texoco Highway project in 1994 and built the Chamapa-Licheria toll road in August of that same year, Tribasa's addition of the Colegio Militar highway completes the western Mexico City ring road system—100 percent owned by Tribasa.

ICA The Concessionaire Division of ICA participated in the construction of 1705-kilometers (1052 mi) of concession toll roads of which 698 kilometers (430 mi) were built directly by ICA. Two highway concessions were awarded to ICA in 1992: the 247-kilometer (152 mi) Saltillo Turnpike and the Plan de Barrancas turnpike extension between Guadalajara and Tepic. They also participated in the Mexico City—Guadalajara toll road project.

Max Liebers Guillen, Manager of Infrastructure Projects for ICA, said that government Requests For Proposals for concession-type toll road projects are similar to other country's RFPs—they require the bidder to provide proof of experience in toll road construction and to display the required financial background and capability to finance and control these kinds of projects. Highway specifications are provided to the bidder along with the proposed routing of the highway. The government purchases the land and all rights-of-way and continues to own the land during the concession period; however, the cost of the land that has been acquired must be reimbursed by the successful bidder.

In most RFPs, the government provides the bidders with projected traffic flows and with the allowable toll rates, but the government will not specify the length of the concession period. Toll rates are generally indexed for inflation and are adjusted every three months. If actual traffic flow is considerably less than that provided to bidders, the government generally allows the concessionaire to extend the term of the concession agreement. Initially it was thought that 12- to 18-year concession periods would be feasible, but now, according to Guillen, the government feels that 30-year concession periods are required to provide a concessionaire with an adequate return on investment. When the government approached another developer and offered to extend its franchise period, the developer replied, "Are you asking me if I want to lose money over a longer period of time?"

Guillen said that private toll road toll charges are at a distinct disadvantage when compared to government-owned tolled highways, and driver's complain that the private turnpike rates are much higher than government-operated systems. When the government prepares its toll road rate schedule, it takes into account only those costs associated with the operation and maintenance of the roadway exclusive of the initial cost to construct the highway. But private operators must include all costs, including construction costs; therefore, their rates, out of necessity, are higher, requiring higher tolls, which is an unfair advantage.

When asked how the Mexican government can help those developers who have participated in the toll road concession program, Guillen said that the government is considering allowing private toll road operators to collect revenues from existing government-owned highways in exchange for assuming responsibility for the operation and maintenance of these roads. Although this may seem to be a likely method by which to provide private developers with an opportunity to increase their revenues, it may just be another trap. Because there are presently no legal limits on the gross weight of trucks, it would appear that many heavily traveled government highways may have been subjected to abuse by overloaded vehicles, and the potential for substantial repair and maintenance costs could be lurking out there for the unsuspecting private developer.

Guillen said that 3000 kilometers (1851 mi) of Mexican federal highways are required to be constructed over the next five years to complete the national highway system, but this program is on hold now for obvious reasons—the high cost of money and the depressed state of the economy. He said that many state governments are looking to the BOT concept to complete their intrastate highway systems, and several adjacent states are teaming up to combine their plans in order to attract private developers.

According to Guillen, when the BOT toll road program is revitalized in Mexico, some form of government participation will be required. The federal government must be willing to assume some risks such as guaranteeing revenue if actual traffic flows fall below acceptable minimums, a system similar to the shadow tolling concept in the United Kingdom. A new law, which is expected to be enacted in late 1995 or early 1996, would establish maximum weight limits for over-the-road truck traffic. This should lower maintenance costs on future highways when the private concession program comes back on stream.

Carlos Pescador Zamora, Business Development Manager for ICA, said that the company has turned its BOT experience toward other areas. ICA recently completed a three-level garage under the Palacio de Belles Artes, a splendid white marble art deco concert hall and art center, which was built under the supervision of the Italian architect Adamo Boari in 1904 and is located in the historic center of Mexico City. Although the government owns the land, ICA has the underground parking concession with a 20-year term. ICA designed, constructed, financed, operates, and maintains this facility and will turn it back to the government at the end of the concession period. The hourly parking rates are established somewhat differently from other lots in the area, and the government had no restriction on the permissable levels of parking rates. Although ICA charges a more or less standard rate of 7 to 8 pesos (U.S.$1 to 1.14) for the first hour, other garages charge for a full hour even if the customer remains one minute past that first hour. ICA divides rate upcharges into 15-minute segments; therefore, on average, they offer less expensive parking rates than their competitors. This is their second BOT parking garage, and they have plans to construct 33 more in Mexico City. Building this three-level structure was no easy task because, among other civil engineering considerations, they had to deal with the high water level that exists under Mexico City. The city was originally built over a lake, and water exists at levels of 1.5 to 2 meters (5 to 6.5 ft) below the surface. Because of the explosive growth of the city over the past several decades, more potable water was pumped from underground. Mexico City has been slowly sinking for years, and aquaduct projects are underway to bring water in from outside the city to slow down the rate of sinking or, with other stabilizing procedures, to stop the sinking.

Preparing for NAFTA

The Mexican government program of highway construction initiated in 1992 included provisions to upgrade many existing roadways. With the expectation of the passage of the North American Free Trade Alliance (NAFTA), the Mexican government envisioned a substantial increase in both imports from the north and exports to the United States and expected to see truck traffic expand accordingly. In mid 1993 NAFTA had been ratified in Canada and Mexico, but environmental concerns remained a barrier between the United States and its southern neighbor.

At a U.S. Council of the Mexico-U.S. Business Committee meeting held in July 1993, the need for an injection of money into Mexico's public sector infrastructure program led to the establishment of the North American Development Bank (NADBANK) whose mission would be endowed with U.S.$1 billion paid in capital and U.S.$4 billion callable capital from the NAFTA signatories. NADBANK would be empowered to make or guarantee loans to fund border infrastructure projects. Although NADBANK would be investing in potable water, waste water, and solid waste projects within 100 kilometer (62 mi) of the U.S.–Mexican border, it was thought that the net result might be to free up public funds for other infrastructure projects like transportation. NADBANK got off to a rocky start when the Mexican economy began to experience meltdown and although callable capital was to be made available by Mexican sources, it appeared in early 1995 that any callable funds required would have to come from the United States.

What Does the Future Hold for Mexican Infrastructure?

The price paid for the BOT tolled roads, an average of U.S.$2.67 million per kilometer (or U.S.$4.32 million per mile), was a large part of the problem, but the high level of toll rates, coupled with a downturned economy, produced traffic flows that were 60 percent less than government projections. What went wrong?

Amidst the euphoria that followed President Salinas' pronouncement of Mexico's ambitious and exciting highway expansion program, government officials hastily provided anxious developers with official traffic projections. Hastily prepared government cost projections and design specifications fell short of actual requirements. Investor groups rushed into the BOT highway business, lured by potential big construction profits and government assurance that their anticipated revenue stream would be supplemented by extended concession periods if traffic flows did not meet expectations.

Just like the commercial real estate boom in the United States in the 1980s, bankers not wishing to miss out on any of the action stepped right up and offered loans without carefully scrutinizing the investor's pro forma statement, weighing anticipated revenue against capital costs, and looking at downside risk. Contractors anxious for work promised to turn these tolled roads back to the government in the shortest period of time, a tune the administration was happy to hear since they convinced the Mexican citizens that there was an urgency to the highway program.

This created a situation where five-year concession periods produced roads where tolls were five to ten times higher than those in the United States for comparable travel distances and where some U.S.$3.75 billion had to be repaid on short-term high-interest loans.

Contractors looking for the "fast buck" on the construction portion of the project could submit "insincere bids," not caring whether the toll road would survive the scrutiny of the marketplace during the short concession period. So what if the venture went bankrupt after two or three years?

The results were inevitable. Traffic flows slowed to a trickle, and motorists preferred to use the older freeways, even though travel time was considerably longer. The Mexican government stepped in to mitigate these problems and extended some of the concession periods from 5 to 10 and from 10 to 15 years. The Mexican highway boom came to a screeching halt after December 1994 when the peso was devalued, leaving highway construction firms with total debts of about U.S.$2.25 billion. The government had to step in and try to effect some sort of bailout.

Federal officials began to restructure the banking system's delinquent highway loan portfolio, and responsibility for this change was vested in the central bank's Fund for Savings Protection (Fobaproa). These restructured highway loans are to be inflation indexed, and borrowers are to be offered extended repayment plans. The World Bank in June 1995 gave Fobaproa U.S.$2.25 billion to help them in this bailout. President Ernesto Zedillo promises deregulation and the streamlining of regulations placed upon small and medium-sized businesses. He is attempting to rid his country of scandal and corruption, a change that can only improve Mexico's image and help the country attract foreign investors. Mexico was attempting to raise U.S.$1.5 billion in international credit markets in late November 1995 to assist in restructuring its debt and to repay some loans it received in an international bailout earlier in the year. To encourage participation in this offering, the government provided investors with an opportunity to invest in 28-day Mexican Treasury Bills, known as *cetes,* which were paying 54.7 percent interest rates in late 1995. Zedillo notes that the rapid paydown of foreign currency debt in 1995 and the beginning of a substantial trade surplus bodes well for the Mexican economy in the years ahead. On December 5, 1995, the government of Mexico reported that 100 foreign companies announced plans to invest U.S.$6.3 billion in the country in 1996, an increase of 53 percent over 1995. With Ernesto Zedillo's leadership, plus the indomitable spirit of the Mexican people, this country will ultimately get back on its feet and forge ahead.

16

WHAT DOES THE FUTURE HOLD?

What will happen in the future? We don't know, but there will certainly be a lot of surprises. Who could have predicted the rapid fragmentation of the former USSR or the Dow Jones punching through 5000 or the Japanese economy in turmoil in 1995 or an Israeli president and PLO leader shaking hands on the White House lawn? Change is inevitable and appears to be taking place with more rapidity than ever before.

The era of globalization has arrived, and with the assistance of computers and satellite telecommunications, the world has become one huge marketplace where investors are constantly shifting funds and investments seeking out their highest returns. Borderless industries search for increasingly cost-effective manufacturing environments and, after moving on to a more attractive one, leave behind a slightly, or sometimes substantially, more developed nation than it first encountered.

We are slowly beginning to recognize that economic growth and infrastructure development go hand in hand. As such, certain trends appear to be emerging.

- Government has begun to realize that it cannot automatically increase taxes and maintain bloated budgets without attracting public scrutiny.
- Government has begun to acknowledge that some of its services can be performed with greater efficiency and, consequently, at less cost by the private sector. The private sector is willing to participate in this new partnership if it is allowed to earn a reasonable return on investment.
- Government has expressed increased concern about the environment and the need to protect it and has heightened public awareness concerning the effect of a deteriorating environment on their future and the future of their children.

THE ROLE INFRASTRUCTURE WILL PLAY IN OUR GLOBAL SOCIETY

If we agree that a healthy infrastructure portends an economically healthy nation, then the converse may be true—a deteriorating and neglected infrastructure portends a declining and deteriorating economy. However, the cost to build and maintain an adequate or superior infrastructure system can be staggering. The United States, one of the world's premier economies, has underfunded its infrastructure for decades. As a percentage of Gross Domestic Product, infrastructure investment in the United States experienced a rapid decline from 1965 to 1989, increasing just slightly above its low peak in the first half of the 1990s (Figure 16.1).

Statistics are always good show stoppers, and a glimpse at the state of American infrastructure is an eye opener.

- Of all the G-7 Countries (Japan, Italy, Germany, France, Canada, United Kingdom, and the United States), the United States had the lowest public infrastructure to Gross Domestic Product ratio.
- The government's General Accounting Office estimates that U.S.$112 billion alone is required over the next three years to bring the country's public school systems up to "good" condition.
- As of 1992, 43 percent of America's lakes, 8 percent of its rivers, and 13 percent of these river estuaries were contaminated with toxic chemicals.
- The U.S. Environmental Protection Agency (EPA) estimates that an additional investment of U.S.$137.2 billion in sewage treatment will be required to meet the needs of the country's growing population into the 21st century.

THIRTY YEARS OF FEDERAL INVESTMENT IN INFRASTRUCTURE—1965–95

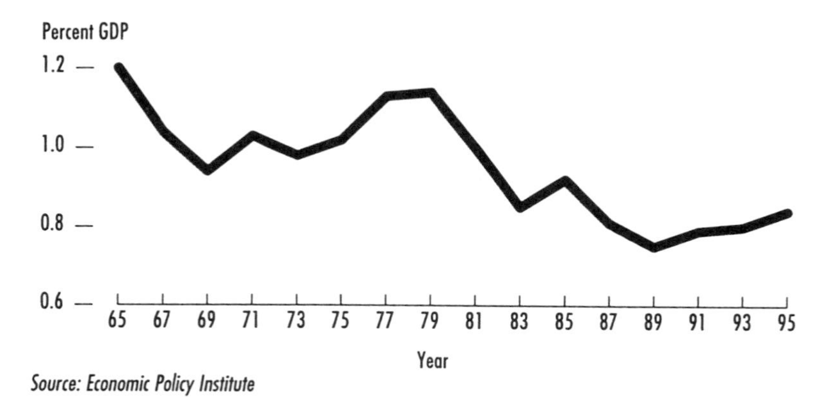

Figure 16.1. U.S. infrastructure investment expressed as percentage of Gross Domestic Product.

- Anyone who lives or works in or near a metropolitan area should not be surprised by the fact that 30.3 percent of these highways are congested.
- Good roads make good cents. According to federal transportation officials, the cost to truck goods in the United States increases 6.3 cents per mile when road conditions drop from "good" to "fair."
- Thirty-two percent of the bridges in the United States require widening or replacement in order to accommodate the existing traffic flows, and U.S.$164.9 billion over the next 20 years is needed to improve these conditions.
- Only two ports on the East Coast are deep enough to handle the latest deep draft ocean-going vessels.
- Municipal drinking water supplies are becoming more susceptable to outbreaks of bacterial infestation, as witness the one that made 400,000 people sick in Milwaukee in 1993.

If the United States of America is one of the world's richest nations and it cannot afford to put its infrastructure back into first-class condition, what is to be expected from less fortunate countries and, in particular, the emerging and developing countries around the world?

Five of the cities with the world's most air pollution are located in Asia—Beijing, Calcutta, Jakarta, Shenyang, and New Delhi. In the last decade of the 20th century, East Asia, with its emphasis on the rapid growth of electrical generating equipment, the prime mover for industrialization, is expected to show a greater increase in carbon dioxide levels and sulfur dioxide emissions than all other regions in the world. The World Bank estimates that by the end of the century Asian countries will need to devote 2 to 3 percent of the region's Gross Domestic Product to deal with environmental problems, an amount equivalent to U.S.$38 billion per year. The Association of Southeast Asian Nations (ASEAN) designated 1995 as the Year of the Environment, and their regional environmental action plan includes three solutions to the problems created by expanding economies in their area:

1. Improve the investment climate of the region in order to attract private participation in many of the infrastructure projects that must be built
2. Provide a political climate where regional governments extend loan guarantees and provide technical assistance for the development of targeted infrastructure projects
3. Develop innovative ecological financial arrangements such as the cancellation of portions of debt obligations if the debtor country agrees to attain certain environmental milestones

Given the financial demands facing most governments around the world today and the constraints placed upon them to satisfy these demands, a stronger partnership among all levels of government and the private sector appears to present a viable alternative to the need to fulfill infrastructure requirements in the most cost-effective and environmentally sound manner. For example, as Europe's Common Market concept coalesces, new transportation systems will be required to link its economic community. The prospect of a Trans-European Highway has prompted the governments of Hungary and Poland to seek private investors to participate in highway expansion programs to match those of their neighbors.

The previous chapters in this book presented case studies of public/private partnerships that worked, some that did not work, and others that were aborted and never had a chance to show what they could do. But is the Build, Operate, Transfer concept the long-term vehicle for these public/private partnerships or is it a flash-in-the pan movement to be supplanted by some other innovative financial chess game in the next century? Does it have enough justification to endure?

The system certainly has its advantages when it is conceived and implemented with sufficient research, resources, and resilience.

- It frees up scarce government funding for other uses.
- It avoids or lessens the need to increase taxes or effect budget cuts in order to build much needed infrastructure projects.
- It should produce projects of higher quality since the concessionaire is obligated to maintain these projects throughout the franchise period.
- It should create a need for less administrative and managerial staff within those agencies of the government that heretofore were charged with the responsibility to supervise construction of these government projects and manage them upon completion.
- It should avoid, or substantially reduce, the cost overruns experienced by most government agencies when they build infrastructure via the more conventional competitive bid method because the contractual responsibility for design and construction rests with the concessionaire.

But the stakes are high for those private investors or consortiums willing to assume the risks that go with the rewards, and the risks are many and varied. They include

- Expropriation of the project once it has been completed. Witness Kumagai Gumi's experience after completing the Second Stage Expressway in Thailand.
- Collapse of a federal economic system or a local government bankruptcy, as happened not halfway around the world, but in Mexico and Orange County, California.
- Uncontrollable and irretrievable cost overruns such as those experienced by the builders and developers of the Channel Tunnel and several Mexican toll roads.
- Stiff citizen resistance to an infrastructure project or the proposed routing of one. Witness the demise of several proposed projects in the State of Washington and Arizona, as well as Tralfagar House's protracted process of gaining public approval for the final routing for the Birmingham Northern Relief Road.
- The ever-present danger of a government reneging on its agreement. Recall Enron Corporation's on-again, off-again Dabhol power plant project in India

and Indonesian President Suharto's decision to drop an American power plant company in favor of a company in which his son had a substantial financial interest.

Setting aside the 1868 Suez Canal project, the Build, Operate, Transfer process is a relatively new project delivery system and is untested over the long term. It may be too soon to tell if the existing BOT projects will survive their concession period and meet their owner's expectations. We cannot know whether the government will receive a valuable asset or an albatross when title to these projects transfers to the appropriate government agency at the expiration period.

By the same token, we cannot know whether some of these projects will be outmoded by new technologies. This may be the case with respect to power generation and water and waste treatment facilities. Power conservation, in some cases, may be substantially less costly than developing new power-generating projects, as some countries in Asia are discovering. The government of Thailand announced a five-year, U.S.$60 million program to save 1400 MW of electricity. For cost effectiveness, consider this cost as opposed to a $1.7 billion, 1300-MW power-generating plant underway in Pakistan or U.S.$1 billion for the proposed 1200-MW Datong generating plant in China.

We also cannot know whether other forms of ground transportation will make some present-day highways less attractive to commuters and travelers. The light rail systems that are being built or expanded in cities around the world may create entirely new people-moving patterns, reducing traffic flows on nearby toll roads to a trickle. Much experience has been gained, however, in assembling complex deals that bring together the technicians (the architects, engineers, and contractors) and the deal makers (the financial institutions, accountants, and lawyers). The size and sophistication of these Build, Operate, Transfer projects to date, by necessity, bring together the large firms in their fields, oftentimes the multinational corporations that roam the globe sniffing out deals and paying allegiance to their stockholders. With the significant costs involved in assembling a BOT proposal and the considerable amount of time it takes to comply with environmental issues and citizen concerns, only those corporations with "deep pockets" can stay the course. As a result, will we begin to see the very thing that we wished to avoid, the monopolizing of the big public/private infrastructure deals where the name of the game is return on investment.

There are possibly no more that two or three dozen players in this field of mega-deals, but we cannot know whether the future will trim them down even further. If competition is substantially reduced, the law of supply and demand may take effect, and the cost of these types of projects could rise to the point where local governments would possibly be able to design and construct them for less money by reverting to the conventional competitive bid process, as they have done in the past. One economist said that the main reason for the success of recently privatized companies is the competitive marketplace. Private corporations, by their nature, can trim down until they obtain optimum productivity and develop a cadre of effective managers. In the public sector, productivity sometimes seeks the lowest common denominator. However, it is conceivable that a federal, state, or local government could prune back its staff, hire the best and the brightest and reward them accordingly, call upon the technical and management expertise for sale in the marketplace, and construct infrastructure projects at a lower cost that the private sector, without the need to turn a profit or show a return on investment.

Given governments' ability to borrow at low interest rates, coupled with its access to a vast pool of experts and information, and with the addition of a little "fire in the belly" attitude, they could become the infrastructure builder of choice in the next century. Already government is beginning to look at other innovative ways to invite private sector participation in these infrastructure projects. The U.S. Department of Transportation, late in 1995, announced a plan to build U.S.$3 billion of new highways by allowing private funding to substitute for state matching dollars. In New York, financier Felix Rohatyn, the force behind New York's Municipal Assistance Corporation (MAC), which helped to pull the city out of its 1970 financial crisis, has proposed a multistate partnership for the purpose of funding new highway construction and other infrastructure projects. Rohatyn's plan involves establishing a corporation to fund U.S.$200 billion projects of this nature over a 10-year period. This multistate-owned corporation would issue long-term bonds backed by 10 cents of the current 18.4 cent per gallon federal gas tax. However, these gasoline tax-backed bonds could be used for nonhighway infrastructure projects such as water and waste treatment plants and even new school construction, a feature that could ultimately kill the plan.

THE PLUSES AND MINUSES OF THE BOT PROCESS

The Pluses

- Government does not need to increase taxes to pay for infrastructure projects because they can be funded by private concerns.
- The administrative staff that handles these types of projects can be either eliminated or transferred to another agency that needs added staff.
- The profit motive, the prime mover in a private venture, will generally tend to put an infrastructure project in place more quickly and more efficiently.
- The private sector is apt to seek a more cost effective approach to the design, financing, and construction of an infrastructure project, thereby reducing the overall cost of the project.
- Because the consortium, in most cases, will operate and maintain the project during the extended concession period, it will employ the latest technologies and strive to achieve the highest-quality levels.
- The private sector, although still subjected to risks during the design and construction phases, is better equipped to deal with these risks on their own than to pursue the adversarial approach common with conventional lump sum government contracts.

- More and better infrastructure projects increase the quality of life and act as an economic stimulus in the area(s) in which they are constructed.
- Particularly in developing or emerging nations, the large design/construction firms, through their interaction with local firms, or by osmosis, effect some degree of technology transfer.

The Minuses

- Government's approach to an infrastructure project does not necessarily depend upon returning a profit, overseeing the construction, or operating and maintaining the project. As a result, all things being equal, the cost of their projects could be less than that in the private sector.
- Project failures and defaults do occur. If one of these projects happens to be a long-term BOT project, the government either will be saddled with its operational and maintenance costs or will be forced to close it down; in either case, a drain on public funds will take place.
- The incentives that governments sometimes provide investors, in the form of tax concessions or some other form of subsidy, may, in the end, be more costly to the taxpayer.
- Citizens may balk at having to pay for what was once a free service or, because of government subsidies, to pay more for equivalent services provided by the private developer.

RESISTANCE TO CHANGE

Since infrastructure projects are locale oriented and can have a profound and immediate effect on a community, changing from a government-created project to a public/private partnership may be viewed as an institutional change. Typically, institutional changes require a rather lengthy gestation period before being accepted. Citizens have always looked to government to provide roads, bridges, water, and waste treatment plants and may be leery of transferring these responsibilities to businessmen whose primary function is to turn a profit.

The case for public/private partnerships in infrastructure projects is strong. If we can learn from our successes and failures, the future for public/private partnerships may be brighter. By looking at what went right and what went wrong, we can determine several ways in which the public/private partnership can be strengthened.

- The method by which government Requests For Proposal from the private sector needs to be changed so as to reduce the cost of proposal preparation, thereby allowing more firms to compete in the process. Partial reimbursement of unsuccessful qualified bidder's proposals might be a start. Each bidder could be instructed, as part of the bid documents, to commit a small amount to a pool with the government matching funds. All qualified bidders, with the exception of the successful one, would share in this "kitty."

- Environmental impact studies and permitting are time consuming and costly endeavors. The government could assume responsibility for this task and hand the successful bidder the necessary permits. In fact, a timeframe could be established for this process. If it exceeds the allotted time, the government would pay the developer interest on its substantiated investment until the permits have been secured. This should speed up the process.
- The concessionaire should be required to provide training sessions for selected government officials during the design, construction, operation, and maintenance stages of the project so that, when their contract expires, there will be a smooth transition of ownership. This will ensure technology and management skill transfer.
- When the concession-type infrastructure project has a positive effect on the local economy, the government might consider rewarding the concessionaire with cash dividends or tax reductions. This would provide another incentive to investors.
- Developers are always required to post bonds in case of default. Why not hold government to the same standard? If they default on their portion of the agreement, the developer could evoke the provisions of the bond.

The Build, Operate, Transfer concept is a sound one; government and industry can work together harmoniously and to each other's mutual benefit. After all, governments seeking private sector assistance in solving some of their economic problems is certainly not a new phenomenon. Back in 1492, King Ferdinand and Queen Isabella of Spain hired a private contractor to map out a new route to India, a journey that turned out to be rather successful.

INDEX